SCRAMBLING TECHNIQUES FOR CDMA COMMUNICATIONS

THE KLUWER INTERNATIONAL SERIES
IN ENGINEERING AND COMPUTER SCIENCE

Distributors for North, Central and South America:
Kluwer Academic Publishers
101 Philip Drive
Assinippi Park
Norwell, Massachusetts 02061 USA
Telephone (781) 871-6600
Fax (781) 681-9045
E-Mail < kluwer@wkap.com >

Distributors for all other countries:
Kluwer Academic Publishers Group
Distribution Centre
Post Office Box 322
3300 AH Dordrecht, THE NETHERLANDS
Telephone 31 78 6392 392
Fax 31 78 6546 474
E-Mail < services@wkap.nl >

 Electronic Services < http://www.wkap.nl >

Library of Congress Cataloging-in-Publication Data

A C.I.P. Catalogue record for this book is available
from the Library of Congress.

Printed on acid-free paper.   Printed in the United States of America

SCRAMBLING TECHNIQUES FOR CDMA COMMUNICATIONS

by

Byeong Gi Lee
Seoul National University

Byoung-Hoon Kim
GCT Semiconductor, Inc.

KLUWER ACADEMIC PUBLISHERS
Boston / Dordrecht / London

Contents

Preface

With the advent of IMT-2000, CDMA has emerged at the focal point of interest in wireless communications. Now it has become impossible to discuss wireless communications without knowing the CDMA technologies. There are a number of books readily published on the CDMA technologies, but they are mostly dealing with the traditional spread-spectrum technologies and the IS-95 based CDMA systems. As a large number of novel and interesting technologies have been newly developed throughout the IMT-2000 standardization process in very recent years, new reference books are now demanding that address the diverse spectrum of the new CDMA technologies.

Spreading, Scrambling and Synchronization, collectively, is a key component of the CDMA technologies necessary for the initialization of all types of CDMA communications. It is a technology unique to the CDMA communications, and thus understanding of the spreading and scrambling techniques is essential for a complete understanding of the CDMA systems. Research of the spreading/scrambling techniques is closely related to that of the code synchronization and identification techniques, and the structure of a CDMA system takes substantially different form depending on the adopted spreading/scrambling methods.

The IMT-2000 standardization has brought about two different types of CDMA technologies which require different forms of spreading/scrambling techniques - - *cdma2000* system and *wideband CDMA* (W-CDMA) system. The fundamental distinction of the two systems is that the cdma2000 is an inter-cell *synchronous* system, whereas the W-CDMA is an inter-cell *asynchronous* system. In the case of the inter-cell synchronous DS/CDMA systems whose earlier example was the IS-95 system, every cell in the cellular system employs a common scrambling sequence with each cell being distinguished by the phase offset of the common sequence, which inevitably necessitates some kind of external timing references for the coordination among the cells. In the case of the inter-cell asynchronous DS/CDMA systems, however, each cell employs different scrambling codes, thereby eliminating the dependency on the external timing references. Cell search and synchronization in this environment becomes a very challenging problem, as allocating different codes to different cells imposes big burden in terms of cell search speed and the related circuit complexity.

There have been reported a considerable amount of works on developing efficient cell search (or scrambling code and timing acquisition) techniques in very recent years, including *three-stage search* technique adopted as the W-CDMA standard. However, this technique has much room for improvement, and researches are still on-going to develop more efficient cell search methods. The *distributed sample-based acquisition* (DSA) technique among the newly emerging techniques may prove to be a highly potential candidate in the future due to its rapid and robust acquisition capability.

This book is intended to deal with the scrambling techniques for use in the CDMA systems, including the synchronous and asynchronous IMT-2000 CDMA systems and those yet to come beyond. It provides some background and fundamentals on sequences and shift register generators in the beginning, and then focuses on various acquisition techniques in the primitive and advanced levels. Much stress is put on spreading and scrambling of the synchronous and asynchronous IMT-2000 systems and other recent code synchronization techniques. Above all, this book has the unique feature that it introduces, comprehensively and thoroughly, the novel acquisition technique DSA and its family, invented by the authors themselves.

It is hoped that the contents of this book appear valuable to wireless communication engineers, especially to those involved in theoretical and design works on spreading, scrambling and synchronization of the CDMA communication systems.

Acknowledgments

This book may be called a sequel of the book *Scrambling Techniques for Digital Transmission* authored by one of us (BGL) and Dr. Seok Chang Kim (published by Springer-Verlag in 1994) in the sense that the theoretical foundation established in the digital transmission environment is applied to CDMA communications in this book: The DSA technique, which is a main theme of this book, is firmly rooted on the DSS theory addressed in the previous book. We recollect the unequaled talents of the deceased coauthor (SCK) who established a solid theoretical foundation of the scrambling techniques for digital transmission, with gratitude and admiration.

This publication is made possible thanks to the helps of several graduate students at Telecommunications and Signal Processing (TSP) Laboratory of Seoul National University and the comfortable research environments provided by Seoul National University, Seoul, Korea, and George Washington University, Washington, DC, USA.

We gratefully acknowledge the contributions of Myung-Kwang Byun, Daeyoung Park, Byeong-Kook Jeong, and Hanbyul Seo in summarizing various references to help writing Chapters 3 and 4. In particular, we are indebted to Daeyoung Park who helped word-processing the text, drawing the figures, formatting the tables, and compositioning the whole contents.

We thank Professors Yong Hwan Lee and Kwang Bok Lee at Seoul National University and Professor Dong In Kim at University of Seoul for kindly reviewing some chapters of the book. Also we are thankful to GCT Semiconductor, Inc. for allowing a grace period to one of us (BHK) to complete drafting the main body of the text.

We would like to thank our wives, Hyeon Soon Kang and Ji-Young Lee, whose love and support at home encouraged and enabled us to concentrate on authoring the book and the related researches at work.

BYEONG GI LEE AND BYOUNG-HOON KIM

Chapter 1

INTRODUCTION

Wireless communication systems can be classified to *frequency division multiple access* (FDMA), *time division multiple access* (TDMA), and *code division multiple access* (CDMA) systems in terms of the employed medium access technology. In the wireless cellular communications arena, FDMA was mainly employed in analog wireless systems such as AMPS, NMT, and TACS, whereas TDMA and has become dominant in digital wireless systems such as GSM, USDC (IS-54), and PDC. CDMA was first applied to commercial use in 1996 through the IS-95 system and has become the standard medium access technology of the IMT-2000 systems [1] (see Fig. 1.1).[2] CDMA is expected to be the major medium access technology in the future public land mobile systems owing to its potential capacity enhancement and the robustness in the multipath fading channel environment.

CDMA or SSMA Communications

CDMA is uniquely featured by its spectrum-spreading randomization process employing a *pseudo-noise* (PN) sequence, thus is often called the *spread spectrum multiple access* (SSMA). As different CDMA users take different PN sequences, each CDMA receiver can discriminate and detect its own signal, by regarding the signals transmitted by other users as noise-like interferences. Fig. 1.2 depicts the block diagram of the generic CDMA (SSMA) communi-

---

[1] UWC-136 is a TDMA based IMT-2000 standard system, while other systems are mostly CDMA based.
[2] AMPS is an abbreviation for Advanced Mobile Phone System; NMT for Nordic Mobile Telephone; TACS for Total Access Communication System; GSM for Global System for Mobile Communications; USDC for United States Digital Cellular; IS for Interim Standard; PDC for Personal Digital Cellular; IMT for International Mobile Telecommunications [1, 2].

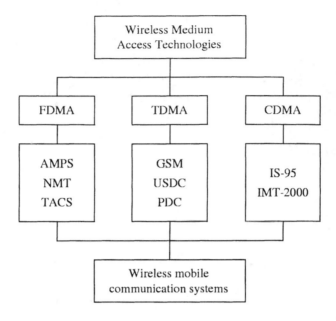

Figure 1.1.   Wireless medium access technologies and wireless mobile communication systems.

cation systems. In the figure, the transmitter and the receiver contain the same pseudo-noise generator.

While the user capacity, defined as the maximum number of simultaneously communicating users, is hard-fixed by the number of allocated frequencies or time slots in the case of the FDMA or TDMA systems, the CDMA user capacity is soft in that the maximum number of users can vary over a wide range depending on the propagation channel characteristics, signal transmission power level, data traffic activity, and others. This soft capacity stems from the unique medium access feature of the CDMA system that the entire common medium is shared by all the active users, without being split into multiple slots with each slot exclusively occupied by different user, as is the case in FDMA or TDMA systems. As each CDMA user randomizes its signal into a random noise, the signal collision among CDMA users can be avoided even though they share the same medium and the same spectrum band. Therefore, each CDMA user can fully exploit the entire medium until the interference level proportional to the number of randomized multi-user signals disables the information conveyance over the medium. This medium sharing method brings forth potential enhancement of user capacity over the medium split-and-dedication method owing to the statistical multiplexing gain principle [3]. It is especially beneficial when the data rate is time-varying, which is commonplace in practical communication applications.

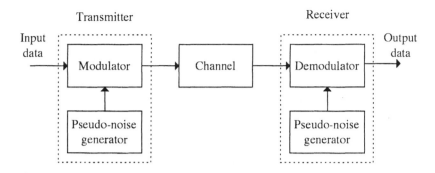

Figure 1.2.   Block diagram of the generic CDMA communication system.

The CDMA systems are further subdivided into the *direct-sequence* (DS) CDMA, the *frequency-hopping* (FH) CDMA, and the *multi-carrier* (MC) CDMA in the respect of spectrum-spreading method [4, 5]. In the DS/CDMA systems, a user-specific PN sequence with a high chip rate is directly multiplied to a low symbol rate data sequence to expand the data spectrum. In the FH/CDMA systems, a user-specific frequency pattern spread over a wide spectrum range adjusts the carrier frequency of the transmitted user signal in every hopping interval, making the wide spectrum fully exploited during the communication. In the MC/CDMA systems, a user signal is copied and transmitted over multiple modulated sub-carriers arranged in order in the wide spectrum, where each user employs a unique amplitude coefficient sequence to modulate the sub-carriers.

The spectrum-spreading feature of the CDMA makes the communication system very robust in the typical multipath fading environment. As the signal bandwidth increases, the probability of deep fading for the entire signal band becomes very low. Thus the receiving side can detect the CDMA signal correctly by using the partial information transmitted over good frequency bands for most of the communication time. This partial spectrum attenuation phenomenon caused by time-varying multipath channel is called frequency-selective fading, and each CDMA receiver mitigates it effectively by employing some unique signal processing methods. For example, the DS/CDMA receiver can intelligently combine the individual multipath signals by employing a RAKE system, thereby maximizing the signal-to-noise ratio at the detector front-end.

In addition to the aforementioned features, the DS/CDMA has several advantages over the other multiple access technologies, which include soft hand-off capability, high information security, simple frequency planning, high frequency reuse efficiency [6], and high transmission power efficiency. All these advantages are rooted on the spectrum-spreading and scrambling processes that employ PN sequences.

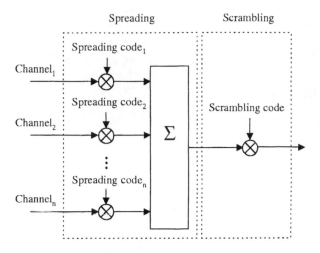

Figure 1.3.   Scrambling and spreading structure in the DS/CDMA system.

Spreading and Scrambling in DS/CDMA Systems

Spectrum-spreading and data randomization can be performed concurrently in the DS/CDMA transmitter through the processing that multiplies a high-rate (or, wide-band) PN chip sequence to a low-rate (or, narrow-band) data sequence. The ratio of the chip rate to the data rate is called the *processing gain* or the *spreading factor* of the resulting wide-band DS/CDMA signal. This corresponds to the spectrum bandwidth expansion ratio of the data signal during the conversion to the DS/CDMA signal.

However, in typical cellular applications, a DS/CDMA signal is generated through two steps - - spreading and scrambling (see Fig. 1.3.): First, an orthogonal spreading code, or, channelization code, is multiplied to the data sequence, which expands the signal bandwidth and makes each user's signal orthogonal to those of the other users (or the other data channels). In the downlink of the IS-95 system, for example, the 64-long orthogonal Walsh codes are employed for the spectrum-spreading and orthogonal channelization, and in the IMT-2000 W-CDMA and cdma2000 systems, the *orthogonal variable spreading factor* (OVSF) codes take the role both in the downlink and in the uplink [7, 8, 9].

Secondly, the channelized data signal is randomized through multiplication of a PN code, typically, of the same rate, which is the chip scrambling process. The employed PN code is called the scrambling code of the signal. As the transmitted DS/CDMA signal usually arrives at the receiving side via multiple propagation paths with different delays, the orthogonality among data channels imposed by the channelization processing cannot often be maintained at the receiver front-end. Furthermore, as the auto- and cross-correlation property of the orthogonal channelization codes is very poor, the interference resulting from the multipath propagation can critically degrade the data detection performance unless another counter-action is taken. Therefore, the scrambling processing that randomizes the user signal while keeping a good correlation property is essential in wireless DS/CDMA communications. In the downlink, a cell-specific scrambling sequence is assigned to each cell, which makes both the neighboring cell interferences and the camping cell multipath inteferences appear like random noises at the front-end of the mobile station receiver. On the other hand, in the uplink, a user-specific scrambling sequence is assigned to each individual user, as the timing alignment among different users is not guaranteed. When a group of users can control their transmission timings such that their signals arrive at the receiver of the base station at the same time, they may employ a common scrambling sequence and user-specific orthogonal codes even in the uplink to get the benefit of interference minimization.

The scrambling code generation methods differ between the inter-cell synchronous and the inter-cell asynchronous DS/CDMA systems. In the inter-cell synchronous systems such as the IS-95 and the cdma2000 [9, 10], all base and mobile stations can align their timers to that of the external timing reference. Thus, for the sake of acquisition circuit simplicity and system coordination facilitation, a common scrambling code is employed in all the base and mobile stations in the system, with each station employing a unique phase shift of the common scrambling code in order to prevent signal collision. On the other hand, in the inter-cell asynchronous systems such as IMT2000 W-CDMA [8, 11], each base station has its own timing reference, which disables employing a common scrambling code over the entire cellular system. Each of base and mobile stations employs a unique scrambling code such that its signal does not collide with those of the others in the system for any phase shift. As a result, the scrambling code acquisition complexity increases significantly, as will be illustrated in the next section.

Scrambling Code Acquisition Techniques

As the first-step signal processing in the DS/CDMA receiver the locally generated scrambling sequence is synchronized to the one superimposed on the received waveform. In general, this synchronization process is accomplished in two steps - - *code acquisition*, which is a coarse alignment process bringing the two scrambling sequences within one chip interval, and *code tracking*, which is a fine tuning and synchronization-maintaining process [4].

In scrambling code acquisition, fast acquisition is one of the most important goals, for which a considerable amount of research has been exerted during the past decades [4, 12, 13]. Fig. 1.4 demonstrates a diverse set of scrambling code

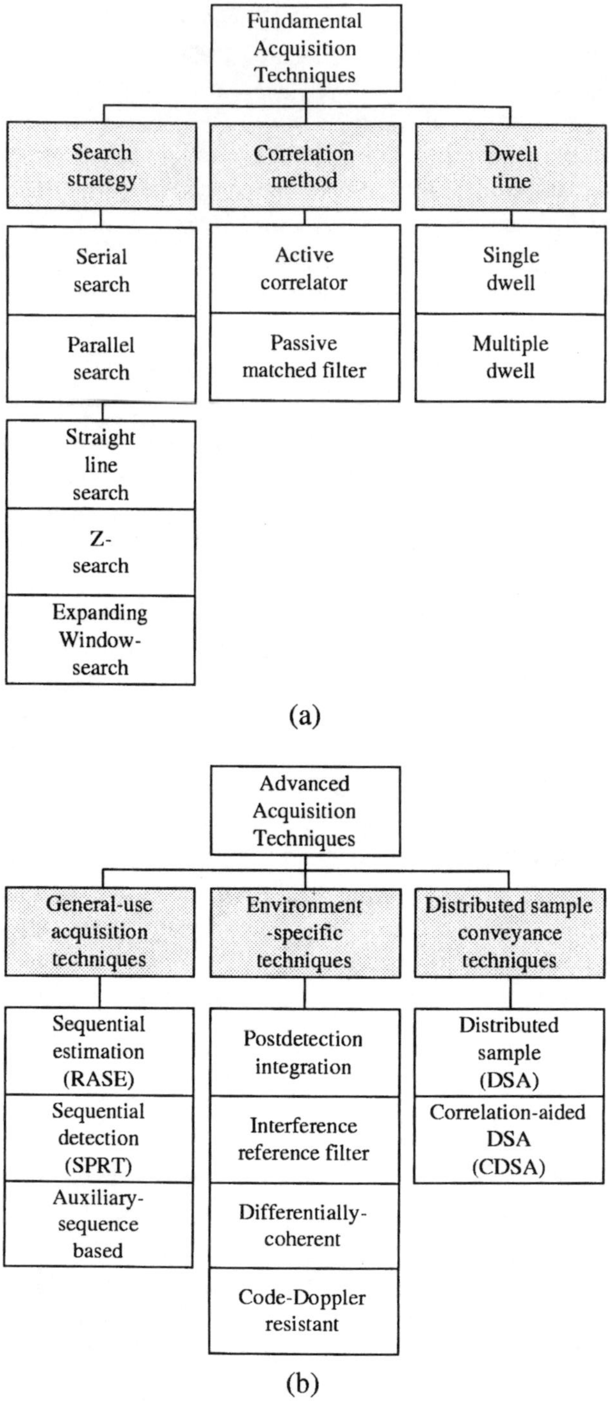

```
                    Fundamental
                    Acquisition
                    Techniques
        ┌───────────────┼───────────────┐
   Search          Correlation          Dwell
   strategy          method              time
   ┌─────────┐     ┌─────────────┐     ┌─────────┐
   │ Serial  │     │   Active    │     │  Single │
   │ search  │     │ correlator  │     │  dwell  │
   ├─────────┤     ├─────────────┤     ├─────────┤
   │Parallel │     │   Passive   │     │Multiple │
   │ search  │     │matched filter│    │  dwell  │
   ├─────────┤     └─────────────┘     └─────────┘
   │Straight │
   │  line   │
   │ search  │
   ├─────────┤
   │   Z-    │
   │ search  │
   ├─────────┤
   │Expanding│
   │ Window- │
   │ search  │
   └─────────┘
```

(a)

```
                    Advanced
                    Acquisition
                    Techniques
        ┌───────────────┼───────────────┐
 General-use        Environment      Distributed sample
 acquisition         -specific         conveyance
 techniques         techniques         techniques
┌─────────────┐   ┌─────────────┐   ┌───────────────┐
│ Sequential  │   │             │   │  Distributed  │
│ estimation  │   │Postdetection│   │    sample     │
│   (RASE)    │   │ integration │   │    (DSA)      │
├─────────────┤   ├─────────────┤   ├───────────────┤
│ Sequential  │   │             │   │Correlation-aided│
│ detection   │   │ Interference│   │     DSA       │
│   (SPRT)    │   │reference filter│ │    (CDSA)     │
├─────────────┤   ├─────────────┤   └───────────────┘
│ Auxiliary-  │   │             │
│  sequence   │   │Differentially-│
│   based     │   │  coherent   │
└─────────────┘   ├─────────────┤
                  │             │
                  │Code-Doppler │
                  │  resistant  │
                  └─────────────┘
```

(b)

Figure 1.4.   Various code acquisition techniques: (a) Fundamental, (b) advanced.

acquisition techniques that have been introduced to date, from fundamental techniques to advanced techniques. [3]

The conventional serial search scheme [12] has the advantage of simple hardware, but the acquisition time is very long for a long-period scrambling sequence because its mean acquisition time is directly proportional to the period of the scrambling sequence employed[14]. Several fast acquisition schemes have been developed at the cost of increased complexity. For example, multiple-dwell approaches were introduced [15, 16] which usually employ a fast decision-rate *matched filter* (MF) for initial searching and a conventional active correlator for verification. This multiple-dwell acquisition process enabled to reduce the mean acquisition time by several factors [4]. For the case when the *a priori* probability for the code phase uncertainty is not uniformly distributed, several time reduction strategies that resort to search pattern variations were proposed, whose examples are Z-search, expanding window search, offset Z-search, and others [4, 17, 18, 19]. Also several sequential test approaches, combined with two threshold comparators and variable dwell time, were introduced for the initial code acquisition process, which could reduce the mean acquisition time further [20, 21, 22, 23].

Unfortunately, those fast acquisition schemes cannot provide the desired level of fast acquisition for very long period scrambling sequences, as they sequentially searched for the valid phase among all candidate local sequence phases, which outnumber the order of the code period. For such a long sequence case the parallel acquisition scheme [24] might render a solution but the hardware complexity, that is, the number of active correlators or matched filters, would increase to the order of code period. To get around this problem, serial-parallel hybrid schemes were proposed for practical use [25, 26, 27, 28], in which a long code sequence of period N was divided into M subsequences, each of which having length N/M, and the acquisition circuit was composed of M parallel matched filters.

Another fascinating trial which can reduce the acquisition time tremendously with a small hardware increase can be found in estimating the state of the involved *shift register generator* (SRG) directly, instead of aligning the code phase based on the correlation value. In principle, it is possible to accomplish code acquisition in about L time units if the current sequence of the transmitter SRG of length L is available, which would take 2^L-1 time units otherwise by employing conventional serial search schemes. Some earlier schemes called the *sequential estimation* [29, 30, 31] made L consecutive hard decisions on the incoming code chips using a chip-matched filter and then loaded them to

---

[3]Refer to Chapters 3 and 4 for detailed descriptions of the scrambling code acquisition techniques. Chapter 3 deals with the fundamental acquisition techniques, whereas Chapter 4 deals with the other advanced techniques.

the receiver SRG as the current SRG states. They were successful in speeding up the acquisition process by additionally employing some proper verification logics. However, the performance degradation, caused by poor estimation of each chip, made it difficult to apply them to the CDMA environment where the average chip-SNR is very low [4]. Even in high SNR environment, in fact, they are not suitable for practical use because, to estimate the chip values, they require to recover the carrier phase before the acquisition process, which, however, is nearly impossible. [4] Note that code acquisition cannot be achieved in a coherent manner. From this viewpoint, the closed-loop coherent acquisition scheme based on an auxiliary sequence [33, 34] is not practically feasible either, in spite of its theoretical novelty.

In addition to fast acquisition, robust acquisition is another important issue to consider in designing an acquisition system. Code acquisition is often conducted in very poor channel environment, where SNR is extremely low due to the shadowing effect, the channel is rapidly fading due to terminal movement, or a considerable frequency offset exists between the transmitter and receiver oscillator circuitry. Thus a new acquisition system often becomes useless in practical environment unless it is designed to overcome all these obstacles.

As an approach to improve the low-SNR acquisition performance, a differentially-coherent acquisition detector was proposed, which could provide an SNR gain of about 5dB over the conventional noncoherent acquisition detectors without using carrier phase information [35]. To prevent the acquisition performance degradation in fast fading channels, chip-differentially coherent detection scheme was proposed, in which the receiver correlation operation is done based on a differentially-detected scrambling sequence [36]. The scheme is reported to be especially useful for code acquisition in very fast Rayleigh fading or large frequency offset channels. On the other hand, for the inter-satellite or underwater acoustic communications, in which the ratio of the terminal velocity to the wave propagation velocity may become considerably high, thereby causing causes the code-Doppler phenomenon (or, time compression of the scrambling sequence itself), several correlator-bank approaches that jointly detect the code timing and the Doppler-shift were introduced to cope with the code-Doppler effect [37, 38, 39]. Aside from those discussed above, there are several additional factors to consider to improve the acquisition system performances - - threshold-setting strategy [40, 41, 42, 43], *multiple access interference* (MAI) effect [44, 45, 46, 47], postdetection integration [28, 48], multipath and the multiple pilots [48], and so forth.

---

[4]A differentially-coherent sequence estimation approach is introduced in [32], where they apply multiple parity-check trinomial equations to estimate reliable chip values. The approach avoids the coherent acquisition problem but requires relatively high computation complexity for low-SNR range operations. Furthermore, it cannot work well in practical time-varying wireless channels.

Recently, with the introduction of the inter-cell asynchronous systems such as the IMT-2000 W-CDMA, the traditional code acquisition issue has entered a new phase. The single code scheme adopted in the inter-cell synchronous systems enables each mobile station to acquire the cell scrambling codes in relatively short time with simple hardware. On the other hand, the inter-cell asynchronous system assigns different PN sequences to different cell sites, eliminating the dependency on external timing references. Consequently, very sophisticated and complex acquisition schemes are needed to acquire the scrambling codes within the allowed time limit, and the code identity and the time shift have to be searched simultaneously [49, 50, 51].

For a fast acquisition of the scrambling codes in the inter-cell asynchronous systems, the 3GPP three-step synchronization scheme based on the *generalized hierarchical Golay* (GHG) code[52] and the *comma-free* code [53, 54] has been adopted as the IMT-2000 W-CDMA cell search (or, initial code synchronization) standard [8, 55]. The three steps in this scheme refer to slot boundary identification based on matched filtering, code group and frame boundary identification, and cell-specific primary scrambling code identification. For the 1st and the 2nd steps, each base station respectively broadcasts the *primary synchronization code* (PSC) and the *secondary synchronization code* (SSC) of length 256 each through the *synchronization channel* (SCH) at every beginning of the slot. In the 3rd step, the cell-specific primary scrambling code is identified by correlating all candidate scrambling codes belonging to the identified code group with the incoming *primary common pilot channel* (CPICH) sequence. [5]

In very recent years, a new acquisition technique that realizes direct SRG acquisition through distributed SRG state sample transmission has been introduced under the name of *distributed sample-based acquisition* (DSA) and a family of variations including the *correlation-aided DSA* (CDSA) followed [56]–[68]. DSA technique basically features two unique mechanisms - - distributed sampling-correction for synchronization of the SRG and distributed conveyance of the state samples via a short period sequence. More specifically, the state of the main SRG in the transmitter is sampled and conveyed to the receiver in a distributed manner, while the state samples are detected and applied to correct the state of the main SRG in a progressive manner. For conveyance of the distributed state samples, a short-period sequence called the igniter sequence is employed. The DSA turned out to be very effective in speeding up the acquisition process at low complexity and making performance reliable even in very poor channel environment. [6]

---

[5]Refer to Chapters 5 and 6 for details of the three-step synchronization scheme in the 3GPP W-CDMA and TD-CDMA systems. Chapter 7 deals with the spreading and scrambling in the 3GPP2 cdma2000 systems.
[6]Refer to Chapters 8 and 9 for detailed descriptions on the DSA and CDSA techniques.

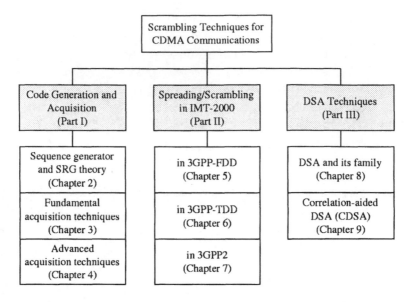

Figure 1.5.   Organization of the book.

Organization of the Book

This book is intended to discuss the scrambling techniques in the wireless communication environment, specifically, in the CDMA communication environment. The contents of the book is arranged into three parts collectively dealing with the scrambling, spreading and synchronization techniques in the DS/CDMA communication systems (see Fig. 1.5).

Part I covers the topics on basic sequences, sequence generation and acquisition. The key features of the popular m-sequences, Gold-sequences, and Kasami-sequences, as well as the embedded SRG theories, are introduced in Chapter 2. Fundamental scrambling code acquisition techniques are briefly addressed in Chapter 3, and advanced code acquisition techniques are discussed in Chapter 4. Included in this selection of acquisition techniques are the rapid acquisition by sequential estimation, the sequential detection based rapid acquisition, the auxiliary sequence based acquisition, postdetection integration techniques, differentially-coherent acquisition, the Doppler-resistant acquisition, and the distributed sample conveying acquisition.

Part II introduces the third-generation DS/CDMA cellular systems, focusing on the scrambling and spreading techniques adopted in the systems. Chapters 5 and 6 deal with the IMT-2000 W-CDMA and TD-CDMA systems, respectively, and Chapter 7 introduces the cdma2000 system as an evolutionary system of the existing IS-95 system. [7]

Part III is dedicated to the DSA technique and its family. Chapter 8 describes the theory, organization, and operation of the DSA, and then extends the discussion to BDSA, PDSA, and D^2SA. Chapter 9 presents the CDSA technique that enhances acquisition robustness in low-SNR fading channels and large frequency-offset environments. Lastly, it discusses the potential applications of the CDSA to the practical inter-cell synchronous (i.e., cdma2000) and asynchronous (i.e., W-CDMA) IMT-2000 systems.

---

[7]These three CDMA systems correspond, respectively, to the 3GPP-FDD, 3GPP-TDD, and 3GPP-2 systems in Fig. 1.5, where 3GPP is an abbreviation for *3rd Generation Partnership Project*, FDD for *Frequency Division Duplex*, and TDD for *Time Division Duplex*.

I

SCRAMBLING CODE GENERATION AND ACQUISITION

Chapter 2

SEQUENCES AND SHIFT REGISTER GENERATORS

PN sequences are used as a means for scrambling and spectrum-spreading modulation in direct sequence systems and as a hopping pattern source in frequency hopping systems. PN sequences are generated using SRGs. PN sequences can be classified into linear sequences and nonlinear sequences depending on the generation method. In this chapter we exclusively deal with binary linear PN sequences as they are practical and widely used in the CDMA communication systems. The reader may refer to [4] for nonlinear sequences.

We use the term *scrambling/spreading code* to refer to the binary code generated by shift-register generator and the term *spreading waveform* to indicate the continuous-time waveform representing the spreading code. The ideal scrambling/spreading code would be an infinite sequence of equally likely random binary digits. The use of an infinite random sequence, however, requires infinitely long storage in both transmitter and receiver. Therefore, for practical use, the periodic *pseudorandom codes* (or *PN codes*) are always employed. In this book we use the term PN code in broad sense to indicate any periodic scrambling/spreading code with noise-like properties.

In order to understand the behaviors of PN generators completely, a strong mathematical basis, including the field concept, is necessary. However, we avoid mathematical discussions in this chapter as it is not the intent of the chapter. The reader may refer to the references [4, 69, 70, 71, 72] for the relevant mathematical descriptions.

In this chapter we briefly discuss the construction and the properties of some key PN sequences such as the maximal-length sequences, Gold sequences, and Kasami sequences, and then introduce the SRG theories that govern the properties and operation of SRGs.

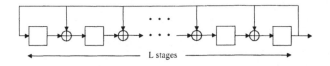

Figure 2.1.   General L-stage shift register with linear feedback.

1.   LINEAR SEQUENCES: M-, GOLD-, AND KASAMI-SEQUENCES

The linear sequences that are most frequently used in the CDMA communications are the maximal-length sequences, Gold sequences, and Kasami sequences. In this section, we review these three sequences in the capacity of sequence generation and the autocorrelation and cross-correlation properties.

1.1   M-SEQUENCES

Fig. 2.1 depicts an L-stage linear shift register generator whose maximum possible period is $N = 2^L - 1$. Shift register sequences having the maximum possible period are called *maximal-length sequences* (or *m-sequences*).

An m-sequence contains one more 1 than 0 in each period, that is, 2^{L-1} 1's and $2^{L-1} - 1$ 0's. It has the *shift-and-add property*, that is, a modulo-2 sum of an m-sequence and any phase shift of the same sequence yields the same m-sequence of different phase. If a window of width L is slid along the sequence for N shifts, every possible L-tuple except for the all-zero L-tuple appears exactly once.

A periodic m-sequence has the two-valued periodic autocorrelation function. The periodic autocorrelation function is defined, in terms of the bipolar sequence,

$$\theta_b(k) = \frac{1}{N} \sum_{n=0}^{N-1} a_n a_{n+k}, \qquad (2.1)$$

where $a_n = (-1)^{b_n}$ and $b_n \in \{0, 1\}$ is the element in the sequence. Ideally, a pseudo-random sequence has the autocorrelation values with the property that $\theta_b(0) = 1$ and $\theta_b(k) = 0$ for $1 \le k \le N - 1$. In the case of m-sequences, however, the periodic autocorrelation values are

$$\theta_b(k) = \begin{cases} 1, & k = lN \\ -\frac{1}{N}, & k \ne lN \end{cases} \qquad (2.2)$$

for an integer l.

For a long m-sequence, having a large value of N, the ratio of the off-peak values of the periodic autocorrelation function to the peak value, $\theta_b(k)/\theta_b(0) =$

$-1/N$, becomes very small. Therefore, m-sequences are nearly ideal in the aspect of autocorrelation function.

The autocorrelation properties of a maximal-length sequence are defined over a complete cycle of the sequence. That is, the two-valued autocorrelation can be guaranteed only when the summation is done over a period N, or equivalently, the integration is over a full period of the continuous-time code waveform $c(t)$. However, in practical spread-spectrum communications, long code synchronization often uses an estimate of the correlation between the received code and the receiver despreading code taken over a partial period so as to reduce the synchronization time. Consequently, the practical long code correlation estimate follows the partial autocorrelation properties of the code.

The *partial autocorrelation function* of the spreading waveform $c(t)$ is defined by

$$R_c(\tau, t, T_w) = \frac{1}{T_w} \int_t^{t+T_w} c(\lambda)c(\lambda + \tau)d\lambda \qquad (2.3)$$

where T_w and t are the duration of the correlation and the starting time of the correlation, respectively. The partial autocorrelation function is dependent on the duration and the starting time of the integration. We take $\tau = kT_c + \tau_\varepsilon$, $T_w = WT_c$, and assume that $t = k'T_c$ and $c(t) = \sum_{n=-\infty}^{\infty} a_n p(t - nT_c)$ for the pulse waveform $p(t)$ and the chip duration T_c. Then, we get the discrete-time expression

$$
\begin{aligned}
R_c(\tau_\varepsilon, k, k', W) &= \frac{1}{W} \sum_{m=k'}^{W+k'-1} \left\{ a_m a_{m+k}\left(1 - \frac{\tau_\varepsilon}{T_c}\right) + a_m a_{m+k+1}\frac{\tau_\varepsilon}{T_c} \right\} \\
&= \left(1 - \frac{\tau_\varepsilon}{T_c}\right)\theta_b(k, k', W) + \frac{\tau_\varepsilon}{T_c}\theta_b(k+1, k', W) \quad (2.4)
\end{aligned}
$$

for

$$\theta_b(k, k', W) = \frac{1}{W} \sum_{m=k'}^{k'+W-1} a_m a_{m+k} \qquad (2.5)$$

and $|\tau_\varepsilon| \leq T_c$. The partial autocorrelation function is not as well behaved as the full-period autocorrelation function. It is not two-valued and its variation, which is a function of the window size and the window placement, can cause serious problems unless special cares are taken in the system design stage. Since the modulo-2 sum of an m-sequence and any phase shift of the same sequence yields another phase of the same m-sequence, the discrete partial autocorrelation function takes the expression

$$\theta_b(k, k', W) = \frac{1}{W} \sum_{i=0}^{W-1} a_{i+q+k'} \qquad (2.6)$$

for an integer q, which itself is a function of k.

The mean and the variance of $\theta_b(k, k', W)$ are useful quantities for the spread spectrum system designer. Due to the above properties of the m-sequence, the mean and the variance take the expressions

$$\overline{\theta_b(k, k', W)} = \frac{1}{N} \sum_{k'=0}^{N-1} \frac{1}{W} \sum_{i=0}^{W-1} a_{i+q+k'} = -\frac{1}{N},$$

$$var[\theta_b(k, k', W)] = \frac{1}{W} \left(1 - \frac{W-1}{N} \right) - \frac{1}{N^2}. \tag{2.7}$$

We can observe that for $W = N$ the variance goes to zero as expected. These relations can be used in determining the approximate threshold values for checking the coincidence of the two sequences.

1.2   GOLD SEQUENCES

In some applications, the cross-correlation properties of PN sequences are as important as the autocorrelation properties. For example, in CDMA, each user is assigned a particular PN sequence and every user transmits signal simultaneously using the same frequency band, so the interference of a user employing a different spreading code to the reference user is dictated by the cross-correlation between the two relevant spreading codes.

It is well known that the periodic cross-correlation function between any pair of m-sequences of the same period has relatively large peaks. As the number of m-sequences increases rapidly with the number of stage L, the probability of large cross-correlation peaks becomes high for long m-sequences. However, such high values of cross-correlation are not desirable in the CDMA communications. Thus new PN sequences with better periodic cross-correlation properties were derived from m-sequences by Gold[73].

We consider an m-sequence that is represented by a binary vector $\mathbf{b}$ of length N, and a second sequence $\tilde{\mathbf{b}}$ obtained by decimating every qth symbol of $\mathbf{b}$, i.e., $\tilde{\mathbf{b}} = \mathbf{b}[q]$. When the decimation of an m-sequence does yield another m-sequence, the decimation is called a *proper decimation*. It can be proven that $\tilde{\mathbf{b}} = \mathbf{b}[q]$ has period N if and only if $gcd(N, q) = 1$, where gcd denotes the greatest common divisor. Thus any pair of m-sequences having the same period N can be related by $\tilde{\mathbf{b}} = \mathbf{b}[q]$ for some q.

Gold proved[73] that certain pairs of m-sequences of length N exhibit a three-valued cross-correlation function with the values $\{-1, -t(L), t(L) - 2\}$ for

$$t(L) = \begin{cases} 2^{(L+1)/2} + 1, & \text{for odd } L \\ 2^{(L+2)/2} + 1, & \text{for even } L \end{cases} \tag{2.8}$$

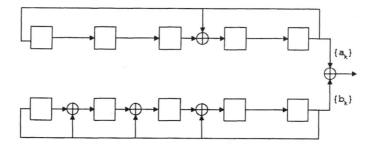

Figure 2.2.   Generation of Gold sequences of length 31.

and the code period $N = 2^L - 1$. Two m-sequences of length N that yield a periodic cross-correlation function taking the possible values $\{-1, -t(L), t(L) - 2\}$ are called *preferred pairs* or *preferred sequences*.

From a pair of preferred sequences, say $\{a_k\}$ and $\{b_k\}$, we construct a set of sequences of length N by taking the modulo-2 sum of $\{a_k\}$ with the N cyclic shifted versions of $\{b_k\}$ or vice versa. Then, we obtain N new periodic sequences with the period $N = 2^L - 1$. We may also include the original sequences $\{a_k\}$ and $\{b_k\}$ to get a set of $N + 2$ sequences. The $N + 2$ sequences constructed in this manner are called *Gold sequences*.

Fig. 2.2 shows a shift register circuit generating two m-sequences and the corresponding Gold sequences for $L = 5$. In this case, there are generated 33 different sequences corresponding to the 33 relative phases of the two m-sequences. With the exception of the sequences $\{a_k\}$ and $\{b_k\}$ themselves, the set of Gold sequences does not include any m-sequences of length N. Hence, their autocorrelation functions are not two-valued. Similar to the case of the cross-correlation function, the off-peak autocorrelation function for a Gold sequence takes the values in the set $\{-1, -t(L), t(L) - 2\}$. Consequently the off-peak values of the autocorrelation function are upper bounded by $t(L)$.

In short, Gold sequences are a family of codes with well-behaved cross-correlation properties that are constructed through a modulo-2 addition of specific relative phases of a preferred pair of m-sequences. The period of any code in the family is N, which is the same as the period of the m-sequences.

1.3   KASAMI SEQUENCES

A procedure similar to that used for generating Gold sequences can generate a smaller set of binary sequences of period $N = 2^L - 1$, for an even L. For a given m-sequence $\{a_k\}$ we derive a binary sequence $\{b_k\}$ by taking every $(2^{L/2} + 1)$th bit of $\{a_k\}$, or by decimating $\{a_k\}$ by $2^{L/2} + 1$. Then the resulting sequence $\{b_k\}$ becomes periodic with the period $2^{L/2} - 1$. We take $N = 2^L - 1$

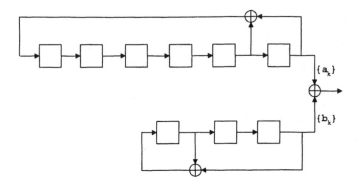

Figure 2.3.   Generation of a small set of Kasami sequences of length 63.

consecutive bits of the sequences $\{a_k\}$ and $\{b_k\}$ to form a new set of sequences by adding, in modulo-2, the bits of $\{a_k\}$ to the bits of $\{b_k\}$ and its $2^{L/2} - 2$ cyclic shifts. If we include $\{a_k\}$ in the set, we obtain a set of $2^{L/2}$ binary sequences of length $N = 2^L - 1$. These are called a *small set* of *Kasami sequences*[74]. Fig. 2.3 depicts a shift register circuit generating a small set of Kasami sequences of length 63.

The autocorrelation function as well as the cross-correlation function of the Kasami sequences take a value in the set $\{-1, -(2^{L/2}+1), 2^{L/2}-1\}$. Hence, the maximum cross-correlation value for any pair of Kasami sequences is

$$\phi_{max} = 2^{L/2} + 1. \qquad (2.9)$$

Welch developed[75] the lower bound of the cross-correlation between any pair of the binary sequences of period N in a set of M sequences

$$\phi_{max} \geq N\sqrt{\frac{M-1}{MN-1}}. \qquad (2.10)$$

The maximum cross-correlation value for the small set of Kasami sequences coincides with this Welch lower bound, and thus it is optimal.

While optimal in cross-correlation properties, a small set of Kasami sequences contains a relatively small number of sequences. So it is difficult to apply such Kasami sequences to the cases when the number of potential communication users is very large and the requested sequence period is short. In order to produce a large number of short-period sequences with good cross-correlation properties, new generation methods have been devised that combine three sequences rather than two. For an m-sequence $\{a_k\}$ with even L, we derive a second sequence $\{b_k\}$ by decimating $\{a_k\}$ by $(2^{(L+2)/2}+1)$, and a third sequence $\{c_k\}$ of a shorter period by decimating $\{a_k\}$ by $(2^{L/2}+1)$. Then

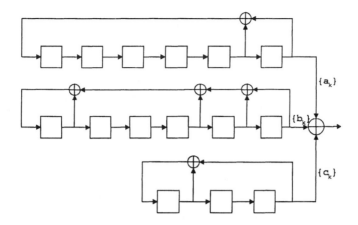

Figure 2.4.   Generation of a large set of Kasami sequences of length 63.

we add, in modulo-2, the three sequences with specific relative phases, thereby obtaining a *large set* of Kasami sequences of period $2^L - 1$. Fig. 2.4 illustrates a sequence generator for a large set of Kasami sequences of period 63 (i.e. $L = 6$). In this case, there are $2^3(2^6 + 1) = 520$ different sequences contained in the large set of Kasami sequences.

The autocorrelation and cross-correlation functions of the sequences both take a value in the set $\{-1, -(2^{(L+2)/2} + 1), 2^{(L+2)/2} - 1, -(2^{L/2} + 1), 2^{L/2} - 1\}$. So the maximum correlation value of the large set of Kasami sequences remains the same as that of the Gold sequences even with the increased number of elements[76].

The Welch bound in (2.10) provides a very loose bound in case $M \geq N$. So, for a large set of Kasami sequences, the lower bound of the cross-correlation is better described by the Sidelnikov bound

$$\phi_{max} > \sqrt{(2N - 2)}. \qquad (2.11)$$

In view of this Sidelnikov bound, the Gold sequence set introduced in the previous section is optimal for odd L, even if not for even L[76].

2.   SHIFT REGISTER GENERATOR THEORY

In scrambling/spreading techniques, there are two core elements that govern the scrambling/spreading behaviors. They are the sequence $\{s_k\}$ to be added to the input data for scrambling/spreading and the SRG which generates the sequence itself. In this section, we introduce the concepts of the sequence space and the SRG space as well as the related theory as a means to rigorously describe the behaviors of sequences and SRGs[69, 77, 78].

2.1   SEQUENCE SPACE.

For a binary-coefficient polynomial $\Psi(x) = \sum_{i=0}^{L} \psi_i x^i$, $\psi_0 = \psi_L = 1$, we define by the *sequence space* $V[\Psi(x)]$ the set[1]

$$V[\Psi(x)] \equiv \left\{ \{s_k, \ k = 0, 1, \cdots \} : \sum_{i=0}^{L} \psi_i s_{k+i} = 0 \text{ for all } k = 0, 1, \cdots \right\}$$
(2.12)

with the sequence addition $\{s_k\} + \{\hat{s}_k\} \equiv \{s_k + \hat{s}_k\}$ and the scalar multiplication $a\{s_k\} \equiv \{as_k\}$ for sequences $\{s_k\}$, $\{\hat{s}_k\}$ in $V[\Psi(x)]$ and a scalar a in the Galois field $GF(2)$.[2] The polynomial $\Psi(x)$ characterizing a sequence space is called the *characteristic polynomial of the sequence space* $V[\Psi(x)]$. Then, a sequence space $V[\Psi(x)]$ becomes a vector space over $GF(2)$.

According to (2.12), for a sequence $\{s_k\}$ in the sequence space $V[\Psi(x)]$ we have the recurrence relation

$$s_k = \sum_{i=1}^{L} \psi_i s_{k+i}, \quad k = 0, 1, \cdots,$$

$$s_{k+L} = \sum_{i=0}^{L-1} \psi_i s_{k+i}, \quad k = 0, 1, \cdots$$
(2.13)

since ψ_0 and ψ_L are both 1. This means that a sequence $\{s_k\}$ in $V[\Psi(x)]$ is completely determined once its L consecutive elements are known. So, if we indicate its first L elements by the *initial vector* $\mathbf{s} \equiv [s_0 \ s_1 \ \cdots _{L-1}]^t$ of the sequence $\{s_k\}$, then all the sequence in $V[\Psi(x)]$ can be determined by inserting 2^L initial vectors to (2.13).

For a sequence space $V[\Psi(x)]$ whose characteristic polynomial $\Psi(x)$ is of degree L, we define by the ith *elementary sequence* $E^i_{\Psi(x)}$ the sequence $\{s_k\}$ in $V[\Psi(x)]$ whose initial vector is $\mathbf{e}_i$, $i = 0, 1, \cdots, L-1$, where $\mathbf{e}_i$ denotes the *basis vector* whose ith element is 1 and the others are all zero. We call the L-vector consisting of the L elementary sequences, $\mathbf{E}_{\Psi(x)} \equiv [E^0_{\Psi(x)} \ E^1_{\Psi(x)} \ \cdots E^{L-1}_{\Psi(x)}]^t$, the *elementary sequence vector*. Then, a sequence in the sequence space $V[\Psi(x)]$ can be expressed in term of the elementary sequence vector as follows [77]:

$$\{s_k\} = \mathbf{s}^t \cdot \mathbf{E}_{\Psi(x)}.$$
(2.14)

This equation means that the L elementary sequences $E^i_{\Psi(x)}, i = 0, 1, \cdots,$ $L-1$, form a basis of the sequence space $V[\Psi(x)]$, and each sequence $\{s_k\}$

---

[1]Addition, or the "+" operation refers to the modulo-2 addition.
[2]$GF(2)$ has two elements 0 and 1, and two operations which are modulo-2 addition and modulo-2 multiplication.

in $V[\Psi(x)]$ is represented by (2.14). We name this the *elementary basis* for the sequence space $V[\Psi(x)]$, and we call the vector **s** the *initial vector for the elementary basis*. The dimension of a sequence space $V[\Psi(x)]$ is identical to the degree of the characteristic polynomial $\Psi(x)$.

For a sequence $\{s_k\}$, the sequence $\{s_{k+m}\}$ is called the *m-delayed sequence*. An m-delayed sequence can be expressed in terms of the elementary sequence vector $\mathbf{E}_{\Psi(x)}$ as

$$\{s_{k+m}\} = \mathbf{s}^t \cdot \mathbf{A}_{\Psi(x)}^m \cdot \mathbf{E}_{\Psi(x)} \tag{2.15}$$

on the elementary basis [77], where $\mathbf{A}_{\Psi(x)}$ is the *companion matrix* for the characteristic polynomial $\Psi(x)$ defined by

$$\mathbf{A}_{\Psi(x)} \equiv \begin{bmatrix} 0 & 0 & \cdots & 0 & \psi_0 \\ 1 & 0 & \cdots & 0 & \psi_1 \\ 0 & 1 & \cdots & 0 & \psi_2 \\ \vdots & \vdots & \ddots & \vdots & \vdots \\ 0 & 0 & \cdots & 1 & \psi_{L-1} \end{bmatrix}. \tag{2.16}$$

As a consequence, we can express the kth element s_k, $k = 0, 1, \cdots$, of sequence $\{s_k\}$ as

$$s_k = \mathbf{s}^t \cdot \mathbf{A}_{\Psi(x)}^k \cdot \mathbf{e}_0. \tag{2.17}$$

For a sequence space $V[\Psi(x)]$ of dimension L, we define the ith *primary sequence* $P_{\Psi(x)}^i$, $i = 0, 1, \cdots, L-1$, to be the $(i+1)$-delayed sequence of the 0th elementary sequence $E_{\Psi(x)}^0$. We call the L-vector consisting of the L primary sequences, $\mathbf{P}_{\Psi(x)} \equiv [P_{\Psi(x)}^0 \ P_{\Psi(x)}^1 \ \cdots \ P_{\Psi(x)}^{L-1}]^t$, the *primary sequence vector*. Then the relations between the elementary and the primary sequence vectors become

$$\mathbf{P}_{\Psi(x)} = \mathbf{B}_{\Psi(x)} \cdot \mathbf{E}_{\Psi(x)},$$
$$\mathbf{E}_{\Psi(x)} = \mathbf{C}_{\Psi(x)} \cdot \mathbf{P}_{\Psi(x)} \tag{2.18}$$

where

$$\mathbf{B}_{\Psi(x)} \equiv \mathbf{C}_{\Psi(x)}^{-1} \equiv \begin{bmatrix} \mathbf{e}_0^t \cdot \mathbf{A}_{\Psi(x)} \\ \mathbf{e}_0^t \cdot \mathbf{A}_{\Psi(x)}^2 \\ \vdots \\ \mathbf{e}_0^t \cdot \mathbf{A}_{\Psi(x)}^L \end{bmatrix}. \tag{2.19}$$

Therefore the L primary sequences $P_{\Psi(x)}^i$, $i = 0, 1, \cdots, L-1$, form a basis of the sequence space $V[\Psi(x)]$ of dimension L, and a sequence $\{s_k\}$ in the sequence space $V[\Psi(x)]$ can be represented by

$$\{s_k\} = \mathbf{p}^t \cdot \mathbf{P}_{\Psi(x)}, \tag{2.20}$$

where

$$\mathbf{p} = \mathbf{C}_{\Psi(x)} \cdot \mathbf{s}. \tag{2.21}$$

We name the basis formed by $P_{\Psi(x)}^i, i = 0, 1, \cdots, L-1$, the *primary basis* for the sequence space $V[\Psi(x)]$, and we call the vector $\mathbf{p}$ the *initial vector for the primary basis*.

An m-delayed sequence $\{s_{k+m}\}$ of a sequence $\{s_k\}$ in $V[\Psi(x)]$ can be equally expressed in terms of primary sequence vector $\mathbf{P}_{\Psi(x)}$ as

$$\{s_{k+m}\} = \mathbf{p}^t \cdot (\mathbf{A}_{\Psi(x)}^t)^m \cdot \mathbf{P}_{\Psi(x)}. \tag{2.22}$$

on the primary basis [77], and, consequently, the kth element $s_k, k = 0, 1, \cdots,$ of the sequence $\{s_k\}$ takes the expression

$$s_k = \mathbf{p}^t \cdot (\mathbf{A}_{\Psi(x)}^t)^t \cdot \mathbf{e}_{L-1}. \tag{2.23}$$

2.2   SRG SPACES

For an SRG of length L, we define the kth *state vector* $\mathbf{d}_k$ to be an L-vector representing the state of the shift registers in the SRG at time k, that is,

$$\mathbf{d}_k = [d_{0,k} \ d_{1,k} \ \cdots \ d_{L-1,k}]^t \tag{2.24}$$

where the state $d_{i,k}, i = 0, 1, \cdots, L-1$, denotes the value of the ith shift register in the SRG at time k. In particular, we call the 0th state vector $\mathbf{d}_0$ the *initial state vector*. In addition, we define by the *state transition matrix* $\mathbf{T}$ an $L \times L$ matrix representing the relation between the state vectors $\mathbf{d}_k$ and $\mathbf{d}_{k-1}$, or more specifically,[3]

$$\mathbf{d}_k = \mathbf{T} \cdot \mathbf{d}_{k-1}. \tag{2.25}$$

Then the configuration of an SRG is *uniquely determined* by its state transition matrix $\mathbf{T}$, and the sequence generated by each shift register is *uniquely determined* when the initial state vector $\mathbf{d}_0$ is additionally furnished.

For the SRG of length L, we define by the ith *SRG sequence* $D_i, i = 0, 1, \cdots, L-1$, the sequence generated by the ith shift register, that is, $D_i \equiv \{d_{i,k}\}$, and define by the *SRG sequence vector* $\mathbf{D}$ the L-vector

$$\mathbf{D} \equiv [D_o \ D_1 \ \cdots \ D_{L-1}]^t. \tag{2.26}$$

Then, the SRG sequence vector $\mathbf{D}$ is uniquely determined once its state transition matrix $\mathbf{T}$ and the initial state vector $\mathbf{d}_0$ are determined. A sequence space $V[\Psi(x)]$ is a *common space*[4] for the L SRG sequences $D_i, i = 0, 1, \cdots, L-1$,

---

[3]The state transition matrix $\mathbf{T}$ for an SRG is always nonsingular unless the SRG is a pathological one.

[4]For N sequences $\{s_k^i\}$, $i = 0, 1, \cdots N-1$, a *common space* refers to a sequence space containing the N sequences[69].

generated by the SRG of length L with the state transition matrix $\mathbf{T}$ and the initial state vector $\mathbf{d}_0$, if and only if the characteristic polynomial $\Psi(x)$ meets the condition[77]

$$\Psi(\mathbf{T}) \cdot \mathbf{d}_0 = 0. \tag{2.27}$$

If $\Psi_{\mathbf{T},\mathbf{d}_0}(x)$ denotes the lowest-degree polynomial that meets (2.27), then the sequence space $V[\Psi_{\mathbf{T},\mathbf{d}_0}(x)]$ is the *minimal space* [5] for the SRG sequence $D_i, i = 0, 1, \cdots, L - 1$, and the SRG sequence vector $\mathbf{D}$ takes the expression [77]

$$\mathbf{D} = [\mathbf{d}_0 \ \mathbf{T} \cdot \mathbf{d}_0 \ \cdots \ \mathbf{T}^{\hat{L}-1} \cdot \mathbf{d}_0] \cdot \mathbf{E}_{\Psi_{\mathbf{T},\mathbf{d}_0}}(x) \tag{2.28}$$

on the elementary basis of the minimal space, where $\hat{L}$ denotes the dimension of $V[\Psi_{\mathbf{T},\mathbf{d}_0}(x)]$.

We define the *SRG space* $V[\mathbf{T}, \mathbf{d}_0]$ to be the vector space formed by D_i's. More specifically,

$$V[\mathbf{T}, \mathbf{d}_0] \equiv \left\{ \sum_{i=0}^{L-1} a_i D_i : a_i \in GF(2), i = 0, 1, \cdots, L - 1 \right\}. \tag{2.29}$$

Then the SRG space $V[\mathbf{T}, \mathbf{d}_0]$ is identical to the minimal space $V[\Psi_{\mathbf{T},\mathbf{d}_0}(x)]$ for the SRG sequences $D_i, i = 0, 1, \cdots, L - 1$ [77].

For an SRG with the state transition matrix $\mathbf{T}$, we define by an *SRG maximal space* $V[\mathbf{T}]$ the largest-dimensional SRG space of all SRG spaces $V[\mathbf{T}, \mathbf{d}_0]$'s obtained by varying the initial state vectors $\mathbf{d}_0$'s, and define by a *maximal initial state vector* $\mathbf{d}_0$ an initial state vector that makes the SRG space $V[\mathbf{T}, \mathbf{d}_0]$ identical to the SRG maximal space $V[\mathbf{T}]$. Then the SRG maximal space $V[\mathbf{T}]$ is unique, and is identical to the sequence space $V[\Psi_{\mathbf{T}}(x)]$ for the minimal polynomial $\Psi_{\mathbf{T}}(x)$ of $\mathbf{T}$.[6] Thus the SRG space $V[\mathbf{T}, \mathbf{d}_0]$ for an SRG with the state transition matrix $\mathbf{T}$ and an initial state vector $\mathbf{d}_0$ is a subspace of the SRG maximal space $V[\mathbf{T}]$. If $\hat{L}$ denotes the dimension of the SRG maximal space $V[\mathbf{T}]$, an initial state vector $\mathbf{d}_0$ is maximal initial state vector, if and only if the *discrimination matrix* $\triangle_{\mathbf{T},\mathbf{d}_0}$ defined by

$$\triangle_{\mathbf{T},\mathbf{d}_0} \equiv [\mathbf{d}_0 \ \mathbf{T} \cdot \mathbf{d}_0 \ \cdots \ \mathbf{T}^{\hat{L}-1} \cdot \mathbf{d}_0] \tag{2.30}$$

is of rank $\hat{L}$.

2.3 SSRG AND MSRG

For a sequence space $V[\Psi(x)]$, we define a *basic SRG* (BSRG) to be an SRG of the smallest length whose SRG maximal space is identical to the sequence

---

[5]For N sequences $\{s_k^i\}$, $i = 0, 1, \cdots N - 1$, a *minimal space* refers to a smallest-dimension common space for the N sequences[69].
[6]The minimal polynomial $\Psi_{\mathbf{T}}(x)$ of matrix $\mathbf{T}$ is the lowest-degree polynomial that makes $\Psi_{\mathbf{T}}(\mathbf{T}) = 0$.

space $V[\Psi(x)]$. That is, a BSRG for a sequence space $V[\Psi(x)]$ refers to a smallest-length SRG which can generate the sequence space $V[\Psi(x)]$. Then, the length of a BSRG for a sequence space is the same as the dimension of the sequence space. A matrix $\mathbf{T}$ is the state transition matrix of a BSRG for the sequence space $V[\Psi(x)]$, if and only if it is similar to the companion matrix $\mathbf{A}_{\Psi(x)}$.[7] An initial state vector $\mathbf{d}_0$ is a maximal initial state vector of the BSRG for a sequence space $V[\Psi(x)]$, if and only if the corresponding discrimination matrix $\triangle_{\mathbf{T},\mathbf{d}_0}$ is nonsingular for the state transition matrix $\mathbf{T}$ taken similar to $\mathbf{A}_{\Psi(x)}$[77].[8]

In fact, for a nonsingular matrix $\mathbf{Q}$, an SRG with the state transition matrix

$$\mathbf{T} = \mathbf{Q} \cdot \mathbf{A}_{\Psi(x)} \cdot \mathbf{Q}^{-1} \qquad (2.31a)$$

is a BSRG for the sequence space $V[\Psi(x)]$; the initial state vector of the form

$$\mathbf{d}_0 = \mathbf{Q} \cdot \mathbf{e}_0 \qquad (2.31b)$$

is a maximal initial state vector for this BSRG; and the BSRG sequence vector $\mathbf{D}$ for this maximal initial state vector becomes

$$\mathbf{D} = \mathbf{Q} \cdot \mathbf{E}_{\Psi(x)}. \qquad (2.31c)$$

Note that the nonsingular matrix $\mathbf{Q}$ itself now becomes the discrimination matrix $\triangle_{\mathbf{T},\mathbf{d}_0}$ for those $\mathbf{T}$ and $\mathbf{d}_0$, and therefore no separate singularity test is necessary on it.

The *simple SRG* (SSRG) and the *modular SRG* (MSRG) [79] are two simple types of SRG's which have the state transition matrices $\mathbf{A}_{\Psi(x)}^t$ and $\mathbf{A}_{\Psi(x)}$, respectively. Since $\mathbf{A}_{\Psi(x)}^t = \mathbf{B}_{\Psi(x)} \cdot \mathbf{A}_{\Psi(x)} \cdot \mathbf{C}_{\Psi(x)}$ for the transformation matrices $\mathbf{B}_{\Psi(x)}$ and $\mathbf{C}_{\Psi(x)} \equiv \mathbf{B}_{\Psi(x)}^{-1}$ in (2.19), $\mathbf{A}_{\Psi(x)}^t$ is similar to $\mathbf{A}_{\Psi(x)}$, and hence the SSRG and the MSRG are both BSRG's. As the companion matrix $\mathbf{A}_{\Psi(x)}$ is uniquely determined for a given characteristic polynomial $\Psi(x)$, so are the SSRG and the MSRG uniquely determined for a given sequence space $V[\Psi(x)]$.

Conventionally, an SSRG is characterized by the so-called *characteristic polynomial* $C(x) = \sum_{i=0}^{L} c_i x^i$, and an MSRG by the so-called *generating polynomial* $G(x) = \sum_{i=0}^{L} g_i x^i$. The configurations of a typical SSRG circuit and a typical MSRG circuit are shown in Fig. 2.5.

The SSRG and MSRG sequences can be represented on the elementary and primary bases, as follows: For the SSRG of the sequence space $V[\Psi(x)]$ of

---

[7] A matrix $\mathbf{M}$ is said *similar* to a matrix $\hat{\mathbf{M}}$ if there exists a nonsingular matrix $\mathbf{Q}$ such that $\mathbf{M} = \mathbf{Q} \cdot \hat{\mathbf{M}} \cdot \mathbf{Q}^{-1}$

[8] Note that in the case of BSRG, $\hat{L}$ in (2.30), which is the dimension of the BSRG maximal space, is identical to the length of the BSRG due to the above properties. Therefore, the discrimination matrix $\triangle_{\mathbf{T},\mathbf{d}_0}$ becomes a square matrix

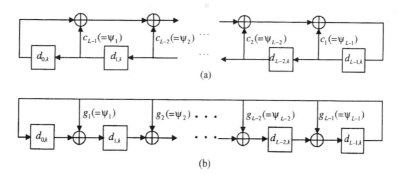

Figure 2.5.   Configurations of (a) SSRG and (b) MSRG.

dimension L, the ith SSRG sequence $D_{S,i}, i = 0, 1, \cdots, L - 1$, for an initial state vector $\mathbf{d}_0$, has the expression

$$D_{S,i} = \mathbf{d}_0^t \cdot \mathbf{A}_{\Psi(x)}^i \cdot \mathbf{E}_{\Psi(x)} \tag{2.32}$$

on the elementary basis of the sequence space $V[\Psi(x)]$. For the MSRG of the sequence space $V[\Psi(x)]$ of dimension L, the ith MSRG sequence $D_{M,i}, i = 0, 1, \cdots, L - 1$, for an initial state vector $\mathbf{d}_0$, has the expression

$$D_{M,i} = \mathbf{d}_0^t \cdot \left(\sum_{j=i+1}^{L} \psi_j (\mathbf{A}_{\Psi(x)}^t)^{j-(i+1)} \right) \cdot \mathbf{P}_{\Psi(x)} \tag{2.33}$$

on the primary basis of the sequence space $V[\Psi(x)]$.

Inserting $i = 0$ and $i = L - 1$, respectively, to (2.32) and (2.33), we obtain the relations

$$D_{S,0} = \mathbf{d}_0^t \cdot \mathbf{E}_{\Psi(x)} \tag{2.34a}$$

$$D_{M,L-1} = \mathbf{d}_0^t \cdot \mathbf{P}_{\Psi(x)}. \tag{2.34b}$$

They are called the *terminating sequences* respectively of the SSRG and the MSRG.

2.4   PRIMITIVE SPACE AND M-SEQUENCE GENERATOR

An irreducible polynomial $\Psi_p(x)$ of degree L is called a *primitive polynomial*, if $\Psi_p(x)$ divides $x^i + 1$ for $i = 2^L - 1$ but not for $i = 0, 1, \cdots, 2^L - 2$. We define *primitive space* $V[\Psi_p(x)]$ to be the sequence space whose characteristic polynomial $\Psi_p(x)$ is a primitive polynomial. Then, a sequence in a primitive space of dimension L is an *m-sequence* of period $2^L - 1$.

An *m-sequence generator* that generates *m*-sequences in the primitive space $V[\Psi_p(x)]$ of dimension L has length L or more. So, for an efficient realization of *m*-sequence generators we need to use a length-L *m*-sequence generator, which we call an *m-sequence minimal generator*. An m-sequence minimal generator has the following properties [77]:

(a) The state transition matrix $\mathbf{T}$ of an *m*-sequence minimal generator for the primitive space $V[\Psi_p(x)]$ is similar to the companion matrix $\mathbf{A}_{\Psi_p(x)}$ for the characteristic polynomial $\Psi_p(x)$.

(b) An arbitrary nonzero initial state vector $\mathbf{d}_0$ is a maximal initial state vector for an *m*-sequence minimal generator.

(c) Each SRG sequence $D_i, i = 0, 1, \cdots, L - 1$, of an *m*-sequence minimal generator becomes an *m*-sequence in the primitive space $V[\Psi_p(x)]$.

We define *generating vector* $\mathbf{h}$ to be an L-vector representing the relation between the *m*-sequence $\{s_k\}$ and the SRG sequence vector $\mathbf{D}$, or more specifically,

$$\{s_k\} = \mathbf{h}^t \cdot \mathbf{D}. \tag{2.35}$$

Then the generating vector $\mathbf{h}$ generating the *m*-sequence $\{s_k\} = \mathbf{s}^t \cdot \mathbf{E}_{\Psi_p(x)}$ is [78]

$$\mathbf{h} = (\triangle_{\mathbf{T},h_0}^t)^{-1} \cdot \mathbf{s}. \tag{2.36}$$

As a special case, an *m*-sequence minimal generator can be realized based on the SSRG or the MSRG as follows [77]:

(a) An *m*-sequence with the expression $\{s_k\} = \mathbf{s}^t \cdot \mathbf{E}_{\Psi_p(x)}$ on the elementary basis can be generated by the SSRG with the initial state vector $\mathbf{d}_0 = \mathbf{s}$ and the generating vector $\mathbf{h} = \mathbf{e}_0$.

(b) An *m*-sequence with the expression $\{s_k\} = \mathbf{p}^t \cdot \mathbf{P}_{\Psi_p(x)}$ on the primary basis can be generated by the MSRG with the initial state vector $\mathbf{d}_0 = \mathbf{p}$ and the generating vector $\mathbf{h} = \mathbf{e}_{L-1}$.

Chapter 3

FUNDAMENTAL CODE ACQUISITION TECHNIQUES

In the DS/CDMA communications, one of the primary functions of the receiver is to despread the received PN code. This is accomplished by generating a local replica of the PN code in the receiver and then synchronizing it to the one superimposed on the received waveform. In general, this synchronization process is accomplished in two steps - - *code acquisition*, which is a coarse alignment process bringing the two PN sequences within one chip interval, and *code tracking*, which is a fine tuning and synchronization-maintaining process [4].

In this chapter, we focus on the fundamental code acquisition techniques. To begin with, we compare the correlator-based integration method with the matched-filter-based one, and compare the serial search method with the parallel one. Next, we briefly discuss the Z- and the expanding-window search methods, and the long and the short code acquisition issues.

1.   CORRELATOR VS. MATCHED FILTER

As the first step of the code acquisition process, the received signal is either correlated by a locally generated PN signal or filtered by a matched filter. The former is called the *active correlator* technique, and the latter the *passive matched filter* (MF) technique.

In the active correlator system, the received signal $r(t)$, which is composed of the PN signal $s(t)$ and noise $n(t)$, is first multiplied by the local PN code reference, and subsequently bandpass-filtered (BPF) and square-law envelope-detected, thereby dropping off the unknown modulated information and unknown carrier phase. The output is then integrated for the duration of τ_d seconds, sampled at interval τ_d, and finally used in making acquisition decision through comparison with threshold. This process is depicted in Fig. 3.1. In

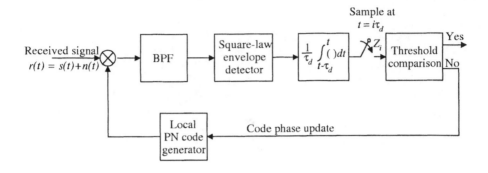

Figure 3.1.   Active correlator acquisition system.

this *multiply-and-integrate* type of correlator/detector structure, the local PN generator runs continuously and a completely *new* set of $M_c \equiv \tau_d/T_c$ chips of the received signal is used for each successive threshold test for the chip interval T_c. This imposes a basic limitation on the search speed as the local PN reference phase can be updated only at the τ_d-second interval. Thus, if the search is conducted in $1/N_s$-chip increments, the search rate R becomes $1/N_s\tau_d$ chip positions per second.

The search rate can be significantly increased by replacing multiply-and-integrate operation with a passive correlator device such as matched filter. Matched filter can be implemented either in continuous-time or discrete-time mode using some state-of-the-art technologies such as charge-coupled devices, surface acoustic wave convolvers, and discrete-time correlators. Fig. 3.2 shows two different types of matched filter implementations in analog form - - one bandpass filter based and the other lowpass filter based. The two implementations are equivalent and produce the identical envelope output $A(t)$ for the identical input $r(t)$. Modern matched-filter synchronization systems are most often implemented digitally. Fig. 3.3 illustrates a digital implementation of the matched filter. In this figure, sampling rate of twice the spreading code chip rate is assumed (i.e., $N_s = 2$), and the coefficients, $\{c_i\}$, $i = 0, \cdots, K$, represent the spreading code sequence.

We consider the synchronizer in Fig. 3.2. We assume that the envelope of the matched filter output is compared with a threshold after each sampling interval T_s. If the sampling interval T_s equals T_c/N_s, the search is conducted at the rate of N_s/T_c sample positions (or, $1/T_c$ chip positions) per second after some initial latency time. However, the matched filter scheme requires more computations than active correlator does. If the sampling rate is N_s/T_c and the correlation length of the matched filter spans M_c chips, the matched filter requires M_cN_s multiplications in every T_c/N_s interval. On the contrary, if we

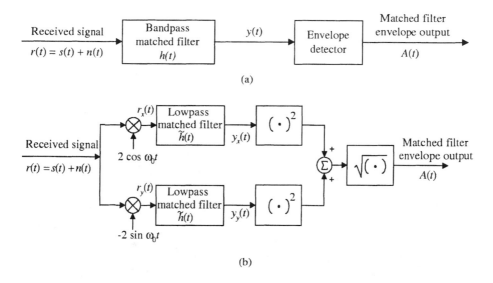

Figure 3.2.   Implementation of matched-filter in analog form : (a) bandpass filter based, (b) lowpass filter based.

use an active correlator, the same number of multiplications are required in every $M_c T_c$ interval. [1]

2.   SERIAL VS. PARALLEL SEARCH

A widely used technique for initial synchronization is *serial search*: All potential code phases and frequencies are searched serially until the correct phase and frequency are identified. Correctness of phase/frequency is determined by attempting to despread the received signal: If the estimated code phase and frequency are both correct, despreading will be properly done and thus a high energy output will be sensed. Otherwise, despreading will not be done properly and the resulting energy will be low. Fig. 3.4 depicts a realization of the maximum-likelihood serial search technique: First, the correlation between the local PN code and the received PN sequence is calculated and stored. Then, the local PN code phase shifts to the next code phase and the same operation is conducted. This process is repeated until all the q cell uncertainty region is examined, or, equivalently, for all $t = i\tau_d$, $i = 1, 2, \cdots, q$. Finally, the code phase that yields the maximum correlation value is selected as the correct one.

A natural extension of this serial technique will be a *parallel search* in which two or more paths search the code phase simultaneously, expecting that by

---

[1] In this complexity calculation, we assumed the case when no chip pulse matched filter is employed.

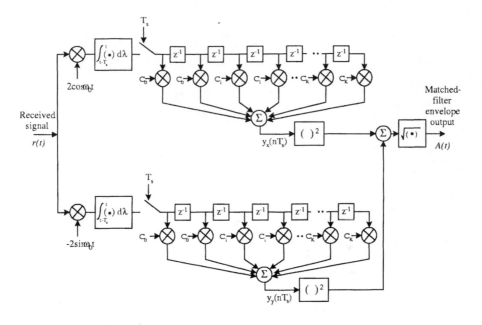

Figure 3.3.   Digital implementation of non-coherent matched filter [72].

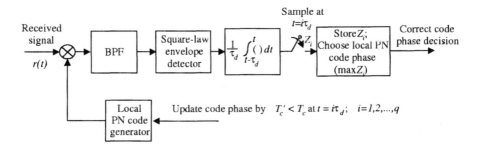

Figure 3.4.   Serial search realization of the maximum-likelihood search technique.

increasing the hardware complexity the acquisition time would decrease in proportion to the number of paths used. Fig. 3.5 shows a parallel search realization in which the entire q cell uncertainty region is subdivided into $N_P (\geq 2)$ identical components, with each responsible for q/N_P code phases. The code phase that yields the maximum among all those despread outputs is determined to be the right code phase in the final stage.

Apparently, the serial search can be implemented with low complexity but at the expense of long acquisition time. On the contrary, the parallel search has more complex hardware, but can achieve a faster acquisition. A mid-

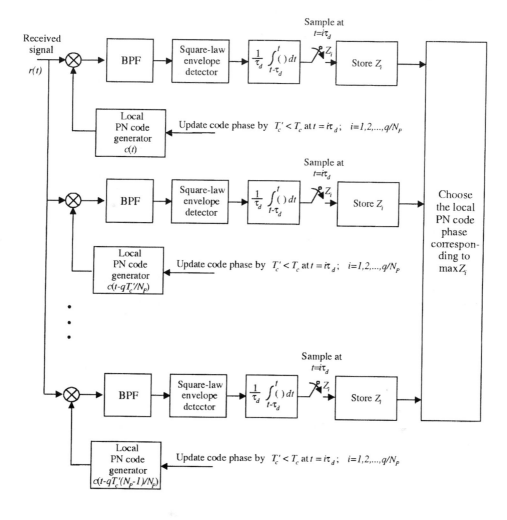

Figure 3.5. Parallel search realization of the maximum-likelihood search technique.

way comprise can be found practically according to the acquisition time and complexity requirement (refer to [25]–[28]).

3.   Z- VS. EXPANDING WINDOW SEARCH

The particular procedure adopted by the receiver in its search through the uncertainty region is called the *search strategy.* The uncertainty is two-dimensional in nature, time and frequency, and the search can be done either *continuously* or in *discrete steps*. In the discrete-step case, the time uncertainty region is quan-

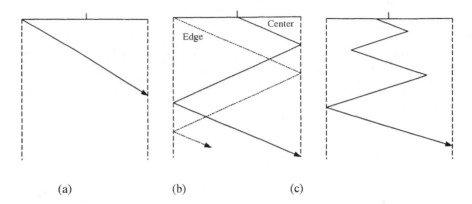

Figure 3.6.   Illustration of three different search strategies: (a) straight line search, (b) Z-search, (c) expanding window search.

tized into a finite number of elements called *cells*, through which the receiver is stepped.

There are several different types of search strategies depending on the nature of the uncertainty region, available prior information, and other factors. Fig. 3.6 illustrates three different types of search strategies - - straight line search, Z-search, and expanding window search.

In case *a priori* information on the correct code-phase is made available through a variety of means such as short preamble, auxiliary reference, and past synchronization history, the distribution of the correct code-phase has been modeled as triangular or truncated Gaussian [80]. In this case, the *Z-search* and the *expanding window (EW) search* strategies appear to achieve a rapid synchronization [19][80]. In the Z-search, the sweep lengths are equal to the number of cells in the whole uncertainty region, while, in the EW search, the sweep length expands from a fraction to the full coverage of the uncertainty region.

In the case of the Z-search strategy, the starting cell position may be specified at the most probable position (center) or the least probable position (edge) and the search path may be made continuous or broken. It is known [17][81] that the broken-center Z-search performs the best while the continuous-center Z performs the worst. The edge Z-search has an intermediate level of performance but does not effectively utilize the available *a priori* information.

In the case of the EW search strategy, when the probability that the received phase is within each phase cell is known, the sweep strategy can be modified to search the most likely phase cells first. For example, if the received phase distribution is Gaussian, a most reasonable search strategy can be found in searching the cells within one standard deviation, or 1σ, of the most likely cell

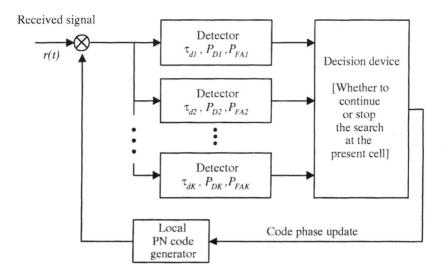

Received signal
$r(t)$

Detector
τ_{d1}, P_{D1}, P_{FA1}

Detector
τ_{d2}, P_{D2}, P_{FA2}

Detector
τ_{dK}, P_{DK}, P_{FAK}

Decision device

[Whether to
continue
or stop
the search
at the
present cell]

Local
PN code
generator

Code phase update

Figure 3.7.   The K-dwell serial synchronization system.

first, expanding to the cells within 2σ and then 3σ. The expanding window may be confined at the 3σ level.

4.   SINGLE VS. MULTIPLE DWELL

In the serial search strategy, *single dwell* system refers to the system in which a single detector is used to examine each possible waveform alignment for a fixed period of time in serial fashion until the correct one is searched.

Fig. 3.1 depicts an example of the single dwell serial PN acquisition system. As we have seen earlier, the received signal is actively correlated with a local replica of the PN code, passed through a band-pass (pre-detection) filter, square-law envelope detected, integrated for a fixed time duration τ_d (which is called the "dwell time"), and finally compared to the preset threshold.

Multiple dwell system is a generalization of the single dwell system. Fig. 3.7 depicts a K-dwell serial synchronization system: The received signal is first multiplied by a locally generated replica of the PN code and then applied to each of K non-coherent detectors. The i-th detector, $i = 1, 2, ..., K$, is characterized by a detection probability P_{Di}, a false alarm probability P_{FAi}, and a dwell time τ_{di}. If the detector dwell times are ordered $\tau_{d1} \leq \tau_{d2} \leq \tau_{d3} \leq ... \leq \tau_{dK}$, the decision whether to continue or to stop the search at the present cell is made by sequentially examining the K detector outputs. If *all* the K detectors indicate that the present cell is correct (i.e., each produces a threshold crossing), then the search stops. If *any* one detector fails to indicate that the present cell is correct

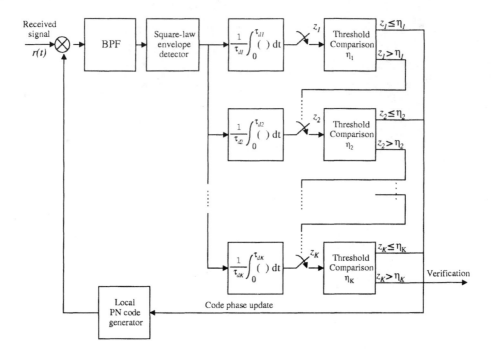

Figure 3.8.   A K-dwell PN acquisition system with non-coherent detection.

(i.e., it fails to produce a threshold crossing), then the search is continued and the time delay τ of the local PN generator is retarded by the amount of the chosen phase update increment. Consequently, once any one detector indicates that the codes are misaligned, the search may move on without waiting for the decision of the remaining detectors.

By virtue of the additional threshold testing, the multiple dwell system does not constrain the examination interval per cell to be constant. Nevertheless, this scheme falls into the class of *fixed dwell* time systems as the variation in integration time is achieved by arranging the examination interval to be a series of fixed short dwell periods, with the decision made after each. The integration time in a given cell examination interval increases toward its maximum value in *discrete steps*. So, the multiple dwell system dismisses each incorrect alignment earlier than the single dwell system which is constrained to integrate over the full examination interval always. Since most of the cells searched indeed correspond to incorrect alignments, this quick elimination capability leads to a considerable reduction in acquisition time, particularly for the long code cases.

The maximum time to search a given cell is τ_{dK}, whereas the minimum time is τ_{d1}. Most cells can be dismissed after a dwell time τ_{dk}; $k \ll K$ in the K-dwell system, whereas every cell must be examined for a time equivalent to

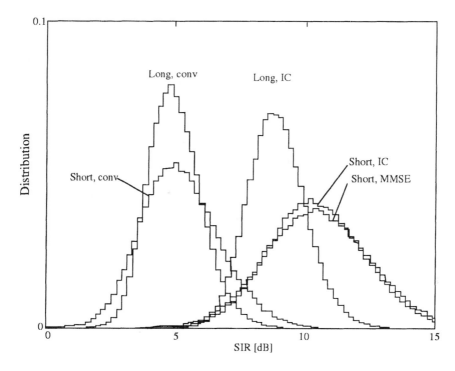

Figure 3.9. Distribution of SIR for the conventional (conv) matched filter receiver, intracell interference cancellation (IC) receiver, and MMSE receiver in nonmultipath environment with perfect power control [82].

τ_{dK} in the single dwell system. Fig. 3.8 depicts a K-dwell time PN acquisition system using non-coherent detection. All the K "integrate-and-dump" blocks begin the integration at the same time instant, with the dumping, however, done at a later time, depending on the magnitude of τ_{dk}

5. SHORT VS. LONG CODE ACQUISITION

The period of a PN sequence may be equal to or greater than the duration of one data bit. In the former case, the PN sequence is called a *short code* sequence and in the latter case, called a *long code* sequence. Short and long code CDMAs are sometimes referred to as *deterministic* CDMA (or D-CDMA) and *random* CDMA (or R-CDMA), respectively, as a longer sequence leads to a more random signal. In a long code system, the crosscorrelation among users changes bit by bit, and consequently the MAI becomes random in time, causing the performance nearly identical among different users. On the other hand, in a short code system, the cross-correlation remains unchanged over time, thus

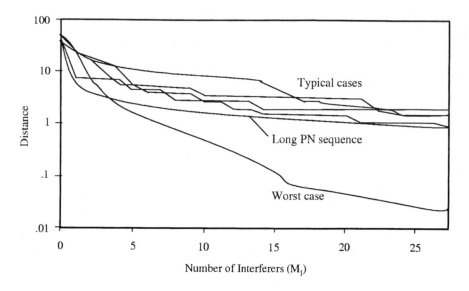

Figure 3.10.   Example of normalized distance measure ρ for a short PN sequence ($N = 63$) [45].

an unfortunate user could be trapped by an inferior performance scenario due to the non-time-varying cross-correlation.

While some interference suppression or cancellation techniques are applicable to both short and long code systems, the complexity, in general, is lower for the short code case due to the cyclostationary interference. For example, an MMSE (minimum mean squared error) type receiver could be used for both short and long sequences, but in practice the short sequence case is better matched because the correlation matrix is estimated by adaptive algorithms under the cyclostationary assumption.

The channel capacity is dictated more by the distribution of the error probability than by its mean value. Fig. 3.9 illustrates a performance comparison of the short and long systems in terms of the average *signal-to-interference ratio* (SIR) [82]. We observe that the variance in SIR is larger for the short code systems than for the corresponding long code cases, with the performance fluctuating in wider range. The intracell *interference cancellation* (IC) receiver of the short code system better performs than the long code system but the variance is higher. In the case of the conventional matched filter detection, average error probability is slightly higher for the long code system than for the short code system, but the variance in performance is larger for the short code system. When fast fading is present, adding more randomization to the interference, the difference in BER diminishes.

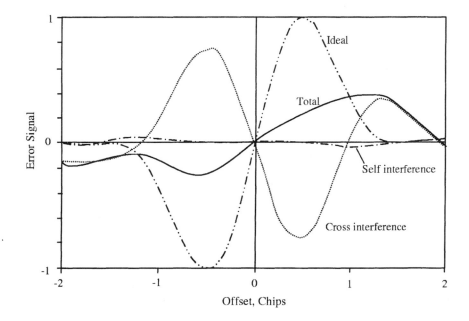

Figure 3.11. An example of S-curve for a short PN sequence ($N = 31$, $M_I = 14$) [45].

The code length also affects the acquisition-based capacity, which refers to the maximum number of simultaneous transmissions supported while maintaining an acceptable acquisition performance [44]. The acquisition window length or the length of the matched filter determines the complexity of acquisition schemes and the acquisition-based capacity depends on the acquisition window length. Unless the receiver knows the delay of the desired user, the probability of interference-caused false alarms increases and the acquisition-based capacity decreases, as the acquisition window length increases. Since the acquisition window length is a linear function of the code length, the acquisition-based capacity, in practice, becomes lower than the value derived based on BER or SNR criteria. This indicates that the acquisition problem limits the capacity of DS/CDMA systems.

The false alarm and detection performance of an acquisition system is determined by the difference of the mean values of the decision statistic in in-sync and out-of-sync states. The decision statistic z for an acquisition with the dwell time, KT_s, is the sum of the K sampled values of the filtered envelope output, for example, $|A(t)|^2$ in Fig. 3.2. A convenient performance indicator is the *normalized distance* ρ defined by [45]

$$\rho = \frac{E\{z|\hat{\tau}_0 = \tau_0\} - E\{z|(|\hat{\tau}_0 - \tau_0|) \gg T_c\}}{\sqrt{K}[\sqrt{var\{z|\hat{\tau}_0 = \tau_0\} + var\{z|(|\hat{\tau}_0 - \tau_0|) \gg T_c\}]}}, \quad (3.1)$$

where τ_0 is the PN chip phase, $\hat{\tau}_0$ the local estimate of the PN chip phase, and T_c a chip interval. Fig. 3.10 plots the normalized distance ρ with respect to the number of interferers, M_I, for a short PN sequence (N=63). Overlaid in the figure is the same performance parameter for a system with a long code spreading sequence. The "worst case" curve happens when the mean value of z at out-of-sync state becomes very large due to interferences. We observe that the normalized distance measure in the worst case is much lower for short sequences than for the long sequence, even if slightly higher in the typical cases. Therefore, in the short code case, the ability to detect the synchronized state may severely degrade in the acquisition mode.

In the tracking mode, the *delay-locked loop* (DLL) normally exhibits an *S-curve* for the expected value of the phase detector output (or error signal) with respect to the phase offset. This curve is determined by calculating the difference of the correlator output corresponding to the the advanced PN code and the the correlator output corresponding to the retarded PN code. Fig. 3.11 shows an example of S-curve for a short PN sequence ($N = 31$). We observe that the interference from other users cancels out a considerable part of the ideal error signal, flattening the S-curve. Consequently, the DLL can become unstable to track. In general, short codes are less random than long codes and this reduced randomness can result in an increased tracking jitter, and possibly leading to a tracking failure.

Chapter 4

ADVANCED CODE ACQUISITION TECHNIQUES

In DS/CDMA communications, code-acquisition (or code-synchronization) is a most essential first-stage processing in the receiver. Code-acquisition refers to the operation of synchronizing the locally generated PN code with the received PN code within one chip interval, beyond which the tracking process takes charge. Consequently, rapid and correct code acquisition is prerequisite to normal operation of the DS/CDMA communications and thus has been a hot issue of research for the past several decades.

Conventionally, code-acquisition has been mainly pursued by taking the correlation of the receiver-generated and the received PN signals and then comparing the result with a threshold, while advancing the phase of the receiver-generated signal by one or a fraction of one chip. However, there also have been introduced a number of advanced code synchronization techniques that take very different approaches, namely, sequential estimation, sequential detection, auxiliary sequence, postdetection integration, interference reference filter, differential-coherence, recirculation loops, distributed samples, or others. This chapter is intended to introduce a variety of such advanced code acquisition techniques one by one.

1. RAPID ACQUISITION BY SEQUENTIAL ESTIMATION (RASE)

The first advanced acquisition technique may be the *rapid acquisition by sequential estimation* (RASE) introduced by Ward [29], which is based on a sequential estimation of the shift register states of the PN sequence generator. The RASE system makes an estimate of the first L received PN code chips, where L is the number of shift registers in the sequence generator, and loads the receiver sequence generator with that estimate. This sets a particular initial

41

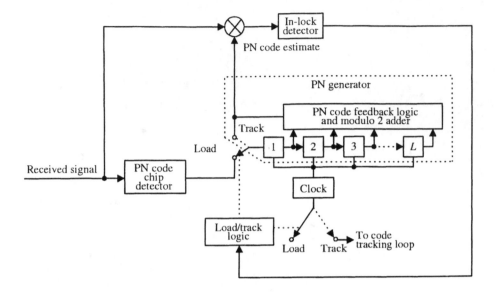

Figure 4.1.   Block diagram of the rapid acquisition by sequential estimation (RASE) technique.
[4]

condition (or starting state) from which the generator begins its operation. Since the next state of a PN sequence generator depends only on the present state, all the subsequent states can be predicted based on the knowledge of the initial condition. The estimation and loading process is repeated periodically until the correct initial state is obtained.

1.1   ORIGINAL RASE TECHNIQUE

Fig. 4.1 depicts the block diagram of the RASE technique. In the figure, the PN code chip detector is the block that makes the estimate of the L received PN code chips. The L chips are loaded to the PN sequence generator in the solid box as the state of the constituent shift registers. The PN sequence generated with the estimated state set is then correlated with the received signal, out of which it is determined whether or not the PN sequence generator is synchronized.

The decision when to stop the estimation and the loading process are based upon a threshold crossing at the in-lock detector. This in-lock detector can make a decision using the test statistic based on the cross-correlation between the received code signal and the locally generated PN code. Once it is determined that a correct estimate is done, the load/track logic inhibits further reloading of the local shift register. This register is then switched to a PN tracking loop which is responsible for maintaining the code phase from that time on.

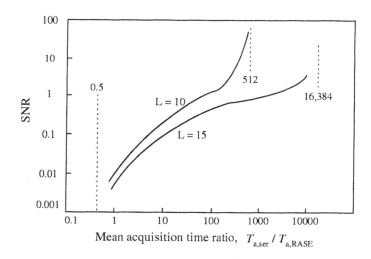

Figure 4.2. Ratio of mean acqusition times of the serial search $(T_{a,ser})$ and the RASE $(T_{a,RASE})$ systems. (AWGN channel) [29]

Fig. 4.2 illustrates the acquisition time performance of the RASE technique in comparison with that of the conventional serial search technique [29]. The figure plots the ratio of the mean acquisition time of the serial search technique, $T_{a,ser}$, and the mean acquisition time of the RASE technique, $T_{a,RASE}$. We observe that the mean acquisition time of the sequential estimation system may be twice as large as that of the serial search system for very small SNR. However, the improvement factor, $T_{a,ser}/T_{a,RASE}$, can increase up to 512 when $L = 10$, and up to 16,384 when $L = 15$.

The RASE system can reduce the acquisition time tremendously at a small hardware increase. This reduction is made possible by estimating the state of the involved SRG directly, instead of aligning the code phase based on the correlation value. In facts, it is possible, in theory, to accomplish code acquisition in L time units if the current sequence of the transmitter SRG of length L is available. This would take 2^L-1 time units if the conventional serial search technique were employed.

Clearly, the success of the RASE scheme depends heavily on the ability to make reliable estimates of the received PN code chips in the presence of noise. For a PN modulated carrier used as the input signal, the estimation process would consist of a simple demodulation and hard-limiting detection of this signal in the same way that one would demodulate and detect any PSK data stream. As such, this scheme falls into the class requiring coherent detection and is thus of limited use in most spread-spectrum applications.

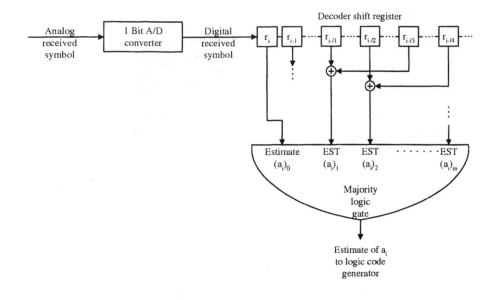

Figure 4.3.   Majority logic acquisition device. [31]

Despite its rapid acquisition capability for moderate SNR, the RASE has drawback of being highly vulnerable to noise and interference signals. This vulnerability stems from the fact that the estimation process is performed on a chip-by-chip basis and, as such, makes no use of the interference rejection capability of PN signals.

1.2   MODIFIED RASE TECHNIQUES

There are several modifications of the Ward's original RASE system that utilize the recursion relation of the PN code to improve the initial n-chip estimates.

Kilgus [31] presented the use of a majority vote which makes multiple independent estimates of each of L chips and then makes a majority logic vote among all the estimates to determine the initial L chips to load to the local shift register. Fig. 4.3 illustrates this arrangement. The number of estimates can vary from one to a large number which is limited only by the hardware complexity. Fig. 4.4 plots the probability to acquire synchronization at one attempt for an SRG of length 13 (or of sequence length 8,191) using a majority logic acquisition device with 4095, 1023, 255, 63, 15, and 1 estimates. We observe that the probability of acquiring an 8,191 lengthed code in one attempt can be made nearly one at -10 dB SNR if 4,095 estimates are made simultaneously.

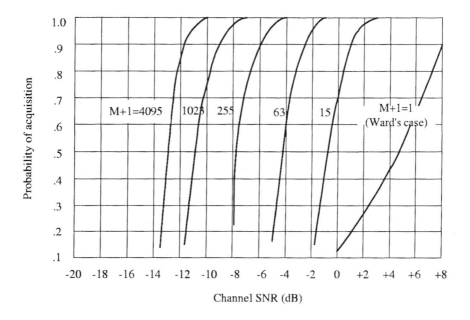

Figure 4.4.   One trial acquisition probability. [31]

Ward and Yiu [30] presented a modified version called *recursion-aided RASE* (RARASE). It utilizes a known recursion relation of the PN sequence to determine if the initial state estimate of the received PN signal is correct and to decide whether or not to make a tracking attempt using that estimate. Then it is possible to discard a large portion of the incorrect initial state estimates using a relatively simple logic. Fig. 4.5 depicts the block diagram of the RARASE technique. The RARASE system has an additional function that sums the estimated code chips according to the recursion relations of the employed PN sequence to generate a *sync-worthiness indicator* (SWI). The SWI is used along with the in-lock indicator to determine when to switch the SRG from the load condition to the tracking condition. Fig. 4.6 illustrates how the SWI works when the switch is held in the "load" position. Fig. 4.6 (a) is a 31-bit m-sequence and Fig. 4.6 (b) is a random series of errors with 25% error probability. These two are combined in Fig. 4.6 (c), representing the sequence flowing into the shift register, and the SWI output is shown in Fig. 4.6 (d). Every "don't try" output corresponds to the case when at least one error is present in the shift register or in the present bit, but every six-bit error-free run appears as a "try sync" case. Consequently, the SWI never excludes the correct initial phases while it does exclude many incorrect phases. "Try sync" outputs also occur when paired errors are present at the SWI input or when errors are present at

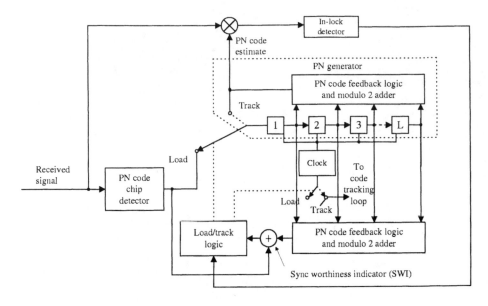

Figure 4.5.   Block diagram of the recursion-aided RASE (RARASE) technique [4].

the shift register outputs that are not checked. So some incorrect phases may result in triggering the tracking mode. Fig. 4.7 illustrates an implementation of the Fig. 4.5 RARASE structure for the SRG with the characteristic polynomial $x^6 + x + 1$. In this implementation the SWI consists of three 3-input mod-2 adders and a 3-input AND gate. Fig. 4.8 plots the ratio of the mean acquisition time of the RARASE system, $T_{a,RARASE}$, to that of the RASE system, $T_{a,RASE}$. We observe that the RARASE system can achieve acquisition time reduction by a factor of 7.5 at -15 dB SNR for an SRG of length 15 (or a PN code of length $2^{15} - 1$).

Barghouthi and Stuber [32] presented a sequence acquisition scheme based on sequential estimation and soft-decision combining, which is applicable to rapidly acquiring long Kasami sequences. It exploits the algebraic properties of Kasami sequences to obtain the estimate of the correct phase of the local shift register by generating and combining multiple statistics for each chip. The mean acquisition time of this scheme is reported to outperform that of the serial search scheme, approaching that of the parallel search schemes [32]. Whereas the parallel schemes achieve the improvement at the expense of hardware complexity, the Barghouthi's scheme does so at the cost of processing complexity.

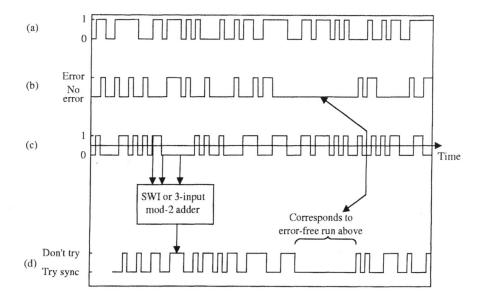

Figure 4.6. Operation of the sync-worthiness indicator (SWI). [30] : (a) error-free sequence, (b) error signal, (c) sequence with errors, (d) SWI output.

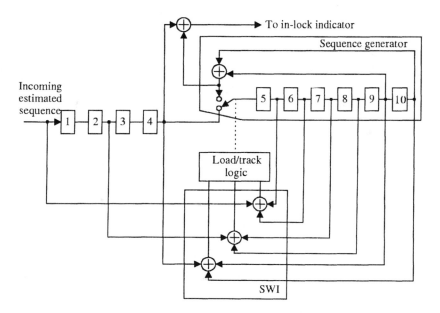

Figure 4.7. Implementation of the RARASE system containing the SRG with the characteristic polynomial $x^6 + x + 1$.

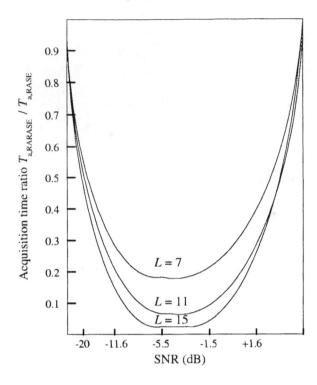

Figure 4.8.   Ratio of mean acquisition times of the RARASE and the RASE systems (AWGN channel) [30].

2.   SEQUENTIAL DETECTION-BASED ACQUISITION

In the aspect of dwell time and verification step size, acquisition schemes can be classified into fixed dwell, multiple dwell, and sequential schemes [4] [14]. The single fixed dwell (or integration) scheme is inefficient as its detector spends as much time for rejecting out-of-sync positions as it does for accepting the in-sync position. In order to minimize the acquisition time, we need such a detector that can quickly dismiss out-of-sync positions but take a long integration for the in-sync cell. The multiple dwell time system introduced in Section 4 of Chapter 2 was an attempt to accomplish the above objective wherein the detector integration time increased in *discrete* steps until the test failed (or any output dropped below the threshold). Thus, at an out-of-sync position, only few steps are needed (i.e., short integration time), whereas all steps are fully needed at the true sync position (i.e., long integration time).

It is also possible to increase the integration time continuously and replace the multiple threshold tests by a continuous test of a single dismissal threshold. Such a variable integration time detector is referred to as a *sequential detector*. The corresponding acquisition system is designed such that the mean time to

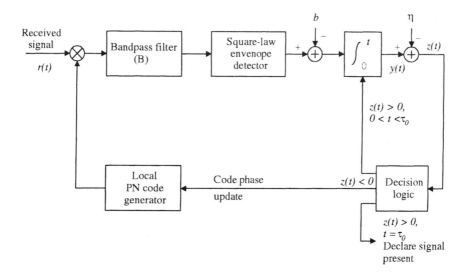

Figure 4.9.   A serial sequential detection PN acquisition system with timeout feature.

dismiss out-of-sync positions is much shorter than the integration time of the single dwell system. As an acquisition process normally spends most time in dismissing out-of-sync positions, the mean acquisition time of the sequential detection based PN acquisition system will become much smaller than that of the single dwell time system.

2.1   SEQUENTIAL DETECTION-BASED SCHEME

Fig. 4.9 depicts a block diagram of a serial sequential detection PN acquisition system. Operation of this system is identical to the single dwell acquisition system up to the square-law envelope detector. Thus, in the absence of the bias voltage b, the output of the continuous time integrator would typically behave as plotted in Fig. 4.10 (a). In particular, the integrator output would follow the integrated mean of the square-law envelope detector output, which will be given by $N_0 Bt$ or $N_0 B(1 + \gamma)t$ for the noise power $N_0 B$ and the pre-detection SNR γ, depending whether the dwelled cell has noise only or signal plus noise. Note that the former corresponds to an out-of-sync cell and the latter corresponds to the in-sync cell. In the presence of bias voltage b with $N_0 B < b < N_0 B(1+\gamma)$, the integrator output waveform changes as shown in Fig. 4.10 (b). In addition, if the threshold η (of negative value) and the follow-up decision logic are applied in such a way that a dismissal occurs whenever the integrator output falls below the threshold then the integrator output waveform changes to the shape shown in Fig. 4.10 (c). It is desirable to choose the threshold value such that it allows for a relatively quick dismissal of out-of-sync position but allows the integrator

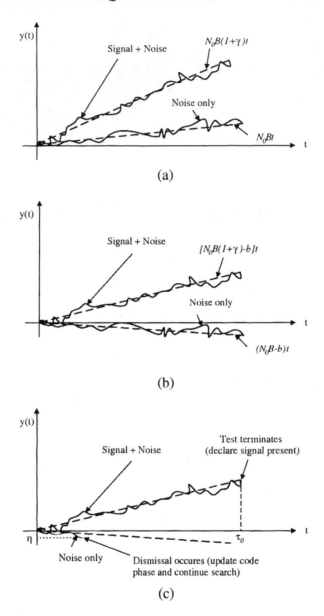

Figure 4.10.   Waveform of the integrator output $y(t)$ [4] : (a) without bias voltage, (b) with bias voltage b, (c) with threshold dismissal and test truncation.

output for the in-sync position to remain above the threshold. If this latter event occurs for a designated interval of time, say τ_0, after which the test is terminated, then the signal is declared present and the cell being searched is declared as the in-sync position. The test truncation time τ_0 is often referred to as the *time-*

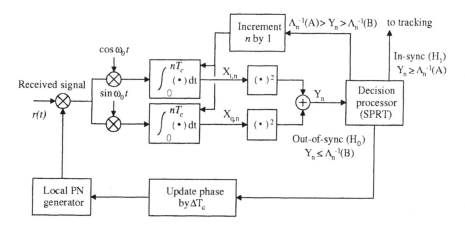

Figure 4.11. Block diagram of the noncoherent serial acquisition scheme employing SPRT [22].

out of the sequential detector. It may be replaced with a test against a second (positive-valued) threshold [21, 22, 23]. In this case, the in-sync declaration would come as soon as the integrator output rises above this positive-valued threshold.

2.2   IMPLEMENTATIONS OF SEQUENTIAL DETECTION-BASED SCHEME

Now we consider noncoherent sequential detection-based acquisition schemes employing the *sequential probability ratio test* (SPRT) or *truncated SPRT* (TSPRT) proposed by Tantaratana, *et al.* [21, 22]. Fig. 4.11 depicts the non-coherent serial acquisition system for this. The SPRT or TSPRT function is embedded in the decision processor. It is assumed that the channel is a slowly-varying fading and additive white Gaussian channel and no data modulation occurs during the acquisition process.

The received signal $r(t)$ is first multiplied by the local PN waveform and then passed through the noncoherent demodulation process. The output Y_n triggers the decision processor to test if the local and the incoming PN waveforms are aligned within $\Delta T_c/2$ interval for $\Delta=1$ or $1/2$. If such a coarse alignment is achieved (i.e., in-sync state), the tracking circuit is initiated. Otherwise, the phase of the local PN waveform is updated by ΔT_c and the acquisition process continues.

We denote the non-alignment and alignment state by H_0 and H_1, respectively. Then, we can write the likelihood ratio as

$$\Lambda_n(y_n) = \frac{f_{Y_n}(y_n|H_1)}{f_{Y_n}(y_n|H_0)}, \tag{4.1}$$

for the conditional probability density functions of Y_n, $f_{Y_n}(y_n|H_i)$, $i = 0, 1$. Based on this, we compare the three decision schemes - - fixed sample size (FSS), SPRT, and TSPRT.

In the FSS (or fixed dwell) decision scheme, the length of the integration is fixed *a priori* and the decision is made based on the resulting test statistic. If the integration time is from 0 to MT_c, the FSS test compares $\Lambda_M(y_M)$ to the given threshold T. Since the likelihood ratio is monotonic, the FSS test may be expressed by

$$y_M \left\{ \begin{array}{c} H_1 \\ \gtreqless \\ H_0 \\ < \end{array} \right. \Lambda_M^{-1}(T). \tag{4.2}$$

In the case of the SPRT, the likelihood ratio $\Lambda_n(y_n)$ is compared with two constant thresholds A and B ($A > B$), for $n = 1, 2, 3, \cdots$, according to the decision logic

$$y_n \left\{ \begin{array}{l} \overset{H_1}{\underset{H_0}{\geq}} \Lambda_n^{-1}(A) \\ \leq \Lambda_n^{-1}(B) \\ \text{otherwise, continue.} \end{array} \right. \tag{4.3}$$

Unless either inequalities is met, the comparison process continues to the next value of n.

One drawback of the SPRT is that there is no upper bound on the test length. Therefore, it can happen that the decision processor is stuck at a particular uncertainty phase for a long period of time. In order to avoid this, the TSPRT puts an upper bound on the test length. More specifically, for an upper bound $\hat{n}$, it takes the decision logic

$$\text{if} \quad n < \hat{n}, \quad y_n \left\{ \begin{array}{l} \overset{H_1}{\underset{H_0}{\geq}} \Lambda_n^{-1}(A) \\ \leq \Lambda_n^{-1}(B) \\ \text{otherwise, continue,} \end{array} \right.$$

$$\text{if} \quad n = \hat{n}, \quad y_{\hat{n}} \left\{ \begin{array}{c} \overset{H_1}{\geq} \Lambda_{\hat{n}}^{-1}(T) \\ \overset{H_0}{<} \Lambda_{\hat{n}}^{-1}(T). \end{array} \right. \tag{4.4}$$

Therefore, the test is truncated at $n = \hat{n}$ if it has not been terminated previously.

Fig. 4.12 plots the performances of the FSS, SPRT, and TSPRT tests in terms of the *average sample number* (ASN) and the power function. The ASN refers to the average number of chips needed for the test to terminate and the power function indicates the probability of accepting H_1. The code phase offset in the horizontal axis is given in units of $T_c/2$: Phase alignment state becomes H_1 for the code phase offset value 0.5 and becomes H_0 for value 2.0 or larger. We observe that the SPRT and the TSPRT are both superior to the FSS test. The SPRT has the smallest ASN among the three at the phse offsets of 0.5 and 2.0, which indicates that it performs the best among the three. In practice, however, the TSPRT is preferable or comparable to SPRT as it is a bounded test.

3.   AUXILIARY SEQUENCE-BASED ACQUISITION

In order to obtain rapid acquisition, it is also possible to utilize an auxiliary signal, instead of the PN signal itself. In this case, the auxiliary signal must be derived from the used PN signal in such a way that the correlation of the auxiliary signal and the PN sequence exhibits some controllable behaviors. Salih and Tantaratana [33] presented an auxiliary signal-based acquisition technique, in which the cross-correlation function of the auxiliary signal and the PN signal has a triangle shape that covers essentially the entire period of the PN signal. Consequently, the correlation can guide the direction to shift the local signal generator for the phase update in the acquisition process.

3.1   ACQUISITION WITH AUXILIARY VCC LOOP

Fig. 4.13 depicts the block diagram of a closed-loop coherent acquisition system with *voltage-controlled-clock* (VCC) loop. The system is composed of two subsystems - - one for the alignment detection and the other for the VCC loop. The VCC loop takes a new local waveform $\alpha(t)$ as the *auxiliary signal*. By taking the auxiliary signal $\alpha(t)$ as a properly weighted sum of shifted versions of the used PN signal $c(t)$, we can make the two signals lightly correlated and their cross-correlation function $R_{c\alpha}(\cdot)$ take the periodical shape shown in Fig. 4.14 [33]. Note that the cross-correlation function has a triangle shape that covers essentially the entire period N of the PN signal. Apparently, this triangular-shaped correlation helps to determine the direction of phase update in the local signal generator.

The VCC loop is for updating the phase of $\alpha(t)$ until it aligns with the phase of the received PN signal. It has a cyclic shift register to store the auxiliary sequence $\{\alpha_k\}$. The shift register output drives the auxiliary signal generator to generate the auxiliary signal $\alpha(t + \frac{T_c}{2})$. The generator also shifts this auxiliary signal by the amount equal to the code phase estimate formed by the VCC (i.e., $\hat{\tau}_m$). The received signal $r(t)$ is correlated with the difference of early and late versions of the local code waveform $\alpha(t)$. The correlated output is lowpass-filtered and then sampled at interval T_c. The real-part of the resulting signal, y_m, is used to control the VCC. Fig. 4.15 plots the averaged VCC input

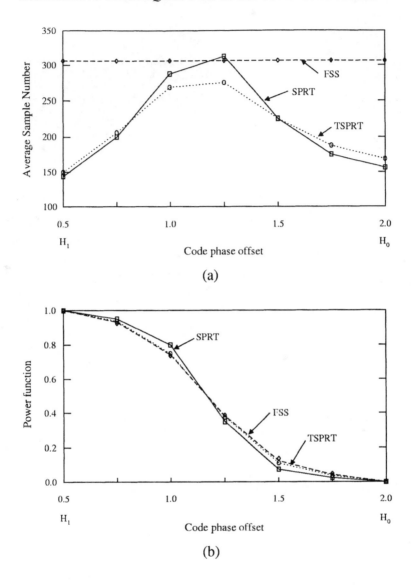

Figure 4.12. Performance of the SPRT, TSPRT, and FSS tests (without fading, SNR=-10dB, $\Delta = 1/2$, A=200, B=0.005) [22]: (a) average sample number (ASN) function, (b) power function.

signal. Note that the average VCC control signal is negative for code phase error e_τ in the interval $(0, \frac{NT_c}{2})$ and positive for e_τ in the interval $(-\frac{NT_c}{2}, 0)$. This enables to acquire synchronization within NT_c wide pull-in range, which is made possible due to the wide-stretching cross-correlation function $R_{c\alpha}(\cdot)$.

On the other hand, the *phase alignment detector* correlates the received signal $r(t)$ with the local PN code waveform $c(t - \hat{\tau}_{ln})$ for a fixed dwell time of nT_c. A

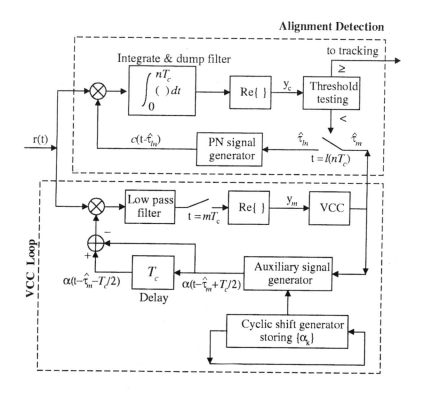

Figure 4.13.   The closed-loop coherent acquisition scheme [33].

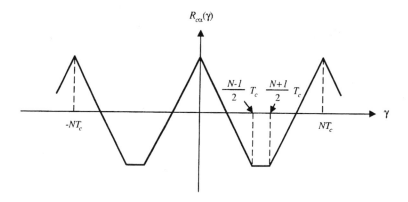

Figure 4.14.   The periodic cross-correlation function of $c(t)$ and $\alpha(t)$

new value of the code phase estimate $\hat{\tau}_m$ is fed from the VCC loop to the detector every nT_c seconds, and is tested for the phase alignment. This continues until the alignment is declared, and then the tracking process is triggered.

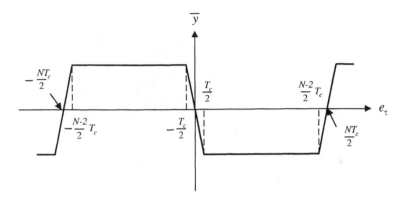

Figure 4.15.   The averaged VCC input signal.

The performance improvement of the VCC loop based auxiliary acquisition system may be examined in terms of mean acquisition time and acquisition time variance. Fig. 4.16 (a) plots the ratio of the mean acquisition time of the VCC auxiliary system, $T_{a,VCC}$, to that of the coherent serial-search single-dwell acquisition system, $T_{a,ser}$. Fig. 4.16 (b) plots the ratio of the acquisition time variance of the VCC auxiliary system, $\sigma^2_{T_{a,VCC}}$, to that of the coherent serial-search single-dwell acquisition system, $\sigma^2_{T_{a,ser}}$. In the figure, P_{fa}, P_d, and K_p denote the false alarm probability, the detection probability, and the penalty time, respectively. We observe from the two figures that the VCC auxiliary system acquires the PN phase at least twice faster than the conventional system, with significantly smaller acquisition time variance.

3.2   ACQUISITION WITH AUXILIARY PRE-LOOP ESTIMATOR

The performance of the VCC loop based auxiliary acquisition technique can be further enhanced by adding a *pre-loop code phase estimator* (PLE) to the system shown in Fig. 4.17 [34]. This PLE-based auxiliary acquisition system operates in the same way as the VCC-based system except that the PLE initially uses the correlation result to estimate the incoming code phase. The operation goes in two steps as follows: At $t = 0$, switch S1 connects the system input to the PLE, switch S2 is closed, and switch S3 connects τ_{in} to the auxiliary signal generator. So, the received signal $r(t)$ is routed to the PLE, and the auxiliary signal generator uses τ_{in} as the initial code phase with the output $\alpha(t - \tau_{in})$ being fed to the PLE. The PLE correlates the received signal $r(t)$ with the auxiliary signal for the duration of $L_I N T_c$. At $t = L_I N T_c$, the PLE gets its code phase estimate and the switches S1, S2, and S3 are switched to the positions shown in the figure. Hence, the VCC loop and the phase alignment

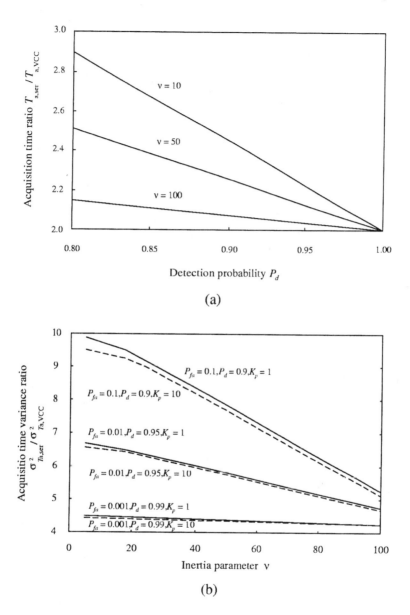

Figure 4.16.   Mean acquisition time performance of the VCC-based auxiliary acquisition system in comparison with the serial search system (ν: system inertia parameter proportional to the VCC loop filter length) [33] : (a) ratio of mean acquisition time, (b) ratio of acquisition time variance.

detector gets activated while the PLE is cut off from the system. Beyond that point, the operation is the same as the VCC loop based system. Note that, in the PLE, the lower correlator branch estimates the code phase error magnitude,

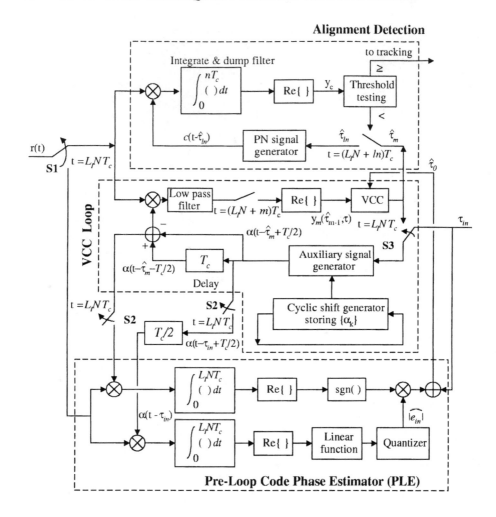

Figure 4.17. The closed-loop coherent acquisition system with a pre-loop estimator [34].

whereas the upper correlator branch provides the direction for the code phase update.

Fig. 4.18 (a) and (b) compare the mean acquisition time of the PLE-based auxiliary acquisition system, $T_{a,PLE}$, with that of the coherent serial acquisition system, $T_{a,ser}$, and that of the VCC-based auxiliary acquisition system, $T_{a,VCC}$, respectively. In the figure, L_I denotes the integration length of the PLE (normalized by NT_c). We observe that the PLE-based acquisition system acquires the incoming code phase faster than both the coherent serial-search and the VCC-based auxiliary systems in general. However, the acquisition speed gain diminishes if L_I becomes very small, which may be due to poor estimates. In the limiting case when L_I becomes zero, the PLE-based system reduces to

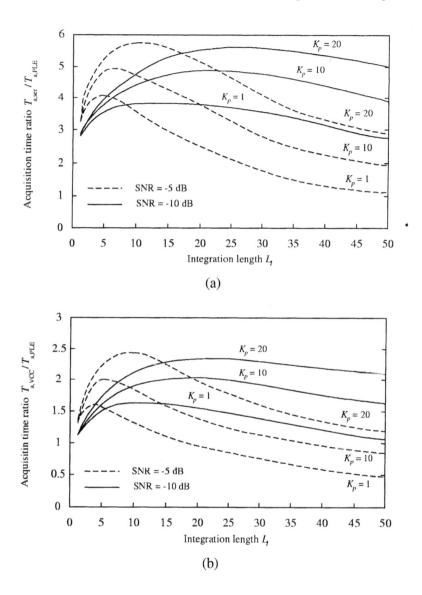

Figure 4.18. Mean acquisition time performance of the PLE-based auxiliary acquisition system [34] : (a) ratio $T_{a,ser}/T_{a.PLE}$, (b) ratio $T_{a,VCC}/T_{a,PLE}$.

the VCC-based system. On the other hand, in good channel condition (i.e., SNR=-5dB), the PLE-based system may perform inferior to the VCC-based scheme if L_I is chosen too large. Fig. 4.19 (a) and (b) plot the ratio of the acquisition time variance in a similar manner.

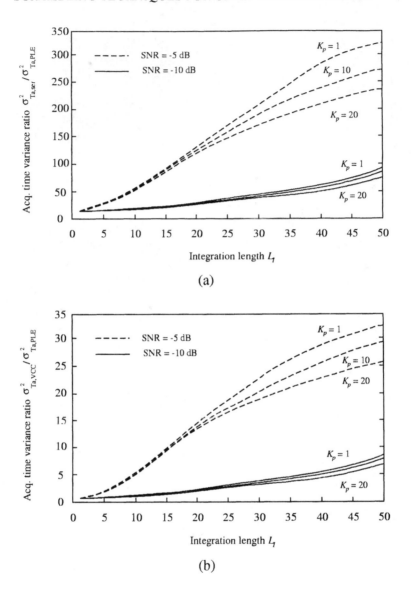

Figure 4.19. Acquisition time variance ratio of the PLE-based auxiliary acquisition system [34] : (a) ratio $\sigma^2_{T_a,ser}/\sigma^2_{T_a,PLE}$, (b) ratio $\sigma^2_{T_a,VCC}/\sigma^2_{T_a,PLE}$.

4.   ACQUISITION BASED ON POSTDETECTION INTEGRATION

In the channel environment with fading or poor SNR condition, it may be desirable to take advantage of multiple observations of information to improve the acquisition performance. In this section, we consider one of such acquisition

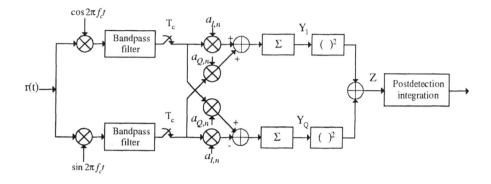

Figure 4.20.   Block diagram of noncoherent QPSK demodulator with postdetection integration.

techniques based on postdetection integration. The postdetection integration receiver combines multiple signal samples observed over multiple time intervals to produce a robust decision statistic in the fading channel. The number of observations should be optimized depending on the channel fading rate for best acquisition performance.

Fig. 4.20 depicts the block diagram of noncoherent QPSK demodulator with the postdetection integration function for code acquisition. In order to improve the detection performance, the observation period is increased by taking a sum of successive demodulator outputs as the decision variable. The postdetection integration process generates the decision variable by taking a summation of M consecutive values of Z, where M represents the number of the postdetection integration.

In general, the pdf of the decision variable depends on the autocorrelation function of the received signal, which represents short-term fading characteristics. Fig. 4.21 shows the effect of the postdetection integration process on the detection probability and mean acquisition time. It can be seen that the detection performance improves as the number of postdetection integration increases. This happens simply because the degree of freedom of the decision variable increases as the number of postdetection integration increases. However, the increase of detection probability becomes marginal as the number of postdetection integration, M, increases. On the other hand, the mean acquisition time decreases for $M = 2$ but begins to increase beyond that number. For a larger number M, a longer observation is required for the decision. For a number of postdetection integrations of up to two or three, the overall detection performance improves because the improvement in detection time through the increased degree of freedom (or diversity order) is dominant over the increase

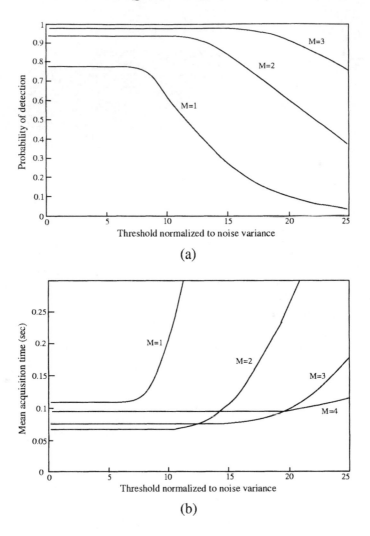

Figure 4.21. Effects of postdetection integration on acquisition. (M: number of postdetection integration, dual antenna, 1.2288 Mcps, $SIR_c = -16dB$, frequency selective Rayleigh fading channel, Doppler spread 50Hz) [28]: (a) probability of detection, (b) mean acquisition time.

in processing time. Beyond that point, however, the improvement in detection is overshadowed by the increase of processing time.

5.   ACQUISITION BASED ON INTERFERENCE REFERENCE FILTER

In an environment where bursty transmissions occur, rapid acquisition is especially important as it must be repeated for every bursty traffic. In this case it is desirable to arrange a synchronization preamble which contains multiple (say,

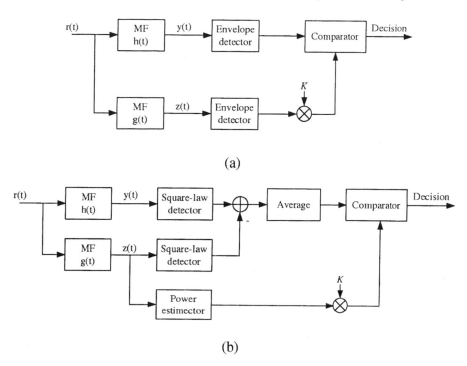

Figure 4.22. Block diagram of reference filter-based acquisition system [42]: (a) hard-decision approach, (b) soft-decision approach.

N) repetitions of the same PN sequence and to arrange the receiver to acquire synchronization based on the number of threshold crossings of the multiple PN sequences. In this arrangement, there are two possible approaches in making decision in the proper alignment of the PN sequence - - *hard-decision* [83, 84] and *soft-decision* [42]. For both approaches, a critical factor that determines the acquisition performance is a threshold setting. The interference reference filter technique is one of practically useful threshold setting methods.

Fig. 4.22 shows the block diagram of the reference filter-based acquisition system. The received signal is passed through two parallel bandpass matched filters $h(t)$ and $g(t)$. The impulse response $h(t)$ is a time-reversed delayed version of the transmitted PN sequence. Matched filter $g(t)$ has the same structure but its PN code is chosen orthogonal to the transmitted PN code. The lower branch is designed to provide a reference for the background interference, whereas the upper branch is for the detection. In the case of the hard-decision approach in Fig. 4.22 (a) the matched filter outputs are envelope-detected for noncoherent acquisition. The output of the lower-branch envelope detector is then multiplied by a gain factor $K (\geq 1)$ chosen to keep the probability of false

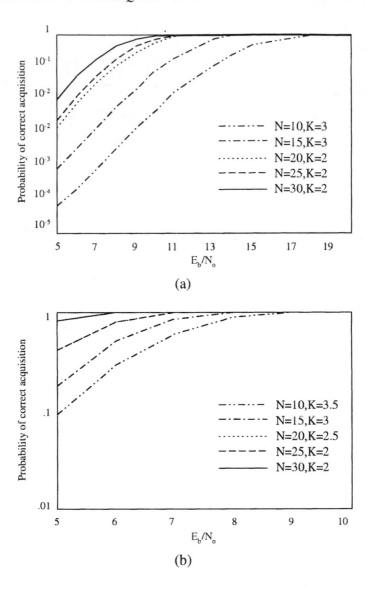

Figure 4.23. Probability of correct acquisition of the reference filter-based acquisition technique in the white Gaussian channel. (N is the number of code repetitions and K is the normalized threshold) [42]: (a) hard-decision approach, (b) soft-decision approach.

acquisition below a certain level. The result provides a reference threshold for the detected output in the upper branch to make the final decision.

In the case of the soft-decision approach shown in Fig. 4.22 (b), the matched filter outputs are square-law detected. Then the difference of the two outputs is averaged for each sampling instant over all the PN code repetitions (i.e.,

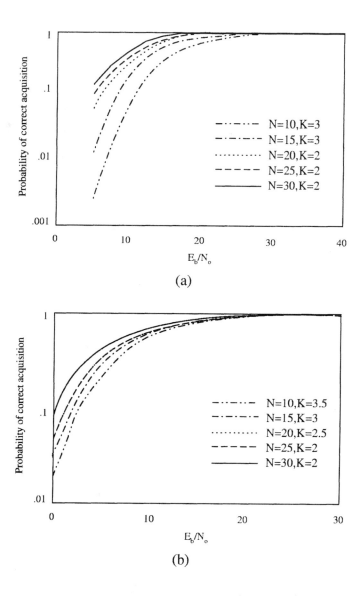

Figure 4.24. Probability of correct acquisition of the reference filter-based acquisition technique in the Rayleigh fading channel (N is the number of code repetitions and K is the normalized threshold) [42]: (a) hard-decision approach, (b) soft-decision approach.

N) within one preamble. The averaged value is then compared, for the final decision to the threshold, which is the product of a gain factor $K (\geq 1)$ and the noise variance at the output of the reference matched filter.

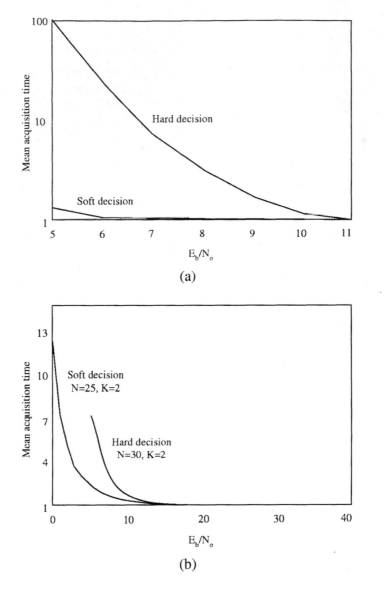

Figure 4.25.   Mean acquisition time in preamble intervals of the reference filter-based acquisition technique ($N=30$ for HD; $N=30$ or 25 for SD) [42]: (a) in white Gaussian channel, (b) in Rayleigh fading channel.

Fig. 4.23 plots the probability of correct acquisition in the white Gaussian channel for the hard-decision (a) and the soft-decision (b) approaches [42]. Fig. 4.24 plots the same probability in the Rayleigh fading channel for the hard-decision (a) and the soft-decision (b) approaches. In both sets of graphs, we observe that the probability of correct acquisition improves significantly as the number of repetition of the PN code within a preamble, N, increases.

Fig. 4.25 plots the mean acquisition time performance in the white Gaussian (a) and Rayleigh fading (b) channels. In both figures, we observe that the soft-decision approach makes much faster acquisition than the hard-decision approach.

6. DIFFERENTIALLY-COHERENT ACQUISITION

The use of differentially-coherent acquisition techniques can provide better performance than the use of noncoherent acquisition. Further, to prevent the performance degradation in the fast fading environment, chip-differentially-coherent acquisition technique was proposed, in which the correlation is made using a differentially-detected PN sequence.

6.1 DIFFERENTIALLY-COHERENT DETECTION

Zarrabizadeh and Sousa [35] presented a differentially-coherent acquisition detector, which could provide some SNR gain over the conventional noncoherent acquisition detectors without using the carrier phase information. Under hypothesis H_0(i.e., out-of-synch state), the decision variable is the product of consecutive uncorrelated samples in the case of differentially-coherent detector, whereas it is the square of a single sample in the case of a noncoherent detector. This distinction brings forth a significant acquisition performance improvement to the differentially-coherent detection scheme over the noncoherent one.

Depending upon the duration of observation, the differentially-coherent acquisition detectors can be categorized into *partial period correlation* (PPC) and *full period correlation* (FPC) detectors. The correlation is performed over a full code period in the FPC scheme, whereas it is done over a segment of the long PN sequence in the PPC scheme [4]. The detection performances of the PPC and the FPC are similar in noncoherent detection, but are different in differentially-coherent detection. In noncoherent detection, the decision variable is the envelope of the noncoherent correlator output. In contrast, in the PPC or the FPC detection, the decision variable is the real part of the differentially-coherent correlator output. With the differential detection scheme, however, the self-noise component of the consecutive samples, whose product forms the decision variable, is the same in the FPC, but is independent random variable in the PPC. Furthermore, unlike in noncoherent detection, the thermal noise components of the two samples multiplied together are statistically independent in both differential schemes. Therefore we can expect that the differential techniques outperform the noncoherent technique in terms of the detection probability, false alarm probability, and mean acquisition time.

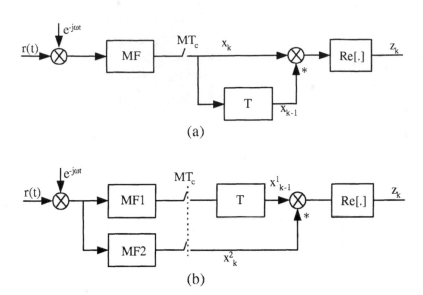

Figure 4.26.   Block diagram of the differentially-coherent acquisition techniques: (a) FPC, (b) PPC.

Fig. 4.26 depicts the receiver structures for the differentially-coherent FPC and PPC acquisition schemes. First, the received signals are demodulated. In the case of the FPC, the demodulated signal is then fed into a matched filter of a full code period. The output sample of the matched filter is multiplied by the conjugate of the T-delayed samples for the period T. In the case of the PPC, the demodulated signal is fed into two different matched filters, whose matching intervals correspond to two different segments of the PN sequence and the matching interval of the second matched filter corresponds to time-advanced version of that of the first matched filter by T. Note that neither the matching interval of the matched filters nor T is the full period of the PN sequence in the case of the PPC. The output sample of the first matched filter is delayed by T and multiplied by the conjugate of the output sample of the second matched filter. In the case of both FPC and PPC, the real part of the multiplied value is taken to be the decision variable. Each decision variable is compared with the given threshold, set by a conventional threshold setting algorithm. If the sample value exceeds the threshold, then the receiver declares in-sync phase and goes to tracking mode(in the single-dwell case) or to the verification mode(in the multiple-dwell case).

Under the hypothesis H_1 for PPC and FPC, the decision variable is given as

$$Z_k = (M\sqrt{E_c T_c} + N_k^R)(M\sqrt{E_c T_c} + N_{k-M}^R) + N_k^I N_{k-M}^I \qquad (4.5)$$

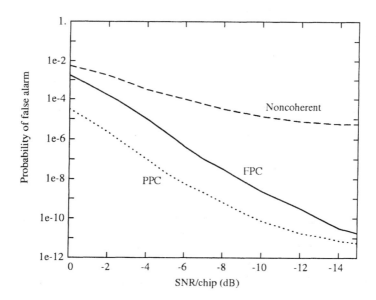

Figure 4.27.   Probabilities of false alarm of differentially-coherent schemes. (AWGN channel, normalized threshold level $V_n = 25$) [35].

where M is the number of integrated chips, E_c the received chip energy, T_c the chip period, and (N_k^R, N_{k-M}^R) and (N_k^I, N_{k-M}^I) are the real and imaginary parts of the thermal noise components of the matched filter, respectively. Under the hypothesis H_1, the detection probabilities for FPC and PPC are the same because the independent thermal noise components are the only random components which are involved in the decision variable and no self-noise component exists.[1] However, under the hypothesis H_0, this is not the case.

Under the hypothesis H_0 for FPC scheme, the decision variable is given as

$$Z_k = (Y_k + N_k^R)(Y_k + N_{k-M}^R) + N_k^I N_{k-M}^I \qquad (4.6)$$

where Y_k is the self-noise term and (N_k^R, N_{k-M}^R) and (N_k^I, N_{k-M}^I) are the real and imaginary parts of the thermal noise components of the matched filter, respectively. For FPC scheme, two self-noise terms are identical. On the other hand, under the hypothesis H_0 for PPC scheme, the decision variable is given as

$$Z_k = (Y_k^2 + N_k^{R,2})(Y_{k-M}^1 + N_{k-M}^{R,1}) + N_k^{I,2} N_{k-M}^{I,1} \qquad (4.7)$$

where Y_{k-M}^1 and Y_k^2 are the self-noise terms, and $(N_k^{R,2}, N_{k-M}^{R,1})$ and $(N_k^{I,2}, N_{k-M}^{I,1})$ are the real and imaginary parts of the thermal noise components of

---

[1]For simplicity, we assume in this section that the fractional chip alignment offset is zero.

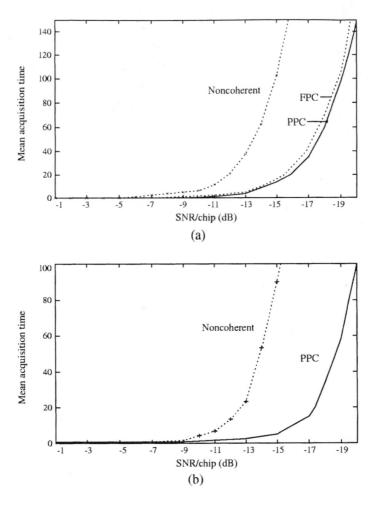

Figure 4.28. Mean acquisition time versus chip SNR (normalized by the uncertainty region length) [35] : (a) single-dwell detector, (b) multiple-dwell detector.

the two different matched filters. For PPC scheme, the self-noise and thermal noise components of the two matched filter outputs are clearly uncorrelated.

Fig. 4.27 plots the false alarm probabilities of the differentially-coherent FPC and PPC schemes in comparison with the noncoherent correlation scheme. We can observe that both FPC and PPC perform better than the noncoherent scheme, with the PPC outperforming the FPC. The superiority of the PPC is mainly due to the independent thermal-noise and self-noise components in the decision variable. At low SNR, the independent thermal-noise term, which is common to both PPC and FPC, is dominant, while at high SNR, the self-noise component dictates the false alarm probability.

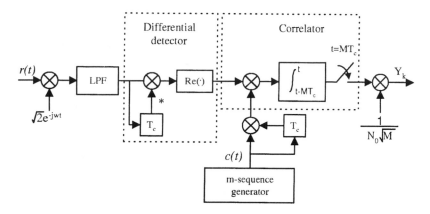

Figure 4.29.   Block diagram of chip-differentially-coherent acquisition technique.

The mean acquisition time performances of the differentially-coherent schemes are plotted in Fig. 4.28 in comparison with the noncoherent scheme. The figure plots the acquisition time of each scheme normalized by the corresponding uncertainty region length. It is assumed that the phase step advances by one chip period and the number of integrated chips, M, is 64. The code period of the FPC scheme is 64 and that of the PPC scheme is $2^{15} - 1$ chips. The penalty time is set to be 100 times greater than the number of integrated chips in the FPC and greater than the verification time in the PPC by the same factor. From the figure, we observe that the differential schemes can acquire synchronization faster than the noncoherent scheme by about 10 times. We also find that the PPC scheme slightly outperforms the FPC scheme in acquisition time also. This trend applies to both the single dwell and multiple dwell detections. The multiple dwell case considers the PPC scheme only, as the FPC scheme is not commonly used along with the multiple dwell detectors.

6.2   CHIP-DIFFERENTIALLY-COHERENT DETECTION

In order to prevent acquisition performance degradation in fast fading channels, Chung [36] proposed a chip-differentially-coherent detection scheme that processes the received signal at the baseband using a differential detector with one-chip time delay. The chip-differentially-coherent detector then performs a coherent partial correlation of the detector output with the product of a local PN code and its one-chip delayed phase to form a test sample. It preserves the pseudo-noise attribute at the output on one hand, and effectively suppresses the fluctuating phase components caused by fading and carrier frequency offset on the other. This enhances the in-sync correlation at the following correlator.

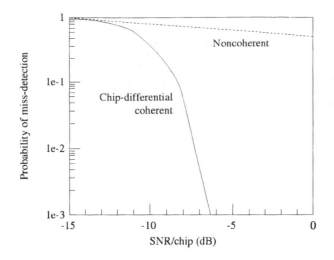

Figure 4.30. Detection performance of chip-differentially-coherent schemes ($P_{fa} = 10^{-2}$, matched filter length $M = 1,024$) [36].

Fig. 4.29 depicts the block diagram of the chip-differentially-coherent detector. We observe that the chip-differential detector first processes the received signal at baseband using a complex differential detector of delay T_c. The detected output is correlated with the product of a local code and its one-chip delayed phase. In case the received and local codes are closely correlated, the correlated output will exceed the threshold, loading the receiver to the in-synch state. There are two reasons for employing a differential detecting preprocessor: First, from the shift-and-add property of m-sequences [4], the product of the DS spreading waveform and its chip-time delayed phase yields another phase. So the differential detector can preserve the pseudo-noise attributes at its output if an m-sequence is used as the embedded spreading code. Secondly, since the differential detector suppresses time-varying phase components caused by fading and frequency offset, it can strongly enhance the in-sync correlation at the next correlator.

Fig. 4.30 plots the miss-detection probability of chip-differentially-coherent acquisition system in comparison with that of the conventional noncoherent parallel I-Q system, where the decision thresholds are selected to maintain false alarm probability $P_{fa} = 10^{-2}$. We observe significant performance improvement of the chip-differentially-coherent detector over the conventional noncoherent I-Q detector.

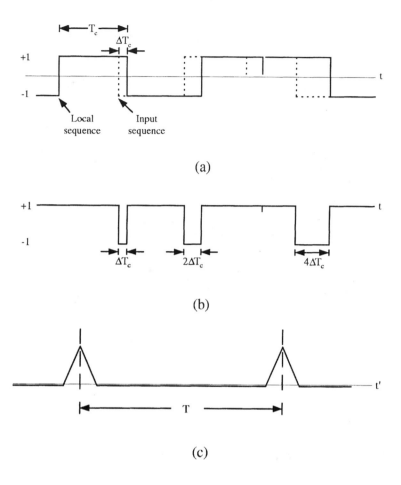

Figure 4.31. Illustration of code-Doppler shift [39]: (a) local sequence and input sequence, (b) correlation product, and (c) sliding correlator output signal.

7. ACQUISITION IN THE PRESENCE OF CODE-DOPPLER SHIFT

In some applications such as high dynamic GPS receiver or satellite communications receiver, acquisition needs to be done in the presence of severe Doppler shift. Doppler shift affects the acquisition performance in terms of carrier-frequency offset and code-frequency offset. In low-speed mobile communications, the code-frequency offset has little effect on the acquisition, so only the carrier-frequency offset may be considered. In high-speed mobile communications, however, the code-frequency offset becomes a critical factor of the performance degradation. In this section, we introduce an acquisition technique proposed by Glisic *et. al.* [39] that handles the code-Doppler problem effectively by employing the recirculation loop approach.

Code-Doppler shift (or code-frequency offset) refers to the deviation of chip interval from the nominal value. Fig. 4.31 illustrates the code-Doppler shift, where Fig. 4.31 (a) overlays the waveform of the input sequence on the waveform of the local sequence and Fig. 4.31 (b) shows their correlation product. If there were no code Doppler, the correlation would take the form in Fig. 4.31 (c) for the correlation peak period $T = N^2 T_c$ (N is the sequence length and T_c is the chip interval). [2] In the presence of code Doppler, however, T_c changes by ΔT_c and the period T changes by $\Delta T = N^2 \Delta T_c$. For example, for the code rate 2Mcps, carrier Doppler 30kHz, code Doppler 30Hz, and sequence length 1,000, we get $T = 0.5$ sec, $\Delta T_c = 7.5$ps, and $\Delta T = 7.5 \mu$s.

Fig. 4.32 depicts the block diagram of the code-Doppler acquisition system that employs single or multiple recirculation loops (or accumulators). Each recirculation loop is intended to improve the SNR of the decision variable tuned to the particular delay $D_i, i = 1, 2, \cdots, n$. In case the code-Doppler shift is unknown but constant, there occurs sliding over the set of possible samples of the triangular correlation function, and consequently, the number of accumulated samples in each accumulator loop is limited. By equipping multiple accumulator loops, the sliding effect can be compensated, since when an accumulator output decreases another accumulator output will increase, thereby making the overall signal level at the achieved output constant.

Fig. 4.33 plots the performance of the recirculation loop based acquisition system in terms of the improvement over the reference acquisition technique which employs sliding correlation, envelope detection, threshold comparison, and correct cell verification functions. For unknown Doppler input, the multiple recirculation loop based acquisition system with a varying number of accumulators and a linear combiner exhibits the improvement factor of the mean acquisition time and variance shown in the figure [39]. We observe that the improvement factor is more significant in lower SNR range. We also observe that the performance improvement is higher for one-accumulator system than for five-accumulator system in this case. It is because only one recirculation loop is accumulating the signal with others contributing to noise accumulation. For the case with a larger range of code-Doppler, however, a larger number of recirculation loops are needed to cover the whole uncertainty region.

The recirculation loop based acquisition scheme contrasts to the FFT based acquisition scheme [37] in the fundamental principle of approach. The FFT scheme partitions the uncertain carrier and code frequency ranges into several subbands and limits the maximum amount of Doppler effect in each subband

---

[2]Full-period correlation is assumed here.

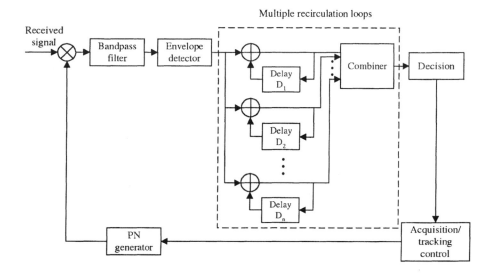

Figure 4.32.   Block diagram of the code-Doppler acquisition technique employing recirculation loops.

during the correlation process. The recirculation loop scheme, in contrast, partitions the period of correlation pulses rather than the code Doppler itself and uses multiple correlators matched to different values of the code Doppler. In essence, a bank of FFT processors in the FFT scheme is converted into a bank of accumulators in the recirculation loop scheme, and consequently the recirculation loop scheme renders a simpler implementation than the FFT scheme does.

8.   ACQUISITION BY DISTRIBUTED SRG STATE SAMPLE CONVEYANCE

Recently, a novel acquisition technique that realizes the direct SRG acquisition through distributed SRG state sample conveyance was introduced under the name of *distributed sample-based acquisition* (DSA) [56], and a family of variations followed [59]–[67]. The DSA basically features two new mechanisms - - distributed sampling-correction for the synchronization of the SRG and distributed conveyance of the state samples via a short-period sequence. It turned out that this combination is very effective in realizing a rapid acquisition scheme at low complexity and making the performance reliable even in very poor channel environments.

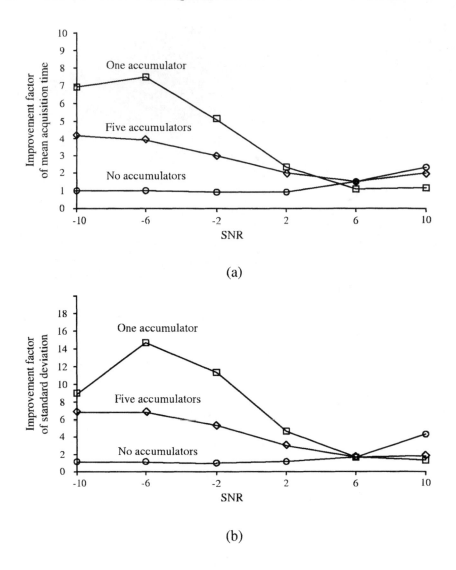

Figure 4.33.   Performance improvement of the recirculation loop based acquisition scheme over the reference system without recirculation loop [39]: (a) Improvement factor of the mean acquisition time, (b) improvement factor of the standard deviation.

In the DSA, the state of the main SRG in the transmitter is sampled and conveyed to the receiver in a *distributed* manner, where the state samples are detected and applied to correct the state of the main SRG in progressive manner. For the conveyance of the distributed state samples a short-period sequence, called *igniter sequence*, is employed.

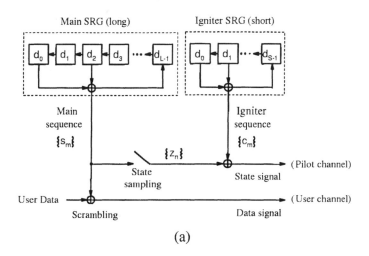

(a)

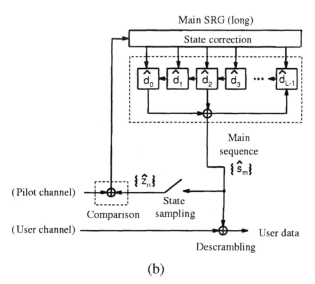

(b)

Figure 4.34. Functional block diagram of the DSA: (a) Transmitter, (b) receiver.

8.1 ORGANIZATION AND OPERATION OF THE DSA

Fig. 4.34 shows the functional block diagram of the DSA system. In the transmitter, the main SRG generates the *main sequence* $\{s_m\}$ of long period $N_M \equiv 2^L - 1$, which is used for data scrambling and whose fast acquisition is our ultimate goal. On the other hand, the igniter SRG generates the *igniter sequence* $\{c_m\}$ of short period $N_I \equiv 2^S$. The time-advanced state sampling block takes the state sample z_n of the main SRG at time $(R + n - 1)N_I$ in advance, for a reference value R. [3] The state samples are scrambled by the igniter sequence and conveyed over the pilot channel, while the user data is scrambled by the main sequence and transmitted over user channel.

In the receiver, the conveyed sample z_n is first passed to the comparator at time $(R + n)N_I$. At the same instant, the state sampling block generates the state sample of the main SRG, $\hat{z}_n$, and feeds to the comparator for comparison with the conveyed counterpart z_n. If $\hat{z}_n$ coincides with z_n, no action takes place, but otherwise the correction block is triggered to correct the state of the main SRG at time $(R + n)N_I + D_c$, with D_c chosen to be in $0 < D_c \leq N_I$.

If the sampling and correction circuits are designed according to the theorems in [56], the receiver main SRG gets synchronized to the transmitter main SRG after L comparison-correction operations provided that no detection error occurs.

For safety purpose against detection error, a verification process is appended to check whether or not the conveyed and receiver-generated state samples coincide V more times after the L comparison-correction operations, for a V chosen to meet the performance target. If all the V state sample sets coincide, then the receiver declares completion of synchronization of the main sequence and the tracking and estimation processes follow. Otherwise the acquisition process is reinitiated.

8.2   DSA BASED DS/CDMA ACQUISITION SYSTEM

Fig. 4.35 depicts the functional block diagram of the DS/CDMA system employing the basic DSA scheme [56, 63], in which the DSA functional blocks in Fig. 4.34 are embedded.

The transmitter part consists of a *DSA-spreader* and a *sample-spreader*, and the receiver part contains their despreading counter parts, that is, *DSA-despreader* and *sample-despreader*. The DSA-spreader/despreader pair take the synchronization function while the sample-spreader/despreader pair take the sample conveyance function. Those two functions are supported by the *main SRG* residing in the DSA-spreader and the *igniter SRG* residing in the sample-spreader, respectively.

In the transmitter (or the *base station* (BS)), the DPSK modulator maps the state sample, generated by the time-advanced sampling block, to the corresponding PSK *state symbol* x_n and produces the DPSK *pilot symbol* f_n by adding the phase of x_n to the phase accumulated up to the previous time slot.[4] The resulting pilot symbol is spread by a period of the igniter sequence and

---

[3]Note that the system description is given on a discrete time basis, with the unit time set to the chip duration T_c.

[4]In this section, we take an example of differential DSA (or DPSK-based DSA) [63] for the illustration of the DSA system.

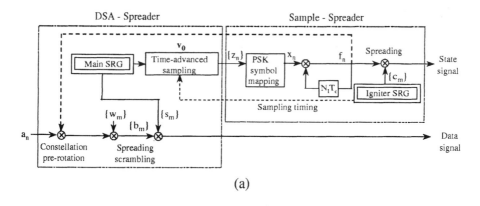

(a)

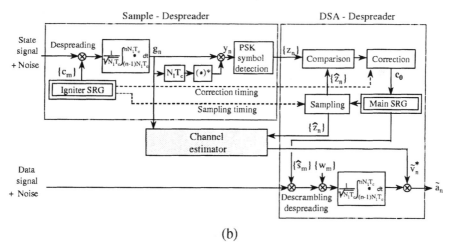

(b)

Figure 4.35. Functional block diagram of the DS/CDMA system employing the DSA scheme: (a) Transmitter, (b) receiver.

transmitted through the pilot channel in the interval $[(R + n - 1)N_I, (R + n)N_I)$. On the other hand, each user's data a_n is multiplied by the pilot symbol f_n, spread by one of the orthogonal channelization sequences $\{w_m\}$ and scrambled by the main sequence $\{s_m\}$, and then transmitted in the interval $[(R + n - 1)N_I, (R + n)N_I)$. The state signal (or, pilot signal) and the data signal are propagated over the same multipath fading channel to arrive at the mobile station.

The receiver (or the *mobile station* (MS)) first acquires the DPSK modulated igniter sequence employing the simple noncoherent threshold detector [4]. We assume that the simple serial search method is applied to the igniter sequence acquisition, while any other search methods may be used to speed up the acquisition process. After the timing synchronization of the igniter sequence, the MS

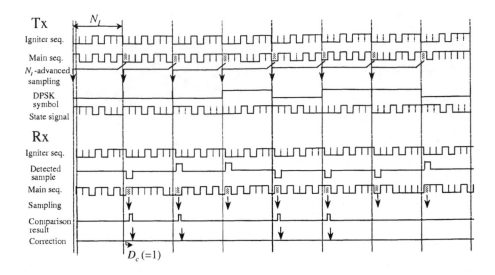

Figure 4.36.   System timing diagram example (N_I=8, D_c=1, DBPSK modulation).

despreads the received state signal and differentially detects the conveyed sample z_n. Then the comparison-correction based SRG synchronization process follows as described above.

Fig. 4.36 depicts the timing relations among various processings in the transmitter and the receiver.

Once the synchronization of the SRG is completed, the receiver despreads the data signal by multiplying the synchronized main sequence and the corresponding channelization code, then coherently demodulates the despread data by using the channel estimate obtained from the pilot channel sequence (or, state signal).

8.3   VARIATIONS OF THE DSA TECHNIQUE

The original DSA scheme was designed on the basis of binary orthogonal state symbol generation [56], in which the state signal was used directly as the channel estimation reference, without employing data constellation pre-rotation shown in Fig. 4.34. The DSA was then extended to the *batch DSA* (BDSA) [59] and the *parallel DSA* (PDSA) [60] that are capable of manipulating multiple samples concurrently, for application to general M-ary signaling or multi-carrier DS/CDMA systems. The acquisition performance degradation that can happen when the number of concurrently manipulated state samples increases was resolved by introducing the *DPSK-signaling DSA* (D^2SA) [63]. For an enhanced robustness of the DSA scheme in the worst-case channel environment, the *correlation-aided DSA* (CDSA) that incorporates the state symbol correlation process was introduced [67].

In the CDSA, a state symbol correlation process is added as an extension to the original comparison-correction based acquisition process. If the initial-stage acquisition, which is based on the comparison-correction process, fails to acquire synchronization within certain time limit, then the state symbol correlation process is activated based on the state symbol sequences that have been collected until that time. In this second-stage acquisition, synchronization is acquired by determining the shifted sequence that produces the maximum correlation energy as the truly transmitted sequence. The state symbol correlation process indeed provides a very high synchronization success probability even in poor channel environments, but the required additional memory or computation complexity is minimal.

8.4   PERFORMANCES AND COMPLEXITY

The mean acquisition time performances of the DSA are discussed in detail in [56]-[66]. In simple approximation, the overall mean acquisition time of the DSA scheme employing an active correlator is about $(N_I + L + V) \times N_I T_c$ (i.e., the sum of the igniter sequence serial search time and the correction/verification time), while that of the conventional serial search scheme is about $(2^L - 1) \times N_I T_c$, asymptotically (at high SNR). So the relative acquisition time of the DSA, normalized to the serial search case, is approximately $(N_I + L + V)/2^L$, which becomes extremely small as L increases. (V is typically set to a value smaller than L.) As a typical example, the relative acquisition time reduces to 0.004 when $N_I = 128$, $L = 15$ and to 0.001 when $N_I = 256$ and $L = 18$.

Fig. 4.37 plots a typical mean acquisition time performance of the synchronous cellular system employing the CDSA (or DSA) scheme in comparison with that employing the conventional *parallel search acquisition* (PSA) scheme operating 4-parallel correlators [65]. The operating chip rate is 1.2288Mcps, and a 100Hz Doppler Rayleigh fading channel is used. The chip-SNR (γ_c) denotes the average ratio of the pilot chip energy to the one-side noise power spectral density. The DSA and the CDSA can complete the cell search about 30 times faster than the PSA having comparable complexity in moderate SNR range. When the SNR becomes very low, the performance of the DSA shifts to the level of the PSA, but the CDSA maintains the same level of acquisition time gain.

For implementation, the DSA necessitates a main sequence generator, an igniter sequence generator, a state symbol generator and a (time-advanced) sampling circuit, a correction circuit and a verification circuit. With all these functions, however, the required hardware and computation is very simple, so implementation complexity is very low.

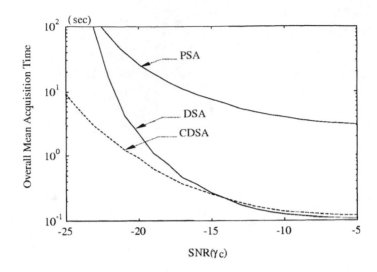

Figure 4.37.   Overall mean acquisition time in 100 Hz Doppler Rayleigh fading channel with respect to chip-SNR(γ_c) [65].

The DSA and its family render a very effective means for rapid and robust acquisition in both inter-cell synchronous and asynchronous DS/CDMA cellular systems. So we will provide more thorough and rigorous discussions on them in separate chapters - - Chapters 8 and 9.

II

SPREADING AND SCRAMBLING IN IMT-2000 DS/CDMA SYSTEMS

Chapter 5

INTER-CELL ASYNCHRONOUS IMT-2000 W-CDMA SYSTEM (3GPP-FDD)

IMT-2000 or the third generation mobile radio system will provide wireless multimedia services with high capacity, quality, and security, taking advantage of the wide spectrum bandwidth, the advanced signal processing, and the intelligent networking technologies. A score of proposals were reviewed as the candidate radio transmission technology before a series of harmonization efforts integrated them to three major IMT-2000 standards: W-CDMA by 3GPP, cdma2000 by 3GPP2, and UWC-136 by UWCC. The UWC-136 standard is based on the advanced TDMA-EDGE technology, so is out of the scope of this book. [1]

The W-CDMA standard is based on the wideband CDMA technology and the GSM network, which yield two different systems called 3GPP-FDD (or UTRA-FDD) and 3GPP-TDD (or UTRA-TDD), respectively. [2] The 3GPP-FDD utilizes dual frequency bands for uplink and downlink communications and operates in the inter-cell asynchronous environment, in general. On the other hand, the 3GPP-TDD utilizes a single frequency band and bi-directional communication is realized through time division duplexing. Furthermore, in the 3GPP-TDD, frame-level inter-cell synchronization is maintained to minimize interference between adjacent base stations or adjacent mobile stations. The cdma2000 standard is an evolutionary outgrowth of the IS-95 technology that utilizes dual frequency bands via FDD for bi-directional communications. Inter-

---

[1] 3GPP and UWC respectively stand for the *3rd Generation Partnership Project* and the *Universal Wireless Communications Consortium*. EDGE stands for the *Enhanced Data rates for GSM (and IS-136) Evolution*.
[2] GSM and UTRA respectively stand for the *Global System for Mobile Communications* and the *Universal Terrestrial Radio Access*. The name W-CDMA often represents the 3GPP-FDD system only, while the 3GPP-TDD system is usually called TD-CDMA system in reflection of its TDMA-CDMA combined multiple access method.

cell synchronization is essentially guaranteed for its operation, which is now serviced by the *global positioning system* (GPS).

In this part, we discuss the spreading and scrambling in the IMT2000 DS/CDMA systems. We divide the discussion into three chapters - - 3GPP-FDD in Chapter 5, 3GPP-TDD in Chapter 6, and cdma2000 in Chapter 7. In the cases of 3GPP-FDD and 3GPP-TDD, we additionally provide descriptions of the relevant physical layer structures and operations before addressing the spreading and scrambling issues, because they are relatively new and thus less known.

In this chapter, we first discuss the structure and operation of the 3GPP-FDD system in detail, and then examine the spreading and scrambling issues in later sections, finally discussing the cell search scheme.

1.   TRANSPORT CHANNELS AND PHYSICAL CHANNELS

Transport channels are the services offered by Layer 1 to the higher layers, while the physical channels are the physical realization of the transport channels which consist of a layered structure of radio frames and time slots. Fig. 5.1 depicts the mapping of transport channels onto physical channels.

1.1   TRANSPORT CHANNELS

Dedicated Channel (DCH): The DCH is a downlink/uplink channel used to carry user or control information between the network and a mobile station. The DCH is composed of *dedicated traffic channel* (DTCH) and *dedicated control channel* (DCCH).

Broadcast Channel (BCH): The BCH is a downlink channel used to broadcast system- and cell-specific information.

Forward Access Channel (FACH): The FACH is a downlink channel used to carry control information to a specific *user equipment* (UE) when the system knows the location cell of the UE.

Paging Channel (PCH): The PCH is a downlink channel used to carry control information to a UE when the system does not know the location cell of the UE.

Random Access Channel (RACH): The RACH is an uplink channel used to carry control information or short user packets from a UE.

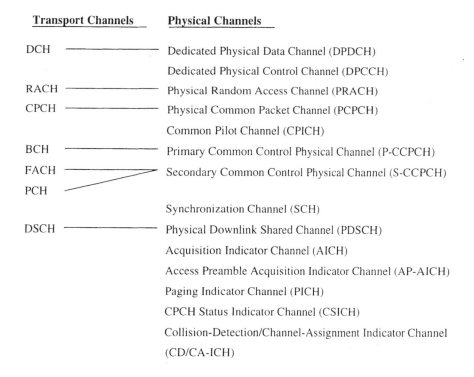

Figure 5.1.   Transport-channel to physical-channel mapping [85].

Common Packet Channel (CPCH): The CPCH is an uplink channel used for high-rate multi-media packet services.

Downlink Shared Channel (DSCH): The DSCH is a downlink channel used for services such as short message service, e-mail applications, *file transfer protocol* (FTP), and multi-media services shared by several UEs. The DSCH is associated with one or several downlink DCHs.

1.2   PHYSICAL CHANNELS

A radio frame is a 10ms processing unit which consists of 15 slots. Each slot corresponds to one power-control period. The length of a radio frame corresponds to 38,400 chips. Thus the length of a slot corresponds to 2,560 chips and the chip rate is 3.84Mcps. The data/control channels are QPSK-modulated both in the uplink and in the downlink but their realizations are different. More specifically, in the uplink the control and the data channels are first BPSK-modulated and then I/Q code-multiplexed, while in the downlink both channels are QPSK-modulated and then time-multiplexed.

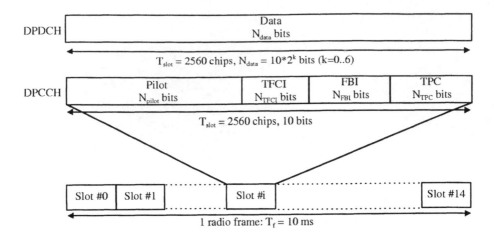

Figure 5.2.   Frame structure for the uplink DPDCH/DPCCH [85].

1.2.1   UPLINK PHYSICAL CHANNELS

Dedicated Physical Data Channel (DPDCH) / Dedicated Physical Control Channel (DPCCH): The DPDCH and the DPCCH are respectively used to carry the DCH transport channel and the control information generated at Layer 1. The Layer 1 information consists of pilot bits for channel estimation, an optional *transport format combination indicator* (TFCI), *feedback information* (FBI) for closed loop mode transmit diversity [D-field] and *site selection diversity transmission* (SSDT) [S-field], and *transmit power control* (TPC) commands. The DPDCH and the DPCCH are I/Q code-multiplexed within each radio frame. Fig. 5.2 shows the frame structure of the uplink dedicated channels, where the parameter k determines the number of bits per uplink DPDCH slot, which is related to the *spreading factor* (SF) of the physical channel as $SF=256/2^k$. We see in the figure that the DPDCH spreading factor may range from 256 down to 4 depending on the data rate, while that of the uplink DPCCH is always 256. The TFCI is used when several simultaneous services are provided and the TPC is used for inner loop power control. The pilot bits have specific patterns optimized for confirmation of the frame synchronization [86] and are used for SIR and channel phase estimations. For the pilot bit patterns corresponding to each physical channels, refer to [85].

Physical Random Access Channel (PRACH): The PRACH is used to carry the RACH. It is based on a slotted ALOHA approach with fast acquisition indication. The UE can start the random-access transmission at the beginning of a number of *access slots*. There are 15 access slots per two frames and they

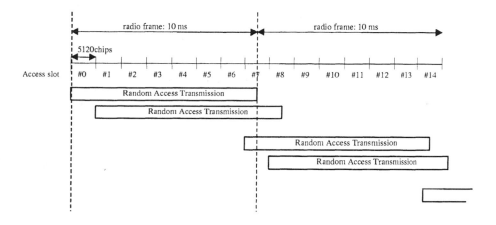

Figure 5.3.   RACH access slot numbers and their spacing [85].

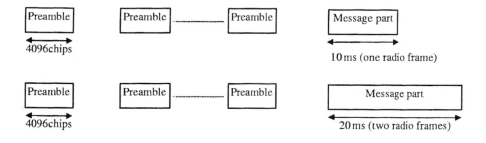

Figure 5.4.   Structure of the random-access transmission [85].

are spaced 5,120 chips apart. Fig. 5.3 depicts the RACH access slot numbers and their spacing. The structure of the random-access transmission is shown in Fig. 5.4. The random-access transmission consists of one of several preambles of length 4,096 chips and a message of length 10ms or 20ms. Each preamble consists of 256 repetitions of a signature of length 16 chips. Fig. 5.5 shows the structure of the 10ms message part radio frame, which consists of a data part to which the RACH transport channel is mapped and a control part that carries Layer 1 control information. The SF of the message data part may range from 256 down to 32 depending on the data rate while that of the control part is fixed at 256. The control part consists of 8 pilot bits and 2 TFCI bits per slot. A 20ms message part consists of two consecutive 10ms message part radio frames. The message part length can be determined from the used signature and/or access slot, as configured by higher layers.

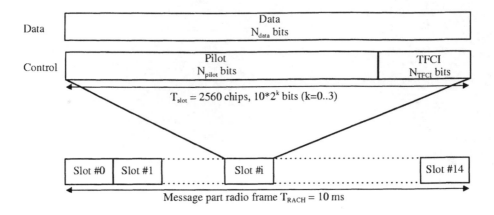

Figure 5.5.   Structure of the random-access message part radio frame [85].

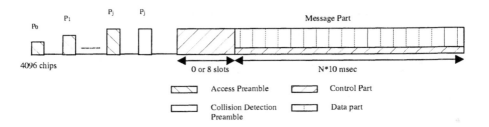

Figure 5.6.   Structure of the CPCH access transmission [85].

Physical Common Packet Channel (PCPCH): The PCPCH is used to carry the CPCH. The CPCH transmission is based on the *digital sense multiple access-collision detection* (DSMA-CD) approach with fast acquisition indication. The access slot timing and structure is identical to those of RACH. The structure of the CPCH access transmission is shown in Fig. 5.6. The PCPCH access transmission consists of one or several *access preambles* (A-P) of length 4,096 chips, one *collision detection preamble* (CD-P) of length 4,096 chips, a *DPCCH power control preamble* (PC-P) which is either 0 or 8 slots in length, and a message of variable length of $N \times 10$ms. The message part frame structure is the same as that of the uplink DPDCH/DPCCH.

1.2.2   DOWNLINK PHYSICAL CHANNELS

Dedicated Physical Channel (DPCH): In the downlink DPCH, the DPDCH and the DPCCH are time-multiplexed. In Fig. 5.7, Data1 and Data2 fields correspond to the DCH while TPC, TFCI (optional), and Pilot fields correspond

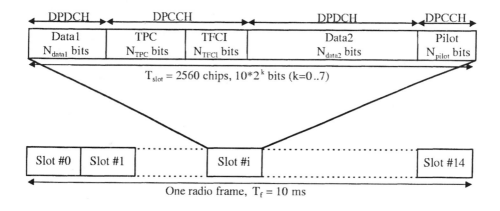

Figure 5.7. Frame structure for the downlink DPCH [85].

to the Layer 1 control information. The parameter k determines the total number of bits per downlink DPCH slot, which is related to the SF of the physical channel in the form $SF=512/2^k$. The SF may thus range from 512 down to 4. In case of multi-code transmission, the DPCCH is transmitted only on the first DPCH.

DL-DPCCH for CPCH: The DL-DPCCH for CPCH is a special case of DPCH used to support CPCH signalling, which has a fixed spreading factor of 512. The slot of the DL-DPCCH for CPCH consists of TPC [2bits], *CPCH control commands* (CCC) [4bits], and pilot bits [4bits]. There are two types of CCC: One is *emergency stop of CPCH transmission* [1111] and the other is *start of message indicator* [1010].

Common Pilot Channel (CPICH): The CPICH is a fixed rate (30 kbps, SF=256) downlink physical channel that carries a pre-defined symbol sequence. There are two types of CPICHs: The *primary CPICH* (P-CPICH) is associated with the 0th channelization code of length 256, scrambled by the primary scrambling code, and broadcast over the entire cell. There is only one P-CPICH per cell and it is the default phase reference for all downlink physical channels. The *secondary CPICH* (S-CPICH) is associated with an arbitrary channelization code of length 256, scrambled by either the primary or a secondary scrambling code, and transmitted over the entire cell or over a part of the cell. There may be zero, one, or several S-CPICHs per cell and the S-CPICH may be the phase reference for the S-CCPCH and the downlink DPCH.

Primary Common Control Physical Channel (P-CCPCH): The P-CCPCH is a fixed rate (30kbps, SF=256) downlink physical channels used to carry the

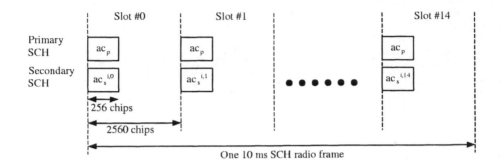

Figure 5.8.   Structure of the synchronization channel (SCH) [85].

BCH transport channel. The P-CCPCH is not transmitted during the first 256 chips of each slot. Instead, *synchronization channel* (SCH) is transmitted during this period. No TPC, no TFCI, and no pilot bits are transmitted over the P-CCPCH.

Secondary Common Control Physical Channel (S-CCPCH): The S-CCPCH is used to carry the FACH and the PCH. The FACH and the PCH may be mapped to the same or to separate S-CCPCHs. The TFCI may be used to support multiple transport format combinations but no TPC is transmitted in S-CCPCHs. The spreading factor may range from 256 down to 4. The S-CCPCH associated with the FACH may be transmitted over a part of the cell.

Synchronization Channel (SCH): The SCH is a downlink signal used for cell search, which consists of two sub channels - the *primary* and the *secondary* SCH. Over the primary SCH, a modulated code of length 256 chips, called *primary synchronization code* (PSC), is transmitted at the beginning of each slot. The PSC is the same for every cell in the system. In parallel with the PSC, a sequence of modulated codes of length 256 chips is transmitted over the secondary SCH, which is called the *secondary synchronization code* (SSC) and has the period of 15. Each SSC is chosen from a set of 16 different codes of length 256. The SCH structure is shown in Fig. 5.8, where C_p and $C_s^{i,k}$ ($i = 0, 1, \cdots, 63$, $k = 0, 1, \cdots, 14$) respectively denote the PSC and the SSC. (The index i denotes the number of the scrambling code group the transmit cell belongs to.) We see in the figure that the PSC and the SSC are modulated by the symbol a, which indicates the presence ($a=+1$) or the absence ($a=-1$) of *space time transmit diversity* (STTD) encoding on the P-CCPCH.

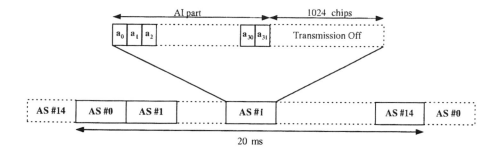

Figure 5.9.   Structure of the acquisition indicator channel (AICH) [85].

Physical Downlink Shared Channel (PDSCH): The PDSCH is used to carry the DSCH and shared by users based on code multiplexing. Each PDSCH radio frame is associated with a downlink DPCH. The PDSCH and the DPCH do not necessarily have the same spreading factors. Furthermore, the PDSCH spreading factor may vary from frame to frame. All relevant Layer 1 control information is transmitted on the DPCCH part of the associated DPCH. The spreading factor may range from 256 down to 4. When a DSCH is mapped to multiple parallel PDSCHs, the parallel PDSCHs operate with the frame synchronized each other and the spreading factor of all PDSCH codes are the same.

Acquisition Indicator Channel (AICH): The AICH is a physical channel used to carry *acquisition indicators* (AI). Acquisition indicator AI_s corresponds to signature s (of length 16) on the PRACH. The AICH structure is depicted in Fig. 5.9. The AICH consists of a repeated sequence of 15 consecutive *access slots* (AS), each of length 40 bit intervals. Each access slot consists of the AI part with 32 real-valued symbols (4,096 chips) $a_0, a_1, \cdots, a_{31}$ and the *transmission-off* part of 1,024 chips. The real-valued symbols $a_0, a_1, \cdots, a_{31}$ are given by

$$a_j = \sum_{s=0}^{15} AI_s \times b_{s,j}, \quad j = 0, 1, \cdots, 31, \tag{5.1}$$

where AI_s takes the value +1 (ACK), -1 (NACK), or 0 (NOT-DETECTED). When +1 has been received, the UE transmits the RACH message part, but when -1 has been received, it reports the status NACK-ON-AICH-RECIEVED to the higher layer and exits the physical random access procedure. When 0 has been received, the UE retransmits the RACH preamble with an increased transmission power and a new access slot and signature combination (if the retransmission counter does not overtake the maximum value). The basis sequences $\{b_{s,2j} : j = 0, 1, \cdots, 15\}$, $s = 0, 1, \cdots, 15$, are the 16 rows of the

16×16 Hadamard matrix taking $+1$ or -1 as the element, and $b_{s,2j+1}$ takes the same value as $b_{s,2j}$ for $j = 0, 1, \cdots, 15$. The spreading factor of the AI part is 256.

CPCH Access Preamble Acquisition Indicator Channel (AP-AICH): The AP-AICH is a physical channel used to carry *AP indicators* (API) of CPCH. API corresponds to AP signature s transmitted by UE. AP-AICH and AICH may use the same or different channelization codes. The AP-AICH structure is the same as that depicted in Fig. 5.9 except that the AI part is replaced with the API part. The real-valued symbols $a_0, a_1, \cdots, a_{31}$ corresponding to the API part are given by

$$a_j = \sum_{s=0}^{15} API_s \times b_{s,j}, \quad j = 0, 1, \cdots, 31, \qquad (5.2)$$

where API_s, the API corresponding to the AP signature s, takes the value $+1$ (ACK), -1 (NACK), or 0 (NOT-DETECTED), and $\{b_{s,j}\}$ are the same Hadamard bases that are used for AICH.

CPCH Collision Detection / Channel Assignment Indicator Channel (CD/CA-ICH): The CD/CA-ICH is a physical channel used to carry *CD indicator* (CDI) if the CA is not active, or *CDI/CA indicator* (CAI) at the same time if the CA is active. The CD/CA-ICH structure is the same as that depicted in Fig. 5.9 except that the AI part is replaced with the CDI/CAI part. The CD/CA-ICH and the AP-AICH may use the same or different channelization codes. In case CA is not active, the real-valued symbols $a_0, a_1, \cdots, a_{31}$ are given by

$$a_j = \sum_{s=0}^{15} CDI_s \times b_{s,j}, \quad j = 0, 1, \cdots, 31, \qquad (5.3)$$

where CDI_s, the CDI corresponding to the CD preamble signature s, takes the value $+1$ (ACK) or 0 (NACK), and $\{b_{s,j}\}$ are the same Hadamard bases that are used for AICH. In case CA is active, the real-valued symbols $a_0, a_1, \cdots, a_{31}$ are given by

$$a_j = \sum_{i=0}^{15} CDI_i \times b_{s_i,j} + \sum_{k=0}^{15} CAI_k \times b_{s_k,j}, \quad j = 0, 1, \cdots, 31, \qquad (5.4)$$

where the subscript s_i and s_k indicate the signature number s depending on the indexes i, k (refer to [85] for the detailed mapping.), and CDI_i and CAI_k, taking the values ± 1 (ACK) and 0 (NACK), are respectively the CDI and the CAI corresponding to the CD preamble i and the assigned channel index k. $\{b_{s,j}\}$ are the same Hadamard bases that are used for AICH.

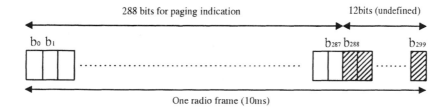

Figure 5.10. Structure of the paging indicator channel (PICH) [85].

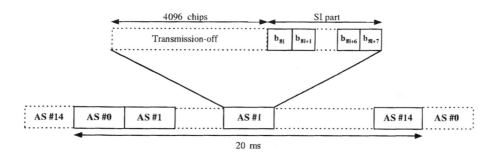

Figure 5.11. Structure of CPCH status indicator channel (CSICH) [85].

Paging Indicator Channel (PICH): The PICH is a fixed rate (SF=256) physical channel used to carry the *paging indicator* (PI). The PICH is always associated with an S-CCPCH to which a PCH transport channel is mapped. Fig. 5.10 shows the frame structure of the PICH. Of 300 bits in a 10ms frame, the first 288 bits are used to carry the PI, while the remaining 12 bits are undefined. N PI's $\{PI_0, PI_1, \cdots, PI_{N-1}\}$ are transmitted in each frame for $N = 18, 36, 72,$ or 144. PI_p can take the value 1 or 0 and is mapped to the PICH by the m times repetition coding $\{b_{mp}, b_{mp+1}, \cdots, b_{mp+m-1}\} = \{(-1)^{PI_p}, (-1)^{PI_p}, \cdots, (-1)^{PI_p}\}$ for $m = 288/N$. When PI_p in a certain frame is set to 1, the UEs associated with this PI should read the corresponding frame of the associated S-CCPCH, but otherwise, they need not.

CPCH Status Indicator Channel (CSICH): The CSICH is a fixed rate (SF=256) physical channel used to carry CPCH status information. Fig. 5.11 depicts the frame structure of the CSICH. The CSICH is always associated with a physical channel used for transmission of CPCH AP-AICH (transmission-off part) and uses the same channelization and scrambling codes. The CSICH frame consists of 15 consecutive access slots of length 5,120 chips. The first

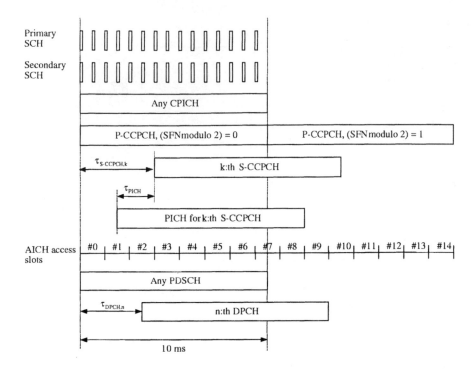

Figure 5.12.   Frame timing and access slot timing of downlink physical channels [85].

4,096 chip period is the transmission-off period (for use of AP-AICH) and a *status indicator* (SI) part consisting of 8 bits is transmitted for the remaining 1,024 chip period. N SI's $\{SI_0, SI_1, \cdots, SI_{N-1}\}$ are transmitted in each frame for $N = 1, 3, 5, 15, 30,$ or 60. SI_p can take the value 1 or 0 and is mapped to the PICH by the m times repetition coding $\{b_{mp}, b_{mp+1}, \cdots, b_{mp+m-1}\} = \{(-1)^{SI_p}, (-1)^{SI_p}, \cdots, (-1)^{SI_p}\}$ for $m = 120/N$. The higher layers shall provide Layer 1 with the mapping between the values of the SI's and the availability of CPCH resources.

2.   TIMING RELATIONS

The P-CCPCH, on which the cell *system frame number* (SFN) is transmitted, is used as the timing reference for all physical channels.

Fig. 5.12 depicts the frame timing of the downlink physical channels. In the downlink, the SCH, CPICH, P-CCPCH, and PDSCH have identical frame timings. The S-CCPCH and DPCH timing may be different for different S-CCPCHs and DPCHs, but the offset from the P-CCPCH frame timing is a

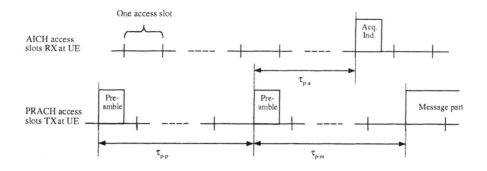

Figure 5.13.   Timing relation between PRACH and AICH as seen at the UE [85].

multiple of 256 chips (or, $\tau_{S-CCPCH,k} = T_k \times 256$ chips, $\tau_{DPCH,n} = T_n \times 256$ chips, $T_k, T_n \in \{0, 1, \cdots, 149\}$). The PICH timing is 3 slots prior to its corresponding S-CCPCH frame timing (or, $\tau_{PICH} = 7,680$ chips). The AICH 0th access slot starts at the same time as the P-CCPCH frames with even-numbered SFN. The PDSCH frame starts anywhere from 3 slots up to 18 slots after the end of the associated DPCH frame. The initial downlink DPCCH/DPDCH frame (or the initial power control preamble if it exists) starts at least 15 slots after the end of the relevant FACH frame transmission. The uplink DPCCH/DPDCH frame is transmitted approximately 1,024 chips after the reception of the first significant path of the corresponding downlink DPCCH/DPDCH frame. This transmission timing offset between the downlink and the uplink enables the transmit power to be adjusted at every slot.

On the other hand, the PRACH/AICH timing relation is depicted in Fig. 5.13, where the preamble-to-preamble distance τ_{p-p} and the preamble-to-message distance τ_{p-m} are at least 3 or 4 access slots (depending on the AICH-TRANSMISSION-TIMING parameter signalled by higher layers), and the preamble-to-AI distance τ_{p-a} corresponds to 1.5 or 2.5 access slots. The random access procedure is as follows [11]:

- The *mobile station* (MS) decodes the BCH to find out the available RACH sub-channels and their scrambling codes and signatures.
- The terminal selects randomly an available signature and one of the *RACH sub-channels* from the group which its *access service class* (ASC) allows to use.
- The downlink power level is measured and the initial RACH power level is set.
- An RACH preamble is sent with the selected signature.

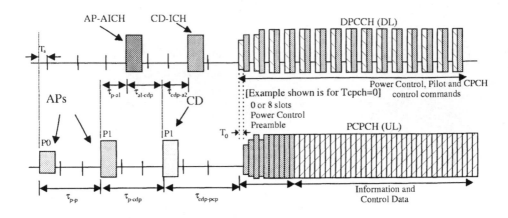

Figure 5.14.   Timing of PCPCH and AICH transmission as seen by the UE, with $T_{cpch} = 0$ [85] ($T_0 = 1,024$ chip length).

- The MS decodes AICH to see whether the *base station* (BS) has detected the preamble.

- In case no AICH is detected, the MS increases the preamble transmission power by a step given by the BS, selects a new signature, and retransmits the preamble in the next available access slot.

- When an AICH transmission is detected from the BS, the MS starts to transmit the message part.

Similar timing relationship and procedure are applied for the PCPCH/AICH, which is depicted in Fig. 5.14, where the relative distance between each channel depends on the CPCH parameter T_{cpch} that takes 0 or 1. The main difference of the CPCH access procedure from that of the RACH lies in incorporating the *collision detection* (CD) operation and a fast power control over the message part. Until the AICH is detected by the MS, the same access procedure as the RACH access procedure is taken. Once the AICH is detected, a CD preamble is sent with another signature, which is randomly chosen from a given signature set. The transmit power level of the CD preamble is kept the same as that of the last *access preamble* (AP). The BS is expected to send the acknowledgement for the signature used in the CD preamble through the CD-ICH. After the MS detects the correct acknowledgement, it starts to transmit the message part which usually lasts over several frames. As the CPCH message part is usually transmitted for a long period, the incorporation of the physical layer collision detection can be an effective means to reduce interference. At the beginning of the CPCH message transmission, an optional power control preamble can be sent over 8 slots to allow the power control process to converge in short time. (Power control step size is 2dB or 3dB.) The power control commands are transmitted over the DL-DPCCH for CPCH. [3]

3.   POWER CONTROL

There are three different power control schemes - - *open loop power control, inner loop power control,* and *outer loop power control.*

The open loop power control is performed by measuring the received power level and adjusting transmit power level based on the measured path loss. It is applied to the initial power setting of the MS in relation to the random access or to the TDD uplink operation where the uplink and the downlink use the same frequency band. Using the measured path loss, in addition to the information from the BS on the interference in the uplink channel and the target SIR, the MS can set its transmit power such that the quality requirements at the BS can be met.

The inner loop power control is a fast power control processing helped by the power control commands that are replied by the receiving side. The inner loop power control has been applied only to the CDMA uplink in order to mitigate the *near-far problem* among the mobile stations, but in the IMT-2000 W-CDMA system (and the cdma2000 system) its use is extended to the downlink as well. The inner loop power control is applied only to the DPCCH/DPDCH and the PCPCH, and (selectively) to the PDSCH. The receiving side measures the SIR level of the incoming data (typically using the pilot part), compares it with a threshold provided on the basis of quality of services (BER, FER, etc.), and then determines the TPC. The power control command returns once per slot (i.e., at the rate of 1,500 Hz) and the transmitting part adjusts its power level conforming to the command.

The transmission timing skew (by 1,024 chips) between the downlink and the uplink slots enables both the BS and the MS to adjust their transmit power only with one slot delay. Fig. 5.15 illustrates the timing relation as well as an exemplary interaction between the BS and the MS for the transmit power control. Nominally three power control step sizes are defined (1dB, 2dB, and 3dB) [55], however, smaller step sizes can be emulated by applying pertinent control algorithms. In soft handover mode, the power control commands transmitted from multiple BSs are first combined applying a specific algorithm and then the power-up or power-down decision is made by the MS. In compressed mode, a larger step size is used for a short period after a compressed frame to make the power level converge to a pertinent level in short time.

---

[3]The CSICH is also associated with the CPCH access procedure, which enables mobile stations to monitor the status of the CPCH usage and get the information of the channel resources.

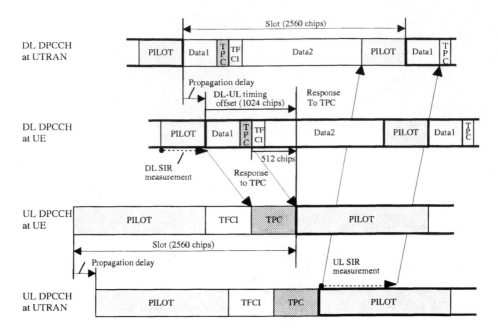

Figure 5.15.   Transmitter power control timing [55].

The outer loop power control is the processing that adjusts the SIR threshold level in complement to the open loop or inner loop power control in order to meet the quality of services (or, FER, BER, etc.) on a long term basis, which is applied relatively at a low rate.

4.   DOWNLINK TRANSMIT DIVERSITY

In the W-CDMA system, along with the conventional reception diversity techniques based on RAKE receivers (in up- and down-links) or multiple receiving antennas (in uplink), *transmit antenna diversity* is adopted in the downlink. Downlink transmit diversity is categorized into two modes - - *open loop transmit diversity* and *closed loop transmit diversity*. The open loop mode consists of *time switched transmit diversity* (TSTD) and *space time block coding based transmit antenna diversity* (STTD), and the closed loop mode consists of transmit antenna phase adjustment (CL mode-1) and transmit antenna phase/amplitude adjustment (CL mode-2). In the closed loop mode, the transmit antenna weight is adjusted based on the FBI information conveyed over the uplink physical control channels. Among the downlink channels, the TSTD is applied only to the SCH, and the STTD is applied on the P-CCPCH, S-CCPCH, DPCH, PICH, PDSCH, AICH, and CSICH. The closed loop transmit diversity

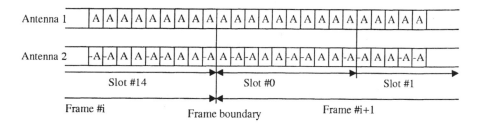

Figure 5.16. Modulation pattern for Common Pilot Channel (with $A = 1 + j$) [85].

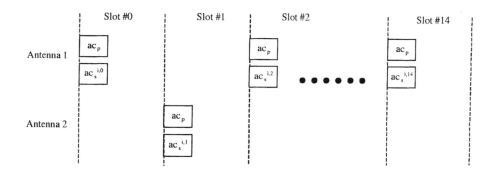

Figure 5.17. Structure of the SCH transmitted by the TSTD scheme [85].

may be applied on the DPCH and PDSCH. If transmit diversity is applied on any of the downlink physical channels, it is also applied on P-CCPCH, SCH, and CPICH. In this case the CPICH is transmitted from both antennas using the different symbol patterns as shown in Fig. 5.16. (In the case of no transmit diversity, the symbol sequence of antenna 1 in the figure is used.) The transmit diversity mode used for a PDSCH frame is the same as that used for the associated DPCH frame. During the duration of the PDSCH frame and within the slot prior to the PDSCH frame, the transmit diversity mode on the associate DPCH may not change. However, it is allowed to change from closed loop mode 1 to mode 2 or vice versa.

Fig. 5.17 illustrates the SCH transmitted by the TSTD scheme. In the even numbered slots both the PSC and the SSC are transmitted on antenna 1, but in the odd numbered slots they are transmitted on antenna 2.

The STTD encoding is applied on blocks of four consecutive channel bits (or, two QPSK symbols). A block diagram of a generic STTD encoder is shown in Fig. 5.18. In the figure, if α_1 and α_2 respectively denote the complex path gains of path 1 and path 2 which are expected to be independent each other, the

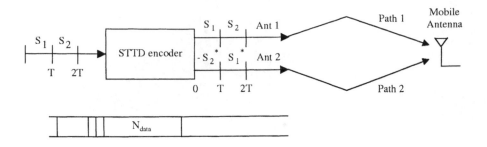

Figure 5.18. Block diagram of the STTD encoder. (S1, S2 are QPSK or discontinuous transmission (DTX) symbols and T denotes the symbol time.) [85]

first and the second received soft symbols R_1 and R_2 can be expressed by

$$R_1 = \alpha_1 S_1 - \alpha_2 S_2^* + N_1, \qquad (5.5a)$$
$$R_2 = \alpha_1 S_2 + \alpha_2 S_1^* + N_2, \qquad (5.5b)$$

where N_1 and N_2 are additive channel noises. The mobile station combines the two consecutive soft symbols such that

$$\hat{S}_1 = \alpha_1^* R_1 + \alpha_2 R_2^* = (|\alpha_1|^2 + |\alpha_2|^2)S_1 + \alpha_1^* N_1 + \alpha_2 N_2^*, \quad (5.6a)$$
$$\hat{S}_2 = -\alpha_2 R_1^* + \alpha_1^* R_2 = (|\alpha_1|^2 + |\alpha_2|^2)S_2 - \alpha_2 N_1^* + \alpha_1^* N_2, (5.6b)$$

and conveys $\hat{S}_1$ and $\hat{S}_2$ to the subsequent decoding block as the soft estimates of the QPSK symbols S_1 and S_2. (For detailed description of the STTD diversity, refer to [87],[88]).

Fig. 5.19 depicts the generic transmitter structure that supports closed loop transmit diversity for DPCH transmission. The complex weight factors (phase adjustments in CL mode 1 and phase/amplitude adjustments in CL mode 2) of the two transmit antennas are determined by the UE such that the UE received power is maximized, and signalled to the BS transceiver using the D-bits of the FBI field of the uplink DPCCH. The UE uses the CPICH to separately estimate the channels seen from each antenna, which is needed for determining the FBI D-bits. The weight update rate is 1,500 Hz (once per slot), and the BS adjusts the antenna weights after calculating the average phase rotations over 2 consecutive slots in CL mode 1 or progressively updates the phase/amplitude weights over 4 consecutive slots in CL mode 2. The use of each mode is controlled via higher layer signalling. (For detailed description of the closed loop transmit diversity, refer to [55].)

Table 5.1 lists the tradeoffs between the open loop and closed loop transmit diversities in terms of link error performance, feedback information, and transmit/receive complexity.

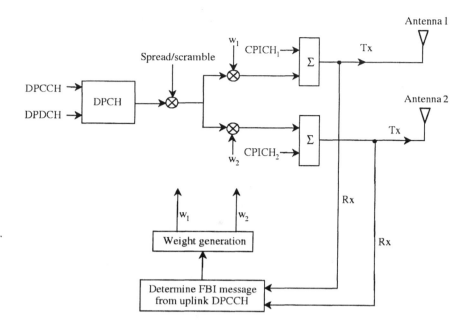

Figure 5.19. Generic downlink transmitter structure to support closed loop mode transmit diversity for the DPCH transmission [55].

Table 5.1. Tradeoffs between open loop and closed loop transmit diversities.

	Open loop	Closed loop
Link error performance	moderate	good
Feedback information	not needed	needed
Tx/Rx complexity	low	high
Examples	STTD, TSTD	CL mode 1, CL mode 2

The *site selection diversity transmission* (SSDT) is an optional macro diversity method in soft handover mode. The main objective of the SSDT is to transmit on the downlink from the primary cell, thus reducing the interference caused by multiple transmissions in soft handover mode. Another objective is to achieve fast site selection without network intervention, thus maintaining the advantage of soft handover. The specific operation is as follows: The UE periodically selects one of the cells from its active set (the cells involved in the soft handover communication) to be *primary* by measuring the *received signal code power* (RSCP) of the CPICHs transmitted by the active cells. (Refer to Section 6.) The cell with the highest CPICH RSCP is selected as the primary cell. In the SSDT mode each cell is assigned a temporary *identification* (ID) and the UE periodically reports the primary cell ID to the connecting cells. The non-primary cells switch off the transmission power for their DPDCHs, only internally adjusting the power according to the power control command. The primary cell ID is delivered by UE to the active cells via uplink FBI S-field. SSDT termination and ID assignment are all carried out by higher layer signalling.

5.   MULTIPLEXING AND CHANNEL CODING

Data stream from/to MAC layer (*transport block / transport block set*) is encoded/decoded to offer transport services over the radio transmission link. Channel coding scheme is a combination of error detection, error correcting, rate matching, interleaving and transport channels mapping onto/splitting from physical channels. Multiple transport channels can be serviced simultaneously through *transport channel multiplexing*. For each *transport channel* (TrCH), data arrives at the coding/multiplexing unit in the form of *transport block sets* once every *transmission time interval* (TTI). The TTI is transport channel specific and takes one of the four values 10, 20, 40, and 80 ms.

5.1   MULTIPLEXING AND CHANNEL CODING PROCEDURES

The overall multiplexing and channel coding structures are depicted in Fig. 5.20 (a) for uplink and Fig. 5.20 (b) for downlink.

First, the error detection is provided on each transport block through a *cyclic redundancy check* (CRC). The CRC length is taken out of 24, 16, 12, 8 or 0 bits and signalled from higher layers. The CRC parity bits are attached to each transport block in reverse order.

All the transport blocks in a TTI are serially concatenated, and then segmented into *code blocks* of the same size (possibly attaching *filler bits* to the beginning of the first code block) in case the number of bits in a TTI is larger than the maximum code block size (504 bits for convolutional coding, and 5,114 bits for turbo coding).

For channel coding one of the three schemes is supported - - no coding, 1/2 or 1/3 convolutional coding with the constraint length $K=9$, and 1/3 turbo coding with $K=4$. The employed turbo encoder structure is shown in Fig. 5.21, which is composed of two constituent recursive systematic convolutional coders and an internal interleaver. In order to realize a pseudo-random permutation which is essential in turbo encoding, a series of processings such as primitive root selection, intra-row permutation, and inter-row permutation are performed in the internal interleaver block [89]. For decoding the turbo coded bit stream in the

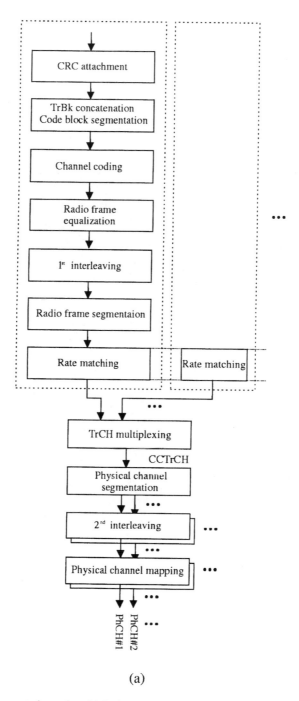

(a)

Figure 5.20.   Transport channel multiplexing structure: (a) for the uplink, (b) for the downlink [89].

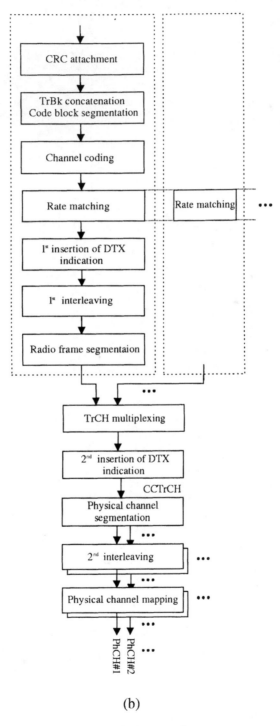

(b)

Figure 5.20. (continued)

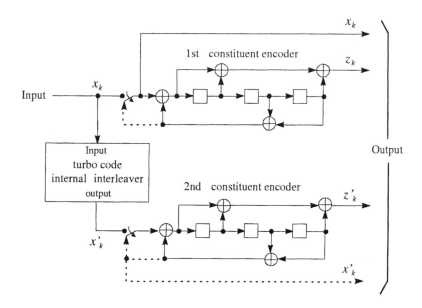

Figure 5.21.   Structure of 1/3 turbo coder. (Dotted lines apply for trellis termination only [89].)

receiving side, one of the two suboptimal *iterative decoding* algorithms, *maximum a posteriori* (MAP) algorithm or *soft output Viterbi algorithm* (SOVA), is practically used. MAP requires more computational complexity than SOVA but it provides a better performance, in general. In order to lower the complexity of the MAP algorithm, *sliding window based Max-Log-MAP algorithms* have been developed with the performance being slightly sacrificed, which is comparable to the SOVA algorithm in terms of complexity. (For the principle and the operation of the Turbo encoder and decoder, refer to [90]–[96].)

The radio frame size equalization is an uplink-unique procedure which possibly pads 0 or 1 bits after the channel coded sequence in order to ensure that the resulting sequence can be segmented into the same sized radio frames. In the downlink the rate matching procedure guarantees the equal radio frame size.

The first interleaving is performed by a block interleaver with inter-column permutations. The number of columns of the block interleaver is the same as that of radio frames in a TTI. Fig. 5.22 illustrates the first interleaving operation. The input bit sequence is written in the matrix row by row. The number of bits at the interleaver input always becomes a multiple of the column numbers owing to the radio frame equalization (for uplink) or the rate matching procedure (for downlink). The indices on top of each column represent the sequence of reading

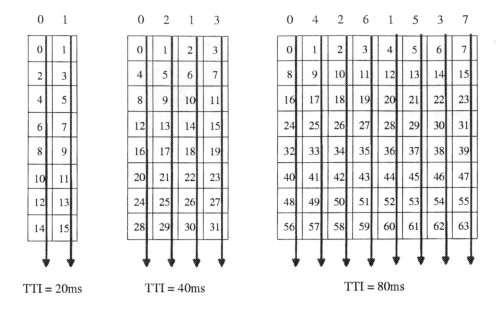

Figure 5.22.   The first interleaving operation.

out. Each column is mapped to its radio frame segment in the TTI. By the first interleaving, the coded bits are sequentially distributed over all the radio frames in the TTI.

Rate matching is a procedure that punctures or repeats the transport channel bits to ensure that the total bit rate of the *coded composite transport channel* (CCTrCH) is identical to the total physical channel bit rate of the allocated physical channel. In other words, rate matching is to fit CCTrCH bits (or, multiplexed transport channel data to be transmitted in one radio frame) into the capacity of the available physical channel(s). (For detailed rate matching procedure, refer to [89].) A remarkable puncturing rule is that the systematic bits (excluding the bits for trellis termination) of turbo encoded transport channel stream are not punctured and all the necessary puncturing is applied only to parity bits and termination bits.

The output of the first interleaving block is segmented into 2, 4, or 8 radio frames if TTI is larger than 10(ms), and then in every 10 ms, one radio frame from each TrCH is delivered to the TrCH multiplexing block. The delivered radio frames are serially multiplexed into a CCTrCH.

In the downlink, spreading factor and the total physical channel bit rate do not change among different TTIs. Rate-matching pattern is calculated on the basis of the transport formats that bring in the maximum CCTrCH bit rate. When the transport format combination (TFC) changes and the new CCTrCH bits

fixed position of TrCHs   flexible position of TrCHs

Figure 5.23. The CCTrCH format after TrCH multiplexing and insertion of DTX indication bits.

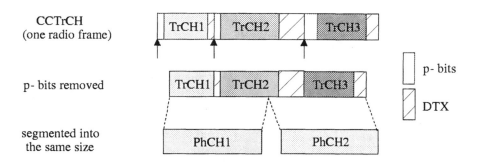

Figure 5.24. The PhCH segmentation (a CCTrCH frame with DTX bits and *p*-bits).

cannot fill all the bit positions of the employed physical channels, *discontinuous transmission* (DTX) indication bits are inserted. DTX indication bits only indicate when the transmission should be turned off. DTX indication bits may be inserted at two different positions of the CCTrCH. First, in case the positions of TrCHs in the radio frame are fixed, DTX indication bits are inserted at the end of each TrCHs. This step is performed before the first interleaving and is called the *1st insertion of DTX indication bits*. Second, when the positions of TrCHs are not fixed but flexible, DTX indication bits are inserted at the end of the CCTrCH. This step is performed right after the TrCH multiplexing and is called the *2nd insertion of DTX indication bits*. One of the two operations is actually performed, depending on the TrCH positioning method. Refer to Fig. 5.23 for the difference of the two cases.

When more than one *physical channel* (PhCH) is used, a CCTrCH can be mapped onto several PhCHs having the same spreading factor. But different CCTrCHs cannot be mapped onto the same PhCH. In compressed mode (refer to the next section for details), by additional puncturing (downlink only), *p*-bits are inserted before the first interleaving. At this stage, these p-bits are removed to make the actual transmission gap. The manipulation of the DTX bits and the *p*-bits are illustrated in Fig. 5.24.

The bits (possibly including the DTX indication bits) in each PhCH are again scattered over all slots through the 2nd interleaving operation. A block interleaver with 30 columns are employed for inter-column permutations. When the input sequence does not fill the interleaver matrix, only the bits originally present in the input sequence are output. The column permutation pattern is given by { 0, 20, 10, 5, 15, 25, 3, 13, 23, 8, 18, 28, 1, 11, 21, 6, 16, 26, 4, 14, 24, 19, 9, 29, 12, 2, 7, 22, 27, 17 }. The interleaver output bits are mapped to the corresponding physical channels and transmitted.

5.2   DIFFERENCE BETWEEN UPLINK AND DOWNLINK PROCEDURES

The number of bits on a transport channel can vary among different TTIs. In the uplink, at every TTI or frame, the physical channels (or the spreading factor) that can accommodate (within puncturing limit) the total bit rate of the CCTrCHs are newly selected and rate matching parameters are newly assigned, in general. Thus the rate matching operation is delayed until the CCTrCH rate is identified. In the downlink, the available physical channel bit rate (or the spreading factor) is fixed by the channelization code(s) assigned by the higher layers and the transmission is turned off if the current bit rate is lower than the maximum available rate, which is the DTX. Thus, rate-matching in the downlink can be performed individually for each TrCH on the basis of the maximum available CCTrCH bit rate. The rate matching parameters are calculated differently for the fixed and flexible TrCH position cases. Furthermore, in the downlink, rate matching is performed before the first interleaving and the rate-matching pattern is applied to the entire bits in a TTI, whereas, in the uplink, rate matching is performed after the first interleaving and is applied to individual radio frame in a TTI. To avoid puncturing adjacent bits in the original bit order (before the first interleaving), initial offset for each radio frame is carefully calculated in the uplink.

Fig. 5.25 (a) and (b) give examples of the uplink and downlink multiplexing, respectively. In the uplink example, the TFC is not changed within the longest TTI (40ms) and the rate-matching parameters remain constant for all radio frames. If the TFC changes, that is, the bit rate of the TrCH with a shorter TTI (TrCH #1 in this example) changes, the rate-matching parameters also change. In the compressed mode, the rate-matching parameters for the compressed radio frames are different because the target bit rate changes. This implies that rate matching for a radio frame can be performed only after the TFC is known for that frame. In other words, though all the frames in the the TrCH of the longest TTI are available at the beginning of the TTI, rate matching cannot be performed in advance in the uplink.

In the downlink, the TFC change of the TrCH with a shorter TTI does not affect the rate matching of the TrCH data with the longest TTI. The rate-matching

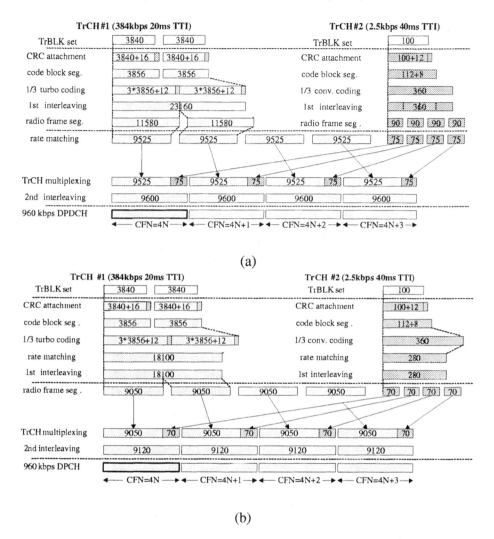

Figure 5.25. Example of TrCH multiplexing: (a) uplink, (b) downlink.

parameters are assigned to each TrCH individually before the current TTI begins.

5.3 TRANSPORT FORMAT COMBINATION INDICATOR

Transport format (TF) is defined as a format offered by the physical layer (or, L1) to the medium access control layer (or, MAC), or vice versa, for the delivery of a transport block set during a TTI on a transport channel [97]. [4] The transport format constitutes two parts - dynamic part and semi-static part. The *dynamic part* contains the attributes of the transport block size (or, number of bits in a transport block) and the transport block set size, and the *semi-static part* contains the attributes of the TTI, type of error protection (convolutional, turbo, no coding), coding rate, static rate matching parameter, puncturing limit (for uplink only), and size of CRC.

Transport format set (TFS) is defined as the set of transport formats associated with a transport channel. The semi-static parts of all transport formats are the same within a transport format set, while the dynamic part may vary among different TTIs effectively realizing variable bit rate services.

Transport format combination (TFC) is defined as the combination of currently valid transport formats which can be simultaneously submitted to the physical layer for transmission on a CCTrCH of an MS, containing one transport format from each transport channel. As was illustrated above, the physical layer multiplexes one or several transport channels and there exists a transport format set applicable for each transport channel. Nevertheless, at a given point of time, not all combinations but only a subset, the TFC, may be submitted to the physical layer.

Transport format combination set (TFCS) is defined as the set of transport format combinations on a CCTrCH. When mapping data onto L1, MAC chooses different transport format combinations given in the TFCS. However, the assignment of the TFCS is done at a higher layer (L3), thus MAC just has a control over the dynamic part of the TFCS at a given point of time.

Transport format indicator (TFI) is a label for a specific transport format within a transport format set. It is used in the inter-layer communication between MAC and L1 each time a transport block set is exchanged between the two layers on a transport channel.

Transport format combination indicator (TFCI) is a representation of the current TFC. The TFCI is used to inform the receiving side of the currently valid TFC, and thus how to decode, demultiplex, and deliver the received data on the appropriate transport channels. MAC indicates the TFI to L1 at each delivery of transport block sets on each transport channel. L1 then builds the TFCI from the TFIs of all parallel transport channels of the MS, processes the transport blocks appropriately and appends the TFCI to the physical control signalling. Through the detection of the TFCI the receiving side is able to identify the TFC. The TFCI signalling may be omitted, instead relying on blind detection (by applying CRC checks).

---

[4]Transport block is the basic unit exchanged between L1 and MAC for the L1 processing. Transport block set is a set of transport blocks, which are exchanged between L1 and MAC at the same time instance using the same transport channel.

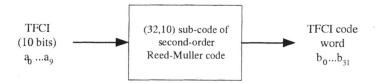

Figure 5.26. Channel coding of TFCI bits [89].

There is a one-to-one correspondence between a certain value of the TFCI and a certain TFC. The TFCI bits are encoded in L1 using a (32, 10) sub-code of the *second-order Reed-Muller* code [98]. The coding procedure is illustrated in Fig. 5.26. If the TFCI consists of less than 10 bits, it is padded with zeros to 10 bits, by setting the most significant bits (a_9, a_8, $\cdots$) to zero. In all uplink channels and the downlink channels with the spreading factor greater than or equal to 128, the bit sequence { b_0, b_1, $\cdots$, b_{29} } is transmitted sequentially in 2 bits per slot in a frame. (b_{30} and b_{31} are not transmitted.) In downlink, when the spreading factor is less than 128, the encoded TFCI codewords are first repeated four times and the resulting sequence of length 128 is sequentially transmitted, yielding 8 encoded TFCI bits per slot in the normal mode and 16 encoded bits in the compressed mode (no TFCI bits are lost even in the compressed mode. See the next section). In this case, the bits from b_0 to b_{23} are transmitted four times while the bits from b_{24} to b_{31} are transmitted three times in a frame. If one of the DCH is associated with a DSCH, the TFCI codeword may be split into two sets, each of which indicates the TFC of the DCH CCTrCH and the TFC of the associated DSCH CCTrCH. In the split mode the TFCI bits are encoded using a (16, 5) *first-order Reed-Muller* (or, bi-orthogonal) code [98].

6. MEASUREMENTS

One of the key services provided by the physical layer is the measurement of various quantities, which are used to trigger or perform a multitude of functions. Both the user equipment side (i.e., MS) and the network side (i.e., BS) are required to perform a variety of measurements. While some of the measurements are critical to the functioning of the network and are mandatory for delivering the basic functionality (e.g., handover measurements, power control measurements), others may be used by the network operators in optimizing the network (e.g., radio environment). Measurements may be made periodical and reported to the upper layers or may be event-triggered (e.g., in the handover or cell re-selection case, a primary CCPCH becomes better than the previous best primary CCPCH). Another reporting strategy may combine the event-triggered and the periodical approaches (e.g., falling of link quality below a certain threshold

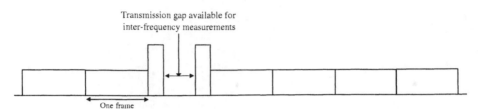

Figure 5.27.   Compressed mode transmission [89].

initiates periodical reporting). The measurements are tightly coupled with the service primitives in that the parameters of the primitives may constitute some of the measurements. The measurements which the physical layer report to the higher layers are as described in the following two subsections that follow the compressed mode section.

6.1   COMPRESSED MODE

Compressed mode is defined as the mechanism by which certain idle periods are created in radio frames so that the MS can perform inter-frequency measurements during these periods. As illustrated in Fig. 5.27, the instantaneous power is increased in the compressed frame in order to keep the quality of service unaffected by the reduced processing gain.   The rate and type of the compressed frames are variable depending on the environment and measurement requirements. When in compressed mode, the information normally transmitted during a frame is compressed in time. The mechanisms used for achieving this compression are puncturing, reduction of the spreading factor by 1/2, and higher layer scheduling. In the downlink, all methods are supported while the puncturing mechanism is not used in the uplink. The maximum transmission gap length is defined to be 7 slots per frame.   The transmission gap pattern structure, position, and repetition are defined with physical channel parameters described in [89].

6.2   USER EQUIPMENT (UE) MEASUREMENT ABILITIES

CPICH received signal code power (RSCP) is the received power on one code measured on the Primary CPICH.

P-CCPCH RSCP is the received power on one code measured on the P-CCPCH from a TDD cell.

SIR is defined as (RSCP/ISCP) × (SF/2). The SIR is measured on the DPCCH after radio link combination. In the calculation, the RSCP indicates the received power on one code measured on the pilot bits, the *interference signal code power* (ISCP) indicates the interference on the received signal measured on the pilot bits, and the SF indicates the spreading factor used.

Received signal strength indicator (RSSI) is the wide-band received power within the relevant channel bandwidth. Measurement is performed on a down-link carrier in UTRA and on the BCCH carrier in GSM.

CPICH Ec/No is the received energy per chip of the P-CPICH divided by the power density in the band. The Ec/No is identical to the ratio RSCP/RSSI.

Transport channel block error rate (BLER) is measured on the basis of the CRC check on each transport block after radio link combination.

UE transmitter power is measured on one carrier.

SFN-CFN observed time difference to cell is defined as $OFF \times 38,400 + T_m$, where $T_m = (T_{UETx} - 1,024) - T_{RxSFN}$, in chip units with the range $[0, 1, \cdots, 38399]$. T_{UETx} is the time when the MS transmits an up-link DPCCH/DPDCH frame, and T_{RxSFN} is the time when the beginning of the neighboring P-CCPCH frame has arrived at the MS most recently before the time instant $(T_{UETx} - 1,024)$. Also, OFF is given by $OFF = (SFN - CFN_{Tx}) \bmod 256$, in number of frames with the range $[0, 1, \cdots, 255]$. CFN_{Tx} is the connection frame number for the MS transmission of an uplink DPCCH/DPDCH frame at the time T_{UETx}. SFN is the system frame number for the neighboring P-CCPCH frame received in the MS at the time T_{RxSFN}.

SFN-SFN observed time difference to cell is defined as $OFF \times 38,400 + T_m$, where $T_m = T_{RxSFN_j} - T_{RxSFN_i}$, in chip units with the range $[0, 1, \cdots, 38399]$ (Type 1). T_{RxSFN_j} is the time when the beginning of a neighboring P-CCPCH frame from cell j has arrived at the MS. T_{RxSFN_i} is the time when the beginning of a neighboring P-CCPCH frame from cell i has arrived at the MS most recently before the time instant T_{RxSFN_j}. Also, OFF is given by $OFF = (SFN_i - SFN_j) \bmod 256$, in number of frames with the range $[0, 1, \cdots, 255]$. SFN_j and SFN_i are respectively the system frame numbers for the P-CCPCH frames received from cell j and i, at the time T_{RxSFN_j} and T_{RxSFN_j}.

For the inter-frequency measurement, the relative timing difference between cell j and cell i is alternatively defined as $T_{CPICHRx_j} - T_{CPICHRx_i}$, where $T_{CPICHRx_j}$ is the time when the MS receives a P-CPICH slot from cell j, and $T_{CPICHRx_i}$ is the time when the MS receives the P-CPICH slot from cell i that is closest in time to the P-CPICH slot received from cell j (Type 2).

UE Rx-Tx time difference is the difference in time between the MS uplink DPCCH/DPDCH frame transmission and the first significant path of the down-link DPCH frame from the measured radio link. Measurements are done for each cell included in the active set.

Aside from those listed above, the observed time difference to the GSM cell and the UE GPS timing of cell frames for location services are additionally measured.

6.3   UTRA NETWORK (UTRAN) MEASUREMENT ABILITIES

RSSI is the wide-band received power within the UTRAN uplink carrier channel bandwidth in an UTRAN access point. [5]

SIR is defined as (RSCP/ISCP) × SF in the uplink.

Transmitted carrier power is the ratio between the total transmitted power and the maximum transmission power. Total transmission power is the mean power on one carrier from one UTRAN access point. Maximum transmission power is the mean power on one carrier from one UTRAN access point when it is transmitted at the configured maximum power for the cell.

Transmitted code power is the transmitted power on one channelization code on one given scrambling code on one given carrier. Measurement is possible on the DPCCH-field of any dedicated radio link transmitted from the UTRAN access point and reflects the power on the pilot bits of the DPCCH-field.

Transport channel BER on the DPDCH data and *physical channel BER* on the DPCCH are measured.

Round trip time (RTT) is defined as $T_{Rx} - T_{Tx}$, where T_{Tx} is the time of transmission of the beginning of a downlink DPCH frame to a UE, and T_{Rx} is the time of reception of the beginning (the first significant path) of the corresponding uplink DPCCH/DPDCH frame from the UE.

Propagation delay is defined as one-way propagation delay as measured during either PRACH or PCPCH access.

Acknowledged PRACH preambles measurement is defined as the total number of acknowledged PRACH preambles per access frame per PRACH. This is equivalent to the number of positive acquisition indicators transmitted per access frame per AICH.

Detected PCPCH access preambles measurement is defined as the total number of detected access preambles per access frame on the PCPCHs belonging to a CPCH set.

Acknowledged PCPCH access preambles mesurement is defined as the total number of acknowledged PCPCH access preambles per access frame on the PCPCHs belonging to a SF. This is equivalent to the number of positive acquisition indicators transmitted for a SF per access frame per AP-AICH.

---

[5]UTRAN access point is defined as a conceptual point within the UTRAN performing radio transmission and reception. A UTRAN access point is associated with one specific cell, i.e., there exists one UTRAN access point for each cell. It is the UTRAN-side end-point of a radio link [99].

In addition, the *UTRAN GPS timing of cell frames for location services* is also measured.

7. UPLINK SPREADING AND SCRAMBLING

Spreading or channelization is a bandwidth-increasing operation that transforms each data symbol into a number of chips. For the channelization operation, data symbols are multiplied by an *orthogonal variable spreading factor* (OVSF) code. Scrambling is a chip randomization operation that multiplies the channelized signals on the I-(real part) and Q-(imaginary part) branches by a complex-valued scrambling code.

Fig. 5.28 depicts the spreading structure of the uplink DPCCH and DPDCHs. The DPCCH and the kth ($1 \leq k \leq 6$) DPDCH, taking the binary values 0 and 1, are mapped to the real values +1 (for 0) and -1 (for 1), and then spread by the channelization codes, C_c and $C_{d,k}$, respectively. One DPCCH and up to six parallel DPDCHs can be transmitted simultaneously, with each BPSK-modulated channel being assigned to I- and Q- branches alternately as shown in the figure. The DPCCH and the DPDCH chips are then weighted by the gain factors β_c and β_d, summed for each branch, and treated as a complex-valued stream of chips. This complex-valued signal is then scrambled by the complex-valued scrambling code $S_{dpch,n}$, for a code index n. The amplitude ratio β_c/β_d is different for each data rate and there are 16 available ratios in all. At every instant in time, at least one of the values β_c and β_d should have the amplitude 1.0.

The spreading and scrambling operation for the PRACH message part as well as the PCPCH message part is the same as that for the DPCCH/DPDCH, but one control channel (Q-branch) and only one data channel (I-channel) can be transmitted simultaneously. We denote the scrambling sequences for the PRACH and the PCPCH message part by $S_{r-msg,n}$ and $S_{c-msg,n}$, respectively.

7.1 UPLINK CHIP MODULATION

In the uplink, the complex-valued chip signal generated by the spreading and scrambling process is QPSK modulated as shown in Fig. 5.29. The real and the imaginary parts of the input complex chip signal S are split and independently pulse-shaped, up-converted by the cosine and sine carriers, and then combined and amplified by the power amplifier for transmission. The employed pulse shaping filter is a *root-raised cosine* (RRC) filter with the roll-off factor α=0.22 [100, 101].

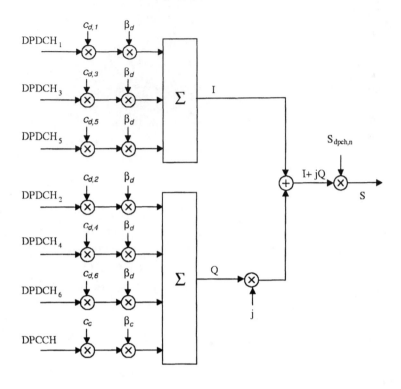

Figure 5.28.   Spreading structure of the uplink DPCCH and DPDCHs [8].

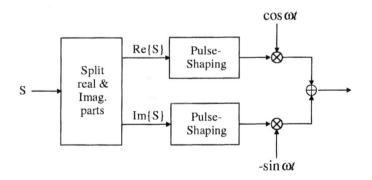

Figure 5.29.   Blockdiagram for modulation [8].

7.2   CHANNELIZATION CODES

The *channelization codes* used for spreading operation are the generalized *Walsh-Hadamard* codes called *OVSF codes,* which preserve the orthogonality

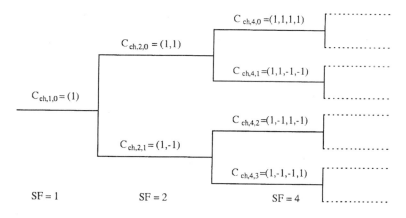

Figure 5.30. Code-tree to generate the OVSF codes for channelization operation [8].

among different physical channels regardless of the spreading factor. The OVSF codes can be defined using the code tree in Fig. 5.30, where $C_{ch,SF,k}$ $(0 \le k \le SF - 1)$ represents the kth channelization of the spreading factor (or, gain) SF. Each level in the code tree corresponds to an individual spreading factor. All codes within the code tree cannot be used simultaneously. A specific code can be used at a moment if and only if no other code on the path from the specific code to the root of the tree or in the sub-tree below the specific code (corresponding to the higher SF) is being used. This means that the number of simultaneously available codes is not fixed but depends on the rate and spreading factor of each physical channel.

In general, the channelization code is generated through the procedure

$$C_{ch,1,0} = 1, \tag{5.7a}$$

$$\begin{bmatrix} C_{ch,2,0} \\ C_{ch,2,1} \end{bmatrix} = \begin{bmatrix} C_{ch,1,0} & C_{ch,1,0} \\ C_{ch,1,0} & -C_{ch,1,0} \end{bmatrix} = \begin{bmatrix} 1 & 1 \\ 1 & -1 \end{bmatrix}, \tag{5.7b}$$

$$\cdots$$

$$\begin{bmatrix} C_{ch,2^{n+1},0} \\ C_{ch,2^{n+1},1} \\ C_{ch,2^{n+1},2} \\ C_{ch,2^{n+1},3} \\ \vdots \\ C_{ch,2^{n+1},2^{n+1}-2} \\ C_{ch,2^{n+1},2^{n+1}-1} \end{bmatrix} = \begin{bmatrix} C_{ch,2^n,0} & C_{ch,2^n,0} \\ C_{ch,2^n,0} & -C_{ch,2^n,0} \\ C_{ch,2^n,1} & C_{ch,2^n,1} \\ C_{ch,2^n,1} & -C_{ch,2^n,1} \\ \vdots \\ C_{ch,2^n,2^n-1} & C_{ch,2^n,2^n-1} \\ C_{ch,2^n,2^n-1} & -C_{ch,2^n,2^n-1} \end{bmatrix}. \tag{5.7c}$$

The DPCCH/DPDCH channelization code allocation rule is as follows: The DPCCH is always spread by the code $C_c = C_{ch,256,0}$. When only one DPDCH

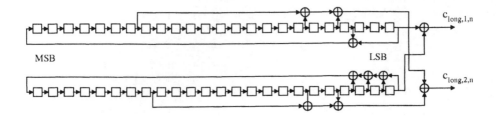

Figure 5.31.   Uplink long scrambling sequence generator [8].

is to be transmitted, the $DPDCH_1$ is spread by $C_{d,1} = C_{ch,SF,SF/4}$ where SF is the spreading factor of the $DPDCH_1$. When plural DPDCHs are to be transmitted, all DPDCHs take the spreading factor of 4. In this case $DPDCH_n$ is spread by $C_{d,n} = C_{ch,4,k}$, where $k = 1$ if $n \in \{1, 2\}$, $k = 3$ if $n \in \{3, 4\}$, and $k = 2$ if $n \in \{5, 6\}$.

The channelization codes for the PRACH message part depend on the preamble signature s ($0 \leq s \leq 15$), which has one-to-one correspondence with each of the 16 nodes in the code-tree of spreading factor 16. The sub-tree of the specified node is used as the spreading code of the message part: More specifically, the control part is spread by the code of length 256, $C_c = C_{ch,256,16 \times s+15}$ (the lowest branch of the sub-tree), and the data part is spread by $C_d = C_{ch,SF,s \times SF/16}$ (the upper-most branch of the sub-tree), where SF is the data part spreading factor ranging from 32 to 256.

The channelization code allocation for the PCPCH message part is the same as that of the DPCCH/DPDCH with only one DPDCH, and the channelization code for the PCPCH power control preamble is the same as that used for the control part of the PCPCH message part.

7.3   SCRAMBLING CODES

The DPCCH/DPDCH may be scrambled by either long or short scrambling codes, while the PRACH message part and the PCPCH message part are scrambled by long scrambling codes.

Long Scrambling Sequence: The uplink long scrambling sequence $C_{long,1,n}$ and $C_{long,2,n}$ are the segments of a set of Gold sequences, which are constructed from position-wise modulo-2 sum of 38,400 chip segments of two binary m-sequences whose characteristic polynomials are $x^{25} + x^3 + 1$ (m-sequence x_n) and $x^{25} + x^3 + x^2 + x + 1$ (m-sequence y). The sequence $C_{long,2,n}$ is a $16,777,232 (= 2^{24} + 16)$ chip shifted version of the sequence $C_{long,1,n}$. The resulting Gold sequence generator configuration is shown in Fig. 5.31. In order to generate different Gold scrambling sequences, different initial state values are loaded in the first (upper) m-sequence generator while all 1's are loaded in the second (lower) generator. More specifically, when we denote by $n_{23}, n_{22}, \cdots, n_0$ the 24 bit binary representation of the scrambling sequence number n with n_0 being the least significant bit, the initial state values of the m-sequence generators are loaded such that

$$x_n(0) = n_0, \; x_n(1) = n_1, \; \cdots, \; x_n(23) = n_{23}, \; x_n(24) = 1, \quad (5.8a)$$
$$y(0) = y(1) = \cdots = y(23) = y(24) = 1, \quad (5.8b)$$

where we have denoted the first m-sequence by x_n, as the initial state of the first m-sequence generator depends on the chosen scrambling sequence number n. The recursion formulae for the subsequent m-sequence generation are

$$
\begin{aligned}
x_n(i+25) &= x_n(i+3) + x_n(i) \; mod \; 2, & (5.9a) \\
y(i+25) &= y(i+3) + y(i+2) + y(i+1) + y(i) \; mod \; 2, & (5.9b) \\
i &= 0, 1, \cdots, 2^{25} - 27,
\end{aligned}
$$

and the binary Gold sequence z_n is generated by

$$z_n(i) = x_n(i) + y(i) \; mod \; 2, \quad i = 0, 1, \cdots, 2^{25} - 2. \quad (5.10)$$

Finally, the real-valued scrambling sequences $C_{long,1,n}$ and $C_{long,2,n}$ are obtained by

$$
\begin{aligned}
C_{long,1,n}(i) &= (-1)^{z_n(i)}, & (5.11a) \\
C_{long,2,n}(i) &= (-1)^{z_n((i+16777232) \; mod \; (2^{25}-1))}, & (5.11b) \\
i &= 0, 1, \cdots, 2^{25} - 2,
\end{aligned}
$$

and the ultimate complex-valued scrambling sequence $C_{long,n}$ is determined, by the *hybrid PSK* (HPSK) construction (refer to the next subsection), to be

$$C_{long,n}(i) = C_{long,1,n}(i)\{1+j(-1)^i C_{long,2,n}(2\lfloor i/2 \rfloor)\}, \quad i = 0, 1, \cdots, 2^{25}-2, \quad (5.12)$$

where $\lfloor x \rfloor$ denotes the largest integer not exceeding x.

Short Scrambling Sequence: The short scrambling sequences $C_{short,1,n}(i)$ and $C_{short,2,n}$ are derived from the family of periodically extended $S(2)$ codes. When we denote by $n_{23}, n_{22}, \cdots, n_0$ the 24 bit representation of the scrambling sequence number n, the quaternary $S(2)$ sequence $z_n(i)$ of length 255 ($0 \le n \le 2^{24} - 1$) is obtained by modulo-4 addition of three sequences, - - a quaternary sequence $a(i)$ and two binary sequences $b(i)$ and $d(i)$ - - according to the relation

$$z_n(i) = a(i) + 2b(i) + 2d(i) \; mod \; 4, \quad i = 0, 1, \cdots, 254, \quad (5.13)$$

Table 5.2.   Mapping from $z_n(i)$ to $C_{short,1,n}(i)$ and $C_{short,2,n}(i)$ [8].

$z_n(i)$	$C_{short,1,n}(i)$	$C_{short,2,n}(i)$
0	+1	+1
1	-1	+1
2	-1	-1
3	+1	-1

where the initial states of the three sequence generators are determined from the sequence number n. The quaternary sequence $a(i)$ is generated by the recursion formula (characteristic polynomial $g_0(x) = x^8 + x^5 + 3x^3 + x^2 + 2x + 1$)

$$a(0) = 2n_0 + 1 \bmod 4, \tag{5.14a}$$

$$a(i) = 2n_i \bmod 4, \quad i = 1, 2, \cdots, 7, \tag{5.14b}$$

$$a(i) = 3a(i-3) + a(i-5) + 3a(i-6) + 2a(i-7) + 3a(i-8) \bmod 4,$$
$$i = 8, 9, \cdots, 254, \tag{5.14c}$$

and the binary sequences $b(i)$ and $d(i)$ are generated by the recursion formulae (characteristic polynomials $g_1(x) = x^8 + x^7 + x^5 + x + 1$ and $g_2(x) = x^8 + x^7 + x^5 + x^4 + 1$)

$$b(i) = n_{8+i} \bmod 2, \quad i = 0, 1, \cdots, 7, \tag{5.15a}$$

$$b(i) = b(i-1) + b(i-3) + b(i-7) + b(i-8) \bmod 2,$$
$$i = 8, 9, \cdots, 254, \tag{5.15b}$$

and

$$d(i) = n_{16+i} \bmod 2, \quad i = 0, 1, \cdots, 7, \tag{5.16a}$$

$$d(i) = d(i-1) + d(i-3) + d(i-4) + d(i-8) \bmod 2,$$
$$i = 8, 9, \cdots, 254. \tag{5.16b}$$

The sequence $z_n(i)$ is extended to the sequence of length 256 by setting $z_n(255) = z_n(0)$. Finally the real-valued binary sequences $C_{short,1,n}(i)$ and $C_{short,2,n}(i)$, $i = 0, 1, \cdots, 255$, are obtained by the mapping relation given in Table 5.2 and the ultimate complex-valued scrambling sequence $C_{short,n}$ is determined, by the HPSK construction

$$C_{short,n}(i) = C_{short,1,n}(i \bmod 256)\{1 + j(-1)^i C_{short,2,n}(2\lfloor (i \bmod 256)/2 \rfloor)\}, \tag{5.17}$$

for $i = 0, 1, 2, \cdots$. The resulting configuration of the uplink short scrambling sequence generator is shown in Fig. 5.32.

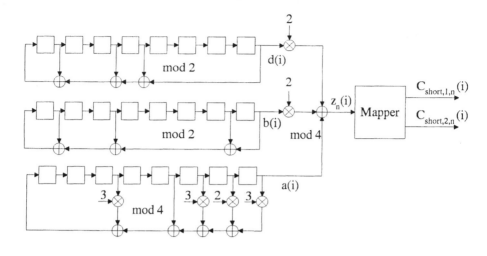

Figure 5.32.   Uplink short scrambling sequence generator [8].

The uplink DPCCH/DPDCH scrambling code may be of either long or short type. When using long [or short] scrambling codes, the nth scrambling code for the DPCCH/DPDCH, $S_{dpch,n}$, is given by $C_{long,n}(i)$ [or $C_{short,n}(i)$], for $i = 0, 1, \cdots, 38399$.

On the other hand, there are 8,192 different PRACH scrambling codes used for the PRACH message part in the system (16 codes per cell), and the nth PRACH message part scrambling code, $S_{r-msg,n}$, is given by $S_{r-msg,n}(i) = C_{long,n}(i + 4,096)$, for $n = 0, 1, \cdots, 8191$ and $i = 0, 1, \cdots, 38399$. There is a one-to-one correspondence between the scrambling codes for the PRACH preamble part and the message part: If the message part scrambling code number is n, the preamble scrambling code number is also n.

The PCPCH scrambling code is cell-specific and has a one-to-one correspondence to the preamble signature sequence and the access sub-channel used by the PCPCH access preamble part. There are 64 PCPCH scrambling codes per cell and 32,768 codes in the system. The PCPCH message part scrambling code may be of either long or short type. When using long [or short] scrambling codes, the nth scrambling code for the PCPCH message part, $S_{c-msg,n}$, is given by $C_{long,n}(i)$ [or $C_{short,n}(i)$], for $n = 8192, 8193, \cdots, 40959$ and $i = 0, 1, \cdots, 38399$. When there are 512 cells in the system, the kth ($k = 16, 17, \cdots, 79$) PCPCH scrambling code in the cell employing the mth ($m = 0, 1, \cdots, 512$) downlink primary scrambling code is defined as $S_{c-msg,n}$ for $n = 64m + k + 8,176$.

The scrambling code for the PCPCH power control preamble is the same as that used for the PCPCH message part. The phase of the scrambling code shall

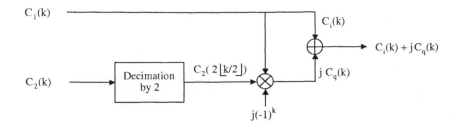

Figure 5.33.   Blockdiagram of HPSK-based scrambling code construction.

be such that the end of the code is aligned with the frame boundary at the end
of the power control preamble.

7.4   HPSK OR OCQPSK FOR SPECTRALLY EFFICIENT SCRAMBLING

The MS *power amplifier* (PA) is typically a nonlinear device that causes
a serious spectral regrowth if the input signal amplitude is not appropriately
controlled to be within the linearly-operating range. It is well known that
the reduction of the *peak-to-average power ratio* (PAR) of the input signal
directly affects the linearity requirement of the PA, which is a critical factor in
the cost and efficiency of the PA device. In order to reduce the PAR and the
spectral regrowth produced by the PA, several modulation methods have been
developed. The *offset QPSK* (OQPSK) [10, 102] and the $\pi/4$-shifted QPSK
[103] techniques are well-known power-efficient modulation techniques and
used in the IS-95 and the U.S. TDMA cellular standards, respectively. However,
these techniques are effective only for single-channel signals with balanced
I/Q-branch power, while in the typical uplink of the IMT-2000 systems multi-
channel signals are simultaneously transmitted with unbalanced I/Q-branch
power after the OVSF channelization shown in Fig. 5.28.

As a spectrally efficient scrambling technique for the multi-channel cases,
introduced was the HPSK [104] shown in Fig. 5.33, which was also proposed independently under the name *orthogonal complex QPSK* (OCQPSK) [105]. For
the HPSK-based scrambling code construction, the two independent binary-
valued (±1) sequences $C_1(k)$ and $C_2(k)$ are first generated at the chip rate,
where $C_1(k)$ is ultimately used as the I-phase scrambling sequence $C_i(k)$.
Then the decimation-by-2 process takes every other chip of the sequence $C_2(k)$
and maintains the same chip value for two chips interval. Finally the decimated chips are multiplied both by $C_1(k)$ and by an alternating binary sequence
$\{+1, -1, +1, -1, \cdots\}$ at the chip rate, producing the Q-phase sequence $C_q(k)$.
The resulting sequences $C_i(k)$ and $C_q(k)$ form the real and imaginary part of
the ultimate complex scrambling sequence $C_i(k) + jC_q(k)$. The phase of the

sequence $C_i(k) + jC_q(k)$ is limited such that at even chip times it is randomly chosen from the four QPSK values of $\pm\pi/4$ and $\pm3\pi/4$, while at odd chip times it changes only by $\pm\pi/2$ from that of the previous even chip time.

If we focus only on the PAR reduction, we had better use the pure $\pi/2$-*shifted BPSK* (which permits only $\pm\pi/2$ transition) for the complex scrambling code generation. However, it is well-known that the BPSK-based scrambling is inferior to the QPSK-based scrambling in terms of the MAI mitigation performance, as the MAI becomes dependent on the carrier phase difference between the desired and the interfering signals in the BPSK-based scrambling [7, 106]. By including QPSK transitions, the performance degradation resulting from the prolonged phase alignment with the MAI signals is mitigated [105]. Therefore, the HPSK may be considered as a hybrid combination of the QPSK for interference mitigation and the $\pi/2$-shifted BPSK for PAR reduction.

As the peak-to-average power statistics of the PA input signal is affected by the low-pass pulse shaping filters and the phase transitions of the input I/Q-data signals as well as the scrambling sequence phase transition, the roll-off factor of the shaping filter and the phase transitions of the input data should also be controlled in a proper way. In general, a sharp filter transition band (or, small roll-off factor) creates a high PAR. Thus the 3GPP W-CDMA system (with roll-off 0.22) will be more advantageous over IS-95 system (with roll-off 0.11) or the cdma2000 system (with roll-off 0.069) in terms of power efficiency. On the other hand, in order not to corrupt the HPSK modulation performance, it is preferable to employ the channelization codes that are constant at least over two chip times.

When the I/Q-data power ratio is 0.25 and the roll-off factor is 0.22, the HPSK provides the PAR reduction of about 1dB over the QPSK or the OQPSK and the performance slightly worse than that of the $\pi/2$-shifted BPSK [104]. Furthermore, in terms of the power amplifier spectral response, the HPSK has the output power gain by about 1dB over the QPSK when an equivalent intermodulation distortion is presumed [104, 105]. In the respect of the MAI rejection performance, the HPSK scrambling is not inferior to the QPSK scrambling unless they are operating under very high signal-to-interference conditions and the number of interfering signals are very small.

7.5   PRACH PREAMBLE CODES

The PRACH preamble code $C_{pre,n,s}$ is a complex-valued sequence constructed by

$$C_{pre,n,s}(i) = S_{r-pre,n}(i) \times C_{sig,s}(i) \times e^{j(\pi/4+\pi i/2)}, \quad i = 0, 1, 2, \cdots, 4095,$$

$$(5.18)$$

Table 5.3.   Preamble Signatures [8].

Preamble	Value of n															
signature	0	1	2	3	4	5	6	7	8	9	10	11	12	13	14	15
$P_0(n)$	1	1	1	1	1	1	1	1	1	1	1	1	1	1	1	1
$P_1(n)$	1	-1	1	-1	1	-1	1	-1	1	-1	1	-1	1	-1	1	-1
$P_2(n)$	1	1	-1	-1	1	1	-1	-1	1	1	-1	-1	1	1	-1	-1
$P_3(n)$	1	-1	-1	1	1	-1	-1	1	1	-1	-1	1	1	-1	-1	1
$P_4(n)$	1	1	1	1	-1	-1	-1	-1	1	1	1	1	-1	-1	-1	-1
$P_5(n)$	1	-1	1	-1	-1	1	-1	1	1	-1	1	-1	-1	1	-1	1
$P_6(n)$	1	1	-1	-1	-1	-1	1	1	1	1	-1	-1	-1	-1	1	1
$P_7(n)$	1	-1	-1	1	-1	1	1	-1	1	-1	-1	1	-1	1	1	-1
$P_8(n)$	1	1	1	1	1	1	1	1	-1	-1	-1	-1	-1	-1	-1	-1
$P_9(n)$	1	-1	1	-1	1	-1	1	-1	-1	1	-1	1	-1	1	-1	1
$P_{10}(n)$	1	1	-1	-1	1	1	-1	-1	-1	-1	1	1	-1	-1	1	1
$P_{11}(n)$	1	-1	-1	1	1	-1	-1	1	-1	1	1	-1	-1	1	1	-1
$P_{12}(n)$	1	1	1	1	-1	-1	-1	-1	-1	-1	-1	-1	1	1	1	1
$P_{13}(n)$	1	-1	1	-1	-1	1	-1	1	-1	1	-1	1	1	-1	1	-1
$P_{14}(n)$	1	1	-1	-1	-1	-1	1	1	-1	-1	1	1	1	1	-1	-1
$P_{15}(n)$	1	-1	-1	1	-1	1	1	-1	-1	1	1	-1	1	-1	-1	1

where $S_{r-pre,n}$ and $C_{sig,s}$ respectively denote the preamble scrambling code and the preamble signature given by

$$S_{r-pre,n}(i) = C_{long,1,n}(i), \quad n = 0, 1, \cdots, 8191, \quad i = 0, 1, \cdots, 4095, \quad (5.19)$$

$$C_{sig,s}(i) = P_s(i \bmod 16), \quad s = 0, 1, \cdots, 15, \quad i = 0, 1, \cdots, 4095, \quad (5.20)$$

for the real-valued long scrambling sequence, $C_{long,1,n}$, and the Hadamard sequence of length 16, $P_s(n)$. (See Table 5.3.)

The 8,192 PRACH preamble scrambling codes are divided into 512 groups with 16 codes in each group, and each group is allocated to each cell of the system according to the mapping given as follows: The kth ($k = 0, 1, \cdots, 15$) PRACH preamble scrambling code within the cell with the downlink primary scrambling code m ($m = 0, 1, \cdots, 511$) is $S_{r-pre,16m+k}$.

7.6   PCPCH PREAMBLE CODES

The PCPCH access preamble code $C_{c-acc,n,s}$ is a complex-valued sequence constructed by

$$C_{c-acc,n,s}(i) = S_{c-acc,n}(i) \times C_{sig,s}(i) \times e^{j(\pi/4 + \pi i/2)}, \quad i = 0, 1, 2, \cdots, 4095, \quad (5.21)$$

where $S_{c-acc,n}$ denotes the access preamble scrambling code given by

$$S_{c-acc,n}(i) = C_{long,1,n}(i), \quad n = 0, 1, \cdots, 40959, \quad i = 0, 1, \cdots, 4095, \quad (5.22)$$

for the real-valued long scrambling sequence $C_{long,1,n}$, and $C_{sig,s}$ $(s = 0, 1, \cdots,$ 15) is the same signature code that is used for the PRACH preamble.

The 40,960 PCPCH access preamble scrambling codes are divided into 512 groups with 80 codes in each group, and each group is allocated to each cell of the system according to the mapping relation given as follows: The kth $(k = 0, 1, \cdots, 79)$ PCPCH access preamble scrambling code within the cell with downlink primary scrambling code m $(m = 0, 1, \cdots, 511)$ is $S_{c-acc,n}$, where $n = 16m + k$ for $k = 0, 1, \cdots, 15$ and $n = 64m + (k - 16) + 8,192$ for $k = 16, 17, \cdots, 79$. The indices $k = 0, 1, \cdots, 15$ may only be used as a PCPCH access preamble part scrambling code if the same code is also used for a PRACH. The indices $k = 16, 17, \cdots, 79$ correspond to PCPCH access preamble scrambling codes which are not shared together with a PRACH. This leads to 32,768 PCPCH specific preamble scrambling codes divided into 512 groups with 64 elements.

The PCPCH CD preamble codes $C_{c-cd,n,s}$, $n = 0, 1, \cdots, 40959$, $s = 0, 1, \cdots, 15$, are constructed by using the same scrambling and signature code set as the PCPCH access preamble codes, i.e.,

$$C_{c-cd,n,s}(i) = S_{c-cd,n}(i) \times C_{sig,s}(i) \times e^{j(\pi/4+\pi i/2)}, \quad i = 0, 1, 2, \cdots, 4095, \tag{5.23}$$

where $S_{c-cd,n} = S_{c-acc,n}$ and $C_{sig,s}$ is a signature code used for the PRACH preamble.

7.7   EFFICIENCY OF THE PRACH AND PCPCH PREAMBLE CODES

The originally proposed PRACH/PCPCH preambles carried one of 16 orthogonal signature codes of length 16, each of which is spread by a cell-specific short orthogonal Gold sequence of length 256 [53]. Thus, as a preamble acquisition procedure, the BS receiver first acquired the short orthogonal Gold sequence, and then despread and detected the signature code symbols one by one. Then, by applying the 16 parallel matched filtering to a collection of 16 successive signature code symbols, it could acquire the 4,096-long preamble boundary and signature. This original preamble code structure enabled a hierarchical acquisition of the PRACH/PCPCH by using spreading code matched filters of length 256 and signature code matched filters of length 16, but brought in several serious problems [107, 108]: First, it caused large cross correlations between preamble codes at the chip lags of multiples of 256. Secondly, it caused large cross correlations between preamble codes at all lags when channel phase rotation is present due to Doppler shifts or BS-MS frequency differences. Thirdly, the performance of frequency offset estimation, which critically affects the coherent detection performance of the message part symbols, degrades severely when there exist interfering preamble transmissions at the chip lags of multiples of 256.

To resolve the problems of the original PRACH/PCPCH preamble code structure, a new structure based on the *long scrambling code* and the *interleaved signature transmission* has been proposed and adopted for the final PRACH/PCPCH preamble scrambling. A key feature of the new preamble code is the use of the long scrambling code of length 4,096 instead of the short spreading code of length 256. The long scrambling code can be generated from the same SRG that is already equipped for the message part scrambling code generation, which was illustrated in the last sections. By adopting the long code as the preamble scrambling code the problem of large cross correlations at the chip lags of multiples of 256 is eliminated. Another new feature is that the 256 chips corresponding to each preamble symbol are interleaved at intervals of 16 over the entire preamble (of length 4,096) while in the original preamble structure all 256 chips are transmitted consecutively. Consequently, in the new structure, a signature block of length 16, which corresponds to one of the 16 Hadamard sequences of length 16, is repeated 256 times to cover the 4,096-long preamble part and the preamble part is scrambled by a long scrambling code. The interleaved signature chip transmission has brought in several advantages over the consecutive chip transmission: In the preamble acquisition procedure the BS receiver can apply partial correlations (down to the 16 chip block units) to the incoming preamble parts without breaking the orthogonality among the preamble signatures, which enables reducing the coherent correlation period down to 16 chip length and accumulating the correlation results noncoherently or differentially to cope with the large BS-MS frequency offset environments. The correlator may be flexibly designed such that the *fast Hadamard transform* (FHT) is applied either to each (descrambled) 16-chip segments or to the 16 partial symbols obtained by accumulating the chips over several segments, with the FHT outputs being combined together in a progressive manner. Furthermore, the frequency offset estimation performance does not degrade severely even when interfering preambles have significant powers [107].

The new preamble code has efficient features in terms of acquisition complexity and power consumption as well. Though ultimate preamble signals take the complex values due to the constellation rotation operation (i.e., multiplication of $e^{j(\pi/4+\pi i/2)}$), the embedded sequences take the real (BPSK-modulated) values, which helps to reduce the BS acquisition circuit complexity, enabling the receiver to despread and descramble both the I-phase and the Q-phase preamble signals with a correlator having real coefficients. The rotated constellation can be reverted by simple sign conversions and I/Q-signal exchanges before the sequence correlation operation. The constellation rotation operation is introduced for the MS transmitter power efficiency: The constellation rotation limits the phase transition between two consecutive chips to $\pm\pi/2$, eliminating the zero crossings and thus reducing the peak-to-average power ratio of the preamble signal. Therefore, the rotation contributes to prolonging the MS battery life without affecting the sequence autocorrelation properties [11].

7.8 COMPARISON OF OQPSK, HPSK, AND PREAMBLE PHASE ROTATION

Note that the OQPSK [10], the HPSK [8, 9], and the preamble constellation rotation (or, the $\pi/2$-shifted BPSK) [8] techniques are the uplink CDMA code construction or modulation techniques adopted to reduce the MS transmitter power consumption by limiting the modulation state transitions, but there exist clear differences in actual realizations: While the phase transition time unit of the HPSK and the preamble constellation rotation is a chip interval, that of the OQPSK is a half chip interval, permitting the I-phase and the Q-phase signals to transit only at every even and odd time unit, respectively. Under the assumption that the I-phase and the Q-phase data channels maintain the same power, when a chip time has elapsed, the OQPSK may take one of the four phase transitions of 0, $\pi/2$, $-\pi/2$, and π, with equal probabilities. On the other hand, the HPSK takes one of four phase transitions of 0, $\pi/2$, $-\pi/2$, and π, with equal probabilities, at every even time unit, but only two phase transitions of $\pm\pi/2$ at every odd time unit. The $\pi/2$-shifted BPSK permits only two phase transitions of $\pm\pi/2$. In terms of zero-crossing frequency, the OQPSK and the $\pi/2$-shifted BPSK permit no zero-crossing, while the HPSK has a zero-crossing probability of $1/4$ at every even chip time. In terms of signal phase randomization which is related to the suppression performance of the MAI, the $\pi/2$-shifted BPSK is less effective than the HPSK and the OQPSK. When the I-phase and the Q-phase data channel powers are not balanced and the pulse shaping is applied, which is typical in the transmission of multi-channel signals, the phase transition and the constellation trajectory become more complicated and the power efficiency of the OQPSK degrades seriously while the $\pi/2$-shifted BPSK and the HPSK maintain high efficiency [104]. Fig 5.34 compares the three modulation schemes in terms of peak-to-average power complementary cumulative distribution functions, using QPSK as the reference.

8. DOWNLINK SPREADING AND SCRAMBLING

Fig. 5.35 depicts the spreading and scrambling operation of the downlink physical channels. The synchronization channels (P-SCH and S-SCH), which are transmitted intermittently with the duty cycle of 10%, are also attached in the figure. Each pair of two consecutive symbols in an input channel are first serial-to-parallel converted and mapped to the I- and Q- branches. Note that the downlink data stream is QPSK-modulated while the uplink data stream is

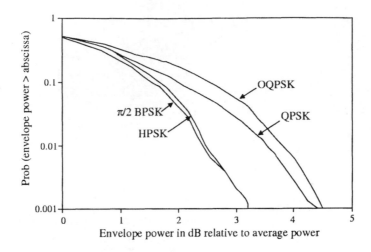

Figure 5.34.   Peak-to-average power complementary cumulative distribution functions (I/Q-data channel power ratio=0.25, root raise cosine filter with roll-off factor 0.22 is assumed.) [104].

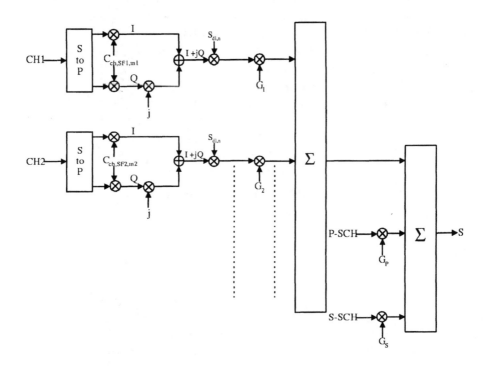

Figure 5.35.   Block diagram for spreading and scrambling of the downlink channels [8].

BPSK-modulated. The I- and Q- branches are then spread to the chip rate by the same real-valued channelization code $C_{ch,SF,m}$. The real-valued chips on the I- and Q- branches are then treated as a single complex-valued chip, which is scrambled by a complex-valued scrambling code $S_{dl,n}$. The scrambling sequence boundary is aligned with the P-CCPCH frame boundary regardless of the applied downlink channels. Each scrambled downlink channel is separately weighted by a weight factor G_i, and all the channels including the P-SCH and the S-SCH are summed together and supplied to the downlink modulation block. Note that the P-SCH and the S-SCH are not scrambled by the cell-specific long code $S_{dl,n}$.

8.1   DOWNLINK CHIP MODULATION

The downlink modulation process is the same as that of the uplink modulation (refer to Fig. 5.29). The complex-valued chip stream generated by the spreading and scrambling process is QPSK modulated. The RRC filter with the roll-off factor 0.22 is also used for downlink pulse shaping.

8.2   CHANNELIZATION CODES

The OVSF codes are employed also for the downlink channelization. The channelization codes of the primary-CPICH and the primary-CCPCH are fixed to $C_{ch,256,0}$ and $C_{ch,256,1}$, respectively. With respect to the spreading factor of 512, when a channelization code $C_{ch,512,2n}$ [or $C_{ch,512,2n+1}$] ($n = 0, 2, 4, \cdots, 510$) is used in the soft handover mode, the channelization code $C_{ch,512,2n+1}$ [or $C_{ch,512,2n}$] is not allocated to the base stations where timing adjustment is used. When the compressed mode is implemented by reducing the spreading factor by 2, the OVSF code used for the compressed frames is $C_{ch,SF/2,\lfloor n/2 \rfloor}$ for an ordinary scrambling code and $C_{ch,SF/2,n \bmod (SF/2)}$ for an alternative scrambling code (refer to the next section), where the channelization code used for the non-compressed mode is assumed to be $C_{ch,SF,n}$. When a PDSCH uses different OVSF codes from frame to frame, the OVSF codes are allocated such that they are taken from the branch of the code tree pointed by the smallest spreading factor used for the connection.

8.3   SCRAMBLING CODES

There are 8,192 ordinary downlink scrambling codes in a system. The downlink scrambling codes are grouped into 512 sets associated with 512 cells of the system, each of which contains a *primary scrambling code* $S_{dl,16m}$ and 15 *secondary scrambling codes* $\{S_{dl,16m+k} : k = 1, 2, \cdots, 15\}$, where

$m = 0, 1, \cdots, 511$. The 512 scrambling code sets are further grouped into 64 scrambling code groups, where the jth code group consists of 8 primary and 8×15 secondary scrambling codes $\{S_{dl,128j+k} : k = 0, 1, \cdots, 127\}$. The scrambling code $S_{dl,n}$ ($n = 0, 1, \cdots, 8191$) is associated with a *left alternative scrambling code* $S_{dl,n+8192}$ and a *right alternative scrambling code* $S_{dl,n+16384}$, which may be used for the compressed frames. If an alternative scrambling code is used in the compressed frames, the left [or the right] alternative scrambling code is used when the index n of the non-compressed frame channelization code $C_{ch,SF,n}$ satisfies $n < SF/2$ [or $n \geq SF/2$]. A CCTrCH may use the mixture of the primary and the secondary scrambling codes, but the CCTrCH for the PDSCH that is associated with a single UE should use a single scrambling code.

The downlink scrambling sequences are the segments of a set of complex Gold sequences, which are constructed from position-wise modulo-2 sum of 38,400 chip segments of two binary m-sequences whose characteristic polynomials are $x^{18} + x^7 + 1$ (m-sequence x) and $x^{18} + x^{10} + x^7 + x^5 + 1$ (m-sequence y). The initial states of the two m-sequence generators are defined as

$$x(0) = 1, x(1) = x(2) = \cdots x(16) = x(17) = 0, \qquad (5.24a)$$
$$y(0) = y(1) = \cdots = y(16) = y(17) = 1. \qquad (5.24b)$$

In order to generate different Gold sequences, different initial state values must be loaded in the first (upper) m-sequence generator when all 1's are loaded in the second (lower) generator. More specifically, when we denote by z_n ($n = 0, 1, \cdots, 2^{18} - 2$) the nth real Gold sequence, $z_n(i)$ is generated by

$$z_n(i) = x\{(i+n) \, mod \, (2^{18} - 1)\} + y(i) \, mod \, 2, \quad i = 0, 1, \cdots, 2^{18} - 2. \quad (5.25)$$

Finally, the nth ultimate complex-valued scrambling sequence $S_{dl,n}$ is generated by

$$S_{dl,n}(i) = (-1)^{z_n(i) + j z_n((i+131072) \, mod \, (2^{18} - 1))}, \quad i = 0, 1, \cdots, 38399. \quad (5.26)$$

The resulting complex Gold sequence generator configuration is shown in Fig. 5.36. The 38,400-long complex scrambling codes are repeated in every 10ms radio frame.

8.4   SYNCHRONIZATION CODES

The *primary synchronization code* (PSC) C_{psc} is a *generalized hierarchical Golay* (GHG) sequence of length 256 having good aperiodic auto-correlation properties. As the name implicates, the GHG sequence is generated by a hierarchical construction using two constituent sequences of length 16, and the

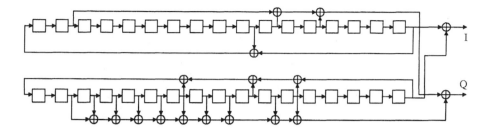

Figure 5.36.   Downlink scrambling sequence generator [8].

constituent sequences themselves are Golay sequences or can be built from shorter Golay sequences. The GHG code satisfies the requirements of the efficient synchronization codes, namely, good correlation performances, low matched-filtering complexity, and mitigation to the frequency discrepancy between the BS and the MS oscillators. The specific code is defined as follows: When we represent the constituent sequence a of length 16 by

$$a = < x_1, x_2, \cdots, x_{16} >$$
$$= < 1, 1, 1, 1, 1, 1, -1, -1, 1, -1, 1, -1, 1, -1, -1, 1 >, \quad (5.27)$$

the PSC is generated by repeating the sequence a modulated by a *Golay complementay* sequence such that

$$C_{psc} = (1+j) \times < a, a, a, -a, -a, a, -a, -a, a, a, a, -a, a, -a, a, a > . \quad (5.28)$$

The efficient implementation of the PSC generator or matched filter will be discussed in detail in Section 9.

The 16 *secondary synchronization codes* (SSC) $\{C_{ssc,1}, C_{ssc,2}, \cdots, C_{ssc,16}\}$ are constructed by position-wise multiplication of a Hadamard sequence and a complementary sequence z defined as

$$z = < b, b, b, -b, b, b, -b, -b, b, -b, b, -b, -b, -b, -b, -b >, \quad (5.29a)$$
$$b = < b_1, b_2 >, \quad (5.29b)$$
$$b_1 = < x_1, x_2, x_3, x_4, x_5, x_6, x_7, x_8 >, \quad (5.29c)$$
$$b_2 = < -x_9, -x_{10}, -x_{11}, -x_{12}, -x_{13}, -x_{14}, -x_{15}, -x_{16} >, \quad (5.29d)$$

where x_i's are defined in (5.28). By taking the constituent sequence b orthogonal to a, the complementary sequence z becomes orthogonal to C_{psc}. The Hadamard sequence h_n is obtained by taking the nth row (from the top) of the Hadamard matrix H_8, which is constructed recursively by

$$H_0 = [1], \quad (5.30a)$$

$$H_m = \begin{bmatrix} H_{m-1} & H_{m-1} \\ H_{m-1} & -H_{m-1} \end{bmatrix}, \quad m \geq 1. \qquad (5.30b)$$

Then the kth SSC, $C_{ssc,k}$, $k = 1, 2, \cdots, 16$, is generated by

$$C_{ssc,k} = (1+j) \times < h_n(0) \times z(0), \ h_n(1) \times z(1), \ \cdots, \ h_n(255) \times z(255) >, \qquad (5.31)$$

for $n = 16(k-1)$. Note that the 16 SSCs are orthogonal themselves due to the modulation by the Hadamard codes. Furthermore, all the SSCs are orthogonal to the PSC as the 16 modulating Hadamard codes do not change their chip values within 16-chip segments and the complementary sequence z is orthogonal to the PSC. The orthogonality helps to minimize the cross-channel interference.

Employing the 16 SSCs as the alphabet of the codeword elements, the 64 S-SCH codewords of length 15 are constructed such that their cyclic-shifts are unique. That is, a cyclic shift from 0 to 14 of any of the 64 S-SCH codewords is not equivalent to any cyclic shifts of the other 63 codewords nor to any other cyclic shifts of itself. Such process is generally called *comma-free* codeword construction. The comma-free codewords embedded in the S-SCH are built from a (15,3) *Reed-Solomon* code over GF(16) [8, 54, 115, 110], which belongs to so-called the *maximum distance separable* (MDS) codes [98] satisfying $d_{min} = n - k + 1$, where d_{min}, $n (= 15)$, and $k (= 3)$ denote the minimum distance between codewords, the codeword length, and the message length, respectively. The number of overall codewords of the 16-ary (15,3) RS code is $16^3 (= 4,096)$. Among them, if we exclude 16 codewords which have the same alphabet in all positions, we get the remaining 4,080 codewords. As the RS code is a cyclic code of length 15, we can divide the 4,080 codewords into 272 codeword sets each of which is composed of 15 different cyclic shifts of a codeword. Therefore, by selecting a codeword from each set, we can construct a Comma-Free RS(15,3) code with 272 codewords and the minimum distance of $d_{min} = 13$. The 64 S-SCH codewords whose indices are listed in Table 5.4 are the elements of a subset of the Comma-Free RS(15,3) code. In the table, if the entry corresponding to Group-l ($l = 0, 1, \cdots, 63$) and Slot-m ($m = 0, 1, \cdots, 14$) is k, $C_{ssc,k}$ is transmitted in the mth time slot of each frame from all the cells belonging to Group-l. The Comma-Free property enables the unique identification of both the scrambling code group and the frame boundary. Furthermore, note that any K length positions of the S-SCH codewords are unique for $3 \leq K \leq 15$ due to the MDS property of the RS code and thus can be used to obtain both the scrambling code group and the frame timing. In fact it can be easily shown that the minimum distance between any such K length positions is $K - 2$, which can effectively be used to reduce the complexity and cell search time in the MS when the SNR is high.

Table 5.4. Allocation of the SSCs for the secondary SCH [8].

Scrambling Code Group	slot number														
	#0	#1	#2	#3	#4	#5	#6	#7	#8	#9	#10	#11	#12	#13	#14
Group 0	1	1	2	8	9	10	15	8	10	16	2	7	15	7	16
Group 1	1	1	5	16	7	3	14	16	3	10	5	12	14	12	10
Group 2	1	2	1	15	5	5	12	16	6	11	2	16	11	15	12
Group 3	1	2	3	1	8	6	5	2	5	8	4	4	6	3	7
Group 4	1	2	16	6	6	11	15	5	12	1	15	12	16	11	2
Group 5	1	3	4	7	4	1	5	5	3	6	2	8	7	6	8
Group 6	1	4	11	3	4	10	9	2	11	2	10	12	12	9	3
Group 7	1	5	6	6	14	9	10	2	13	9	2	5	14	1	13
Group 8	1	6	10	10	4	11	7	13	16	11	13	6	4	1	16
Group 9	1	6	13	2	14	2	6	5	5	13	10	9	1	14	10
Group 10	1	7	8	5	7	2	4	3	8	3	2	6	6	4	5
Group 11	1	7	10	9	16	7	9	15	1	8	16	8	15	2	2
Group 12	1	8	12	9	9	4	13	16	5	1	13	5	12	4	8
Group 13	1	8	14	10	14	1	15	15	8	5	11	4	10	5	4
Group 14	1	9	2	15	15	16	10	7	8	1	10	8	2	16	9
Group 15	1	9	15	6	16	2	13	14	10	11	7	4	5	12	3
Group 16	1	10	9	11	15	7	6	4	16	5	2	12	13	3	14
Group 17	1	11	14	4	13	2	9	10	12	16	8	5	3	15	6
Group 18	1	12	12	13	14	7	2	8	14	2	1	13	11	8	11
Group 19	1	12	15	5	4	14	3	16	7	8	6	2	10	11	13
Group 20	1	15	4	3	7	6	10	13	12	5	14	16	8	2	11
Group 21	1	16	3	12	11	9	13	5	8	2	14	7	4	10	15
Group 22	2	2	5	10	16	11	3	10	11	8	5	13	3	13	8
Group 23	2	2	12	3	15	5	8	3	5	14	12	9	8	9	14
Group 24	2	3	6	16	12	16	3	13	13	6	7	9	2	12	7
Group 25	2	3	8	2	9	15	14	3	14	9	5	5	15	8	12
Group 26	2	4	7	9	5	4	9	11	2	14	5	14	11	16	16
Group 27	2	4	13	12	12	7	15	10	5	2	15	5	13	7	4
Group 28	2	5	9	9	3	12	8	14	15	12	14	5	3	2	15
Group 29	2	5	11	7	2	11	9	4	16	7	16	9	14	14	4
Group 30	2	6	2	13	3	3	12	9	7	16	6	9	16	13	12
Group 31	2	6	9	7	7	16	13	3	12	2	13	12	9	16	6

The detailed cell search procedure in the W-CDMA system is described in the next section. [6]

---

[6]Aside from the physical layer issues we have discussed so far, there are some other channels and advanced schemes under study for inclusion in later releases, which include the DSCH control channel, *fast uplink synchronization channel* (FAUSCH), slow power control, hybrid ARQ, 4-state *serially concatenated convolutional code* (SCCC) based turbo coding, and *opportunity driven multiple access* (ODMA).

Table 5.4. (continued)

Scrambling Code Group	slot number														
	#0	#1	#2	#3	#4	#5	#6	#7	#8	#9	#10	#11	#12	#13	#14
Group 32	2	7	12	15	2	12	4	10	13	15	13	4	5	5	10
Group 33	2	7	14	16	5	9	2	9	16	11	11	5	7	4	14
Group 34	2	8	5	12	5	2	14	14	8	15	3	9	12	15	9
Group 35	2	9	13	4	2	13	8	11	6	4	6	8	15	15	11
Group 36	2	10	3	2	13	16	8	10	8	13	11	11	16	3	5
Group 37	2	11	15	3	11	6	14	10	15	10	6	7	7	14	3
Group 38	2	16	4	5	16	14	7	11	4	11	14	9	9	7	5
Group 39	3	3	4	6	11	12	13	6	12	14	4	5	13	5	14
Group 40	3	3	6	5	16	9	15	5	9	10	6	4	15	4	10
Group 41	3	4	5	14	4	6	12	13	5	13	6	11	11	12	14
Group 42	3	4	9	16	10	4	16	15	3	5	10	5	15	6	6
Group 43	3	4	16	10	5	10	4	9	9	16	15	6	3	5	15
Group 44	3	5	12	11	14	5	11	13	3	6	14	6	13	4	4
Group 45	3	6	4	10	6	5	9	15	4	15	5	16	16	9	10
Group 46	3	7	8	8	16	11	12	4	15	11	4	7	16	3	15
Group 47	3	7	16	11	4	15	3	15	11	12	12	4	7	8	16
Group 48	3	8	7	15	4	8	15	12	3	16	4	16	12	11	11
Group 49	3	8	15	4	16	4	8	7	7	15	12	11	3	16	12
Group 50	3	10	10	15	16	5	4	6	16	4	3	15	9	6	9
Group 51	3	13	11	5	4	12	4	11	6	6	5	3	14	13	12
Group 52	3	14	7	9	14	10	13	8	7	8	10	4	4	13	9
Group 53	5	5	8	14	16	13	6	14	13	7	8	15	6	15	7
Group 54	5	6	11	7	10	8	5	8	7	12	12	10	6	9	11
Group 55	5	6	13	8	13	5	7	7	6	16	14	15	8	16	15
Group 56	5	7	9	10	7	11	6	12	9	12	11	8	8	6	10
Group 57	5	9	6	8	10	9	8	12	5	11	10	11	12	7	7
Group 58	5	10	10	12	8	11	9	7	8	9	5	12	6	7	6
Group 59	5	10	12	6	5	12	8	9	7	6	7	8	11	11	9
Group 60	5	13	15	15	14	8	6	7	16	8	7	13	14	5	16
Group 61	9	10	13	10	11	15	15	9	16	12	14	13	16	14	11
Group 62	9	11	12	15	12	9	13	13	11	14	10	16	15	14	16
Group 63	9	12	10	15	13	14	9	14	15	11	11	13	12	16	10

9.   CELL SEARCH

The code synchronization, or cell search techniques for the inter-cell asynchronous IMT-2000 W-CDMA systems are quite different from the typical technique that have been employed for the inter-cell synchronous systems. As the code/timing uncertainty is very large in the inter-cell asynchronous systems, the traditional correlator-based serial or parallel search acquisition techniques based on the common pilot channel cannot meet the cell search timing and complexity requirements. Therefore, in the W-CDMA systems, two intermittently transmitted synchronization channels, P-SCH and S-SCH, are additionally uti-

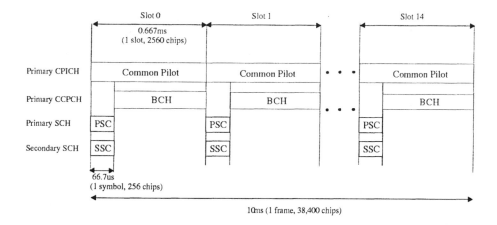

Figure 5.37. Timing relations among the primary CPICH, primary CCPCH, primary SCH, and secondary SCH.

lized in parallel with the continuously transmitted P-CPICH. (Refer to Fig. 5.37 for synchronization related structures.)

The 3GPP-FDD system employs a three-step cell search procedure in order to resolve a large amount of code-timing uncertainty in short time. The three-step cell search consists of slot synchronization, frame synchronization and code-group identification, and scrambling code identification as follows:

(Step 1) Slot Synchronization: In the first step of the cell search procedure, the MS carries out matched filtering on the incoming *primary synchronization code* (PSC) to acquire the slot boundary timing of the BS. As the PSC is common to all cells in the system, the slot boundary can be obtained by detecting the epoch that produces the peaks in the matched filter output.

(Step 2) Frame Synchronization and Code-Group Identification: In the second step, the MS uses the incoming S-SCH to acquire the frame boundary and identify the code group of the cell found in Step 1. In this step all possible *secondary synchronization code* (SSC) codes are correlated with the incoming signal, and the combination of the frame timing and code group number that is maximally correlated with the incoming S-SCH is detected.

(Step 3) Scrambling Code Identification: In the third step, the UE determines the exact primary scrambling code (or, the exact cell number) typically by correlating the incoming P-CPICH with all possible primary scrambling codes associated with the code group identified in Step 2 and detecting the maximally correlated scrambling code. Once the primary scrambling code is identified, the P-CCPCH, which contains the system and cell specific BCH information, can be decoded.

Typically, the three steps are performed in parallel in the form of pipelined processing.

9.1   INITIAL CELL SEARCH

In the initial cell search, the three steps are all carried out sequentially. Under the assumption that the incoming sequence samples are taken at every half chip time, the initial code/timing uncertainty amounts to the determination of a true hypothesis out of $512 \times 38,400 \times 2$ candidates.

In the first step, $2,560 \times 2$ samples are taken per slot period from the GHG matched filter output, which are then usually accumulated non-coherently over several slots to decrease the false acquisition probability. [7] Then the epoch that produces the maximum output energy is chosen as the slot boundary, which reduces the code/timing uncertainty to 512×15.

In the second step, 16 SSCs generated by the corresponding active correlators are correlated with the incoming signal during the first 256-chip period of each slot, and then the resulting 16 correlation output values are accumulated (non-coherently or coherently with the phase reference provided by the PSC) to the 64×15 scrambling code group/timing metrics specified in the Table 5.4. By detecting the maximum out of the 64×15 accumulated metrics, the scrambling code group and the frame boundary are simultaneously acquired, which reduces the code/timing uncertainty to 8×1. The number of slots to be accumulated determining the maximum metric may be adjusted depending on the channel status, but in any case at least 3 slots should be accumulated to differentiate all the group/timing candidates. For details, refer to the distance property of the comma-free codewords in the previous section.

In the third step, 8 candidate scrambling codes generated by active correlators are correlated with the incoming P-CPICH. The maximally correlated code out of the 8 scrambling codes is detected to be the one transmitted from the current cell. At that point all the uncertainty that has remained up to the second step is removed. Even in this last step, in general, the energies taken from the correlator outputs are accumulated over several CPICH symbol periods and used in determining the maximum correlation, thereby reducing the false identification probability (or, the probability that determines a wrong scrambling code out of 8 candidates). An alternative approach can be found in recording

---

[7] The differentially-coherent combining may be applicable for improved performance when there is a constant BS-MS frequency offset and the channel variation rate is very low and TSTD not used. In very slowly varying channels, coherent combining may also be tried.

a scrambling code that produces the maximum output for each symbol period and determining a code as the incoming scrambling code by majority voting on the basis of the observation results over multiple symbol periods. To control the false alarm probability, a threshold detector may be incorporated, which aborts the determined scrambling code if the maximum-taking frequency of the scrambling code does not exceed the threshold over the observed interval [111].

As there may exist a very high frequency offset between the BS and the MS oscillators in the initial cell search stage, the 256-long full correlation with the incoming PSC, SSC, and P-CPICH symbols may not be satisfactory in obtaining the required correlation energy. To avoid such worst-case situation, the MS may reduce the minimum available correlation period to a fraction of the symbol period (or, 256 chips). By reducing the minimum correlation period to $256/N$ chips ($N \geq 2$) and combining successive correlation outputs noncoherently (or coherently after compensating for the phase rotations of each output), the MS can effectively cope with a very high frequency offset or a very fast fading environment. The coherent combining with phase compensation is usually applicable in the SSC correlation step as the current channel phase can be estimated from the PSC that has been synchronized in the first step. In the first and the third steps, noncoherent (or differentially-coherent) combining will be pertinent. However, in the environment where the channel phase varies slowly, the MS may use coherent combining in all steps.

9.2 MULTIPATH PROFILE AND NEIGHBORING CELL SEARCH

For the performance improvement in the multipath channel, the MS usually employs a RAKE receiver that can demodulate multiple radio links (or, paths) simultaneously. The searcher (or, scrambling code acquisition block) finds each path successively after acquiring the strongest path and then assigns each of them a RAKE finger (or, demodulation block). Furthermore, after the MS has camped on one of the cells in the system, it tries to acquire and monitor the radio link status of the neighboring cells to be able to re-select a new cell or perform a soft handover depending on the channel status variations. [8] In the inter-cell synchronous systems, the multipath and the neighboring cell search can be easily done as the timing uncertainty is very small: The MS can easily obtain the information of the relative chip timing differences between the current cell and each neighboring cell, and thus can rapidly complete the neighboring cell search within a small timing uncertainty window (or, timing uncertainty region

---

[8]The measurement control information broadcast in the current cell contains the number of the neighboring cells to monitor. The MS should be able to monitor up to 32 cells given in the measurement control information and record at least 6 strongest cells for cell re-selection or soft handover purposes [112].

due to the propagation delay). However, in the inter-cell asynchronous system, the timing difference between cells varies time to time, so it is impossible to apply the uncertainty window based neighboring cell search. [9] Furthermore, as all the cells use the same PSC of length 256, the MS cannot differentiate, in the first step, the peaks due to the multipaths of the current cell from those due to the neighboring cell signal.

Therefore, searcher control becomes more difficult in the 3GPP-FDD system than in the IS-95 or cdma2000 systems. A practical searcher operation may be done as follows, while several variations may be incorporated for performance improvement: After determining the strongest path of the current cell by the initial cell search procedure, the searcher initiates obtaining the multipath profiles of the current cell. Directly performing the P-CPICH correlation through a timing uncertainty window around the acquired path, this procedure can be completed quickly through a conventional serial or parallel search scheme, as the timing uncertainty window is usually small (e.g., less than 256 chips). [10] After assigning N_m effective multipaths to the RAKE fingers, the MS can decode the control channel which contains the information of the N_n neighboring cell scrambling code identification numbers. Then the searcher initiates obtaining the neighboring cell profiles. In this procedure, it masks out the selected path samples up to that moment and re-select the strongest sample out of the remaining $(2,560 - N_m) \times 2$ sample profiles reported in Step 1. The uncertainty associated with the re-selected sample is not large — 15 slot timing uncertainty and N_n scrambling code uncertainty. Now without resorting to Step 2, the searcher can resolve the uncertainty by performing the P-CPICH correlations in parallel or in serial. (Or by applying Step 2 with reduced accumulation slots, the $15 \times N_n$ uncertainty may be resolved reliably with a few accumulation slots, as the uncertainty itself is not large.) Once the strongest path of the neighboring cell is acquired, the searcher can obtain the delay profile of the cell through the uncertainty window approach as it does for the camping cell. In this way the active set (i.e., the radio links involved in a particular communication service simultaneously at the moment) and the neighboring cell profiles can be successively recorded.

Alternatively, if the cell deployment is planned such that the neighboring cells cannot use the scrambling codes belonging to the same scrambling code group, Step 3 (or P-CPICH identification) may be skipped in the neighboring cell search procedure. More specifically, after re-selecting the strongest sample out of the remaining sample profiles in Step 1, the searcher performs Step 2 to

---

[9]Optionally, the BS in the 3GPP-FDD system may convey the instant timing difference between the current cell and the neighboring cells over the broadcast channel [113]. In this case the uncertainty window based search can be applied.
[10]The uncertainty window size should be designed subject to the deployed cell radius.

resolve the code group and frame timing uncertainty. As the MS already has the information of the N_n candidate scrambling codes (or code groups) available for the neighboring cell, the number of accumulated slots may be reduced significantly without degrading the search performance. Once the code group and frame timing uncertainty is resolved, the P-CPICH correlation with the corresponding scrambling code is performed only for the verification purpose [11].

In practice, the multipath and neighboring cell search procedure should be practically implemented in consideration of the hardware and operational complexity, search algorithm performance, and the involved channel power. In terms of channel power and acquisition performance, it is preferable to exploit the P-CPICH as the nominal BS power allocations to P-SCH, S-SCH, and P-CPICH are 3.15%(-15dB), 3.15%(-15dB), and 10%(-10dB), respectively [101, 112]. Moreover, the P-SCH and the S-SCH are intermittently transmitted, while the P-CPICH is continuously transmitted.

9.3   GENERALIZED HIERARCHICAL GOLAY (GHG) MATCHED FILTER

The *generalized hierarchical Golay* (GHG) code illustrated in the previous section is obtained by combining two proposals, one for the *hierarchical code* [114] and the other for the *Goly complementary code* [115]. In fact, both schemes exhibit good correlation properties and render efficient computations of the correlation in the MS. In the following, we discuss how to generate the two codes and how to combine them to generate the GHG code.

Fig. 5.38 depicts the block diagram of an efficient matched filtering for the hierarchical code of length 256, which is constructed by concatenating the two constituent codes of length 16 each, X_1 and X_2. More specifically, the hierarchical code is generated by spreading each symbol of code X_1 by code X_2. The corresponding hierarchical matched filter consists of two concatenated matched filter blocks as shown in the figure. We can observe that the total number of complex adds needed for each correlation output is (15+15)=30, which is very small compared to that of the typical non-hierarchical matched filter requiring 256 complex adds per output. By selecting proper constituent codes, we can obtain a hierarchical code that exhibits satisfactory correlation performances.

On the other hand, Fig. 5.39 depicts a lattice-type matched filter of the Golay complementary code [116, 117, 118]. The complementary codes [119] are the sequences having the property that the sum of their auto-correlation functions is zero for all time shifts other than zero. Mathematically, complementary

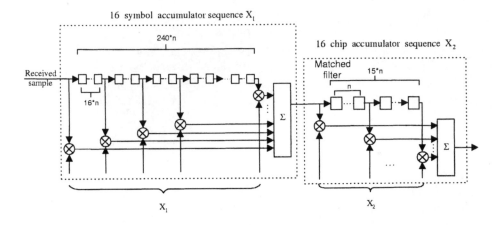

Figure 5.38. An efficient matched filter for the hierarchical code of length 256. (n denotes the chip oversampling ratio.)

sequences $a(i)$ and $b(i)$ of length L can be defined by the relation

$$R_a(k) + R_b(k) = \begin{cases} 1, & k = 0, \\ 0, & k \neq 0, \end{cases} \tag{5.32a}$$

$$R_a(k) = \sum_{i=0}^{L-k-1} a(i)a(i+k)^*, \tag{5.32b}$$

$$R_b(k) = \sum_{i=0}^{L-k-1} b(i)b(i+k)^*, \tag{5.32c}$$

where L is the period of the sequences $a(i)$ and $b(i)$. In general, a polyphase complementary pair of sequences, which contains the Golay complementary codes as a special binary case, can be constructed through the following recursive relation [117]

$$a_0(i) = \delta(i), \tag{5.33a}$$

$$b_0(i) = \delta(i), \tag{5.33b}$$

$$a_n(i) = a_{n-1}(i) + W_n \cdot b_{n-1}(i - D_n), \tag{5.33c}$$

$$b_n(i) = a_{n-1}(i) - W_n \cdot b_{n-1}(i - D_n), \tag{5.33d}$$

for two complementary sequences of length 2^N, a_n and b_n. In the expressions, $\delta(i)$ denotes the Kronecker delta function, i the time index, n the number of iteration ($n \in \{1, 2, \cdots, N\}$), D_n a positive delay chosen as any permutation of $\{1, 2, 2^2, \cdots, 2^{N-1}\}$, and W_n an arbitrary complex number on the unit circle. When each W_n is chosen to be $+1$ or -1, the binary Golay complementary

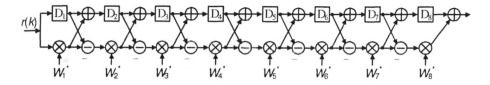

Figure 5.39. Efficient Golay correlator (EGC) for the plain Golay code of length 256.

sequences are generated. It is obvious that the recursive relation yields the lattice structure implementation in Fig. 5.39. It is called the *efficient Golay correlator* (EGC). The EGC requires $2 \log_2 256 = 16$ complex adds for each correlation output, thus lowering the complexity to about a half of that of the hierarchical code generated by Fig. 5.38. Good autocorrelation properties are guaranteed by the relations in Eq. (8.1)

Now by combining the two efficient code generation schemes shown in Figs. 5.38 and 5.39, we can further reduce the correlation complexity, yet maintaining good correlation properties. More specifically, in constructing the ultimate hierarchical code of length 256 from X_1 and X_2 of length 16 each, we take X_2 as a hierarchical code whose constituent sequences are two Golay codes of length 4 each, and X_1 as a Golay code of length 16. Then the resulting GHG code of length 256 coincides with the P-SCH code in Section 9.4 of Chapter 4, which can be efficiently generated by the following methods:

Method 1: The PSC code C_{psc} is hierarchically generated by two constituent sequences using the formula

$$C_{psc}(i) = X_2(i \bmod L_2) \cdot X_1(i \operatorname{div} L_2), \quad i = 0, 1, \cdots, L_1 L_2 - 1, \quad (5.34)$$

where the sequences X_1 and X_2 are length 16 sequences (i.e., $L_1 = L_2 = 16$).

The sequence X_1 is a Golay complementary sequence of length 16 (i.e., $N^{(1)} = 4$) by the delay matrix $D^{(1)} = [8,4,1,2]$ and the weight matrix $W^{(1)} = [1,-1, 1, 1]$. On the other hand, the sequence X_2 is a *generalized hierarchical sequence* constructed by the formula

$$X_2(i) = X_4(i \bmod s + s \cdot (i \operatorname{div} sL_3)) \cdot X_3((i \operatorname{div} s) \bmod L_3),$$
$$i = 0, 1, \cdots, L_3 L_4 - 1, \quad (5.35)$$

for $s=2$. The sequence X_3 and X_4 are two identical Golay complementary sequences of length 4 (i.e., $L_3 = L_4 = 4$, $N^{(3)} = N^{(4)} = 2$) obtained by the delay matrices $D^{(3)} = D^{(4)} = [1, 2]$ and the weight matrices $W^{(3)} = W^{(4)} = [1, 1]$. Note that the generalized hierarchical sequence X_2 is constructed by repeating the first two-chip segment of X_4 by 8 times, the last two-chip segment of X_4 by 8

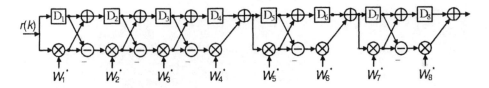

Figure 5.40.   Pruned efficient Golay correlator for the GHG.

times, and then modulating the resulting 16-long sequence by X_3 whose symbol length corresponds to 2 chip intervals. The Golay complementary sequence X_j, $j = 1, 3, 4$, are obtained by taking the sequence $a_{N^{(j)}}$ in the Golay recursion formula

$$a_0(i) = \delta(i), \tag{5.36a}$$

$$b_0(i) = \delta(i), \tag{5.36b}$$

$$a_n(i) = a_{n-1}(i) + W_n^{(j)} \cdot b_{n-1}(i - D_n^{(j)}), \tag{5.36c}$$

$$b_n(i) = a_{n-1}(i) - W_n^{(j)} \cdot b_{n-1}(i - D_n^{(j)}), \tag{5.36d}$$

for $n = 1, 2, \cdots, N^{(j)}$ and $i = 0, 1, \cdots, 2^{N^{(j)}} - 1$. The parameters of each Golay complementary sequence are carefully optimized through computer simulation for the best performances.

Method 2: Alternatively, the PSC code C_{psc} can be viewed as a Golay complementary sequence of length 256 generated with the following parameters applied to the above recursion formula (Note that the ultimate sequence a_8 becomes the desired C_{psc}.):

(a) Let $j = 0$, $N^{(0)} = 8$, $\qquad\qquad\qquad\qquad\qquad$ (5.37a)

(b) $[D_1^{(0)}, D_2^{(0)}, D_3^{(0)}, D_4^{(0)}, D_5^{(0)}, D_6^{(0)}, D_7^{(0)}, D_8^{(0)}]$
    $= [128, 64, 16, 32,\ 8,\ 1,\ 4,\ 2],$ $\qquad\qquad$ (5.37b)

(c) $[W_1^{(0)}, W_2^{(0)}, W_3^{(0)}, W_4^{(0)}, W_5^{(0)}, W_6^{(0)}, W_7^{(0)}, W_8^{(0)}]$
    $= [\,1, -1,\ 1,\ 1,\ 1,\ 1,\ 1,\ 1\,],$ $\qquad\qquad\qquad$ (5.37c)

(d) For $n = 4$ and $n = 6$, set $b_4(i) = a_4(i)$, $b_6(i) = a_6(i)$. (5.37d)

Fig. 5.39 depicts the corresponding matched filter structure for the resulting PSC code, which is called the *pruned efficient Golay correlator*. The delay and weight elements are as given in (5.37b) and (5.37c) above. We observe that the pruned EGC requires only 13 complex adds per correlation output. According to the simulation results reported in [52], the GHG code provides satisfactory slot acquisition performance even in high frequency offset environment, maintaining all benefits of the hierarchical and the Golay complementary codes.

9.4 SSC CORRELATORS

After acquiring the slot boundary, the 16 SSCs are correlated in parallel with the incoming S-SCH signal for the first 256 chip period of each slot. This process produces 16 instant metrics needed for determining the code group as well as the frame boundary. Those instant metrics are added to 64×15 accumulated metrics over K ($K \geq 3$, typically 15) slots (refer to Section 8.4) according to Table 5.4, out of which the maximum value is determined. In accumulating the instant metrics, noncoherent accumulation can be always applied, while the coherent accumulation techniques may be applied selectively for improved performances. The coherent accumulation is applicable after compensating for the phase distortion by using the channel estimates obtained from the synchronized P-SCH.

The 16 SSC active correlators may be efficiently designed by employing a correlator corresponding to the complementary sequence z used for the SSC generation in Section 8.4 and a 16×16 *fast Hadamard transformer* (FHT): In this implementation, the incoming S-SCH signal of length 256 is correlated with the sequence z for the coherent correlation unit of 16 chips, and then the 16 partial-correlation results are successively put in the FHT input storage. If the BS-MS frequency offset is large enough to bring in substantial phase rotation during a 256-chip period, the PSC partial-correlation outputs are also used to compensate the phase rotations of the FHT-input values. [11] Finally, the (channel compensated) input vector of length 16 is FHT-transformed to produce the 16 full-correlation outputs simultaneously.

9.5 COMPUTATIONAL COMPLEXITY

In order to evaluate the computational complexity quantitatively, we assume that two samples are taken per chip from the A/D converter, and the correlation outputs with the coherent correlation period of 256 chips are noncoherently accumulated over a frame (or, 15 slots) in each of the three steps. Under this assumption, the nominal cell search time becomes 30(ms). We examine the number of real additions/multiplications as well as the memory size required for each of the three steps.

---

[11] In case the P-SCH of different cells are time-aligned, the channel estimate obtained through P-SCH correlation is not accurate and causes performance degradation.

In the first step, computational complexity is determined by calculating the total number of operations for obtaining the pruned EGC output I/Q-samples (13×2 additions per I/Q-sample pair), squaring the I/Q samples (2 multiplications per I/Q-sample pair), and then accumulating them to the storage (2 additions per sample). This requires 28 additions and 2 multiplications per half-chip time. For the frame period of 10ms, the number of operations amounts to 2.15×10^6 additions and 1.5×10^5 multiplications. After accumulating the samples over a frame, the 5,120 accumulated values are compared to determine the maximum. The required memory size is $5,120 \times b$ bits when b-bit quantization is assumed.

In the second step, computational complexity is determined by calculating the operations required for the active correlation corresponding to the complementary sequence z (2×256 additions per slot), a 16×16 complex-valued Hadamard transformation ($2 \times 16 \times 4$ additions per slot), the squaring operations of the I/Q transform outputs (2×16 multiplications per slot), and the metric accumulations for RS-decoding ($16 + 64 \times 15$ additions per slot). The total number of operations per frame is 2.4×10^4 additions and 480 multiplications. An the end of each frame, the $960 (= 64 \times 15)$ accumulated metrics are compared to determine the maximum. The required memory size is $960 \times b$ bits for b-bit quantization.

Finally, in the third step, 8 complex Gold codes are correlated with the incoming P-CPICH in parallel, which requires 32 additions per chip period (8×2 additions for correlation and 8×2 additions for accumulation). At the end of each P-CPICH symbol period (or, 256 chips), 2×8 multiplications are needed for obtaining the squared magnitude of the correlation results. The total number of operations per frame amounts to 1.23×10^6 additions and 2,400 multiplications. Finally, 8 metrics are compared to determine the maximum. So the required memory size is negligible.

As is assessed above, the number of operations fluctuate among the three steps. If a multiplication is assumed to be equivalent to 6 additions in terms of computational complexity, the complexity distribution is 70.8% to Step 1, 0.6% to Step 2, and 28.6% to Step 3, respectively. In terms of memory usage, about 80% of the memory is allocated to Step 1, and 20% to Step 2. Moreover, according to [111], the false alarm probability in Step 1 is higher than that of the other steps for the same processing time (1 frame per each step). Consequently, a larger portion of the overall processing time need to be allocated to Step 1 to get improved performances. The complexity distribution among the three steps may change a little depending on the implemental variations (e.g., coherent correlation in Step 2, partial correlation for mitigation of the frequency offset, and parallel manipulation of multiple delay paths in Step 2, and others).

9.6   INITIAL FREQUENCY ACQUISITION

After acquiring the primary scrambling code of the cell, the MS should compensate the BS-MS frequency offset to be able to decode the BCH information in the P-CCPCH. The mass product consumer electronics often employ inexpensive crystal oscillators of the accuracy range of 3-13 ppm, in which case the frequency offset will be in the range of 6-26kHz for the 2GHz operational frequency [111]. Unless the frequency offset is reduced down to an acceptable level (e.g., less than 1kHz), the demodulator cannot track the incoming frequency and thus the communication functions can fail. The frequency acquisition is normally applied to the P-CPICH.

A promising FFT-based frequency acquisition [120] method proposed in [111] is as follows: First, we despread the P-CPICH using a despreading factor 64, thus collecting 40 despread values per slot. Then, we remove the modulation of the despread values using the knowledge of the pilot symbols, and pad 24 zeros to construct a 64-element data vector. Next, we apply a 64-point FFT to the data vector, accumulate the transformed FFT vectors over multiple slots for noise suppression, and perform quadratic interpolation for a refined frequency estimation. Then we can acquire the frequency offset with high resolution by identifying the frequency point producing the peak output energy. If we perform the slot accumulation over 30 slots, we can reduce the frequency offset from 20kHz to below 200Hz with a very high reliability [111].

Another popular frequency acquisition method is presented based on the differential estimation [120] between consecutive despread values. In this case, we multiply each despread P-CPICH value by the complex conjugate of the precedent despread value to get a phase rotation metric. Then we accumulate the metrics over multiple slots for noise suppression, and then convert the accumulated metric to the frequency offset. This differential estimation scheme suffers from noise enhancement, so exhibits poor performances in low SNR ranges and its pull-in range is not wide enough. In overall performance, when the FFT-based schemes and the differential estimation schemes are compared, the FFT-based scheme is reported to outperform the differential estimation scheme by more than 10dB for the P-CPICH SNR below -12dB [111]. However, the computation complexity makes it difficult to apply the FFT scheme directly to the mobile handset.

Chapter 6

INTER-CELL SYNCHRONOUS IMT-2000 TD-CDMA SYSTEM (3GPP-TDD)

The 3GPP-TDD system[1] contrasts to the 3GPP-FDD system in various aspects: The 3GPP-FDD system uses a pair of 5MHz frequency bands for downlink and uplink allocation, but the 3GPP-TDD system uses only a single 5MHz band, with the downlink and the uplink using different time slots on the same frequency band for data transmission. [2] First of all, the 3GPP-FDD base stations are not synchronized each other, while the 3GPP-TDD base stations are operated in frame-level timing-synchronous mode. The 3GPP-FDD system is intended for applications in public macro- and micro-cell environments with the data rates typically going up to 384 kbps with high mobility, while the 3GPP-TDD system is for public micro- and pico-cell environments with high traffic density and indoor coverage, with the data rates stretching up to 2Mbps. The 3GPP-TDD system is particularly suitable for the case when an application requires high data rates and tends to create highly asymmetric traffic such as Internet access [121].

In spite of the above differences in the frequency band usage and the target applications, both systems share much in common in the aspects of basic system parameters (e.g., chip rate, bandwidth, modulation scheme, etc.), channel structure, and physical layer procedure. In this chapter we first discuss the physical layer structures and operations of the 3GPP-TDD system, focusing on its essential differences from the 3GPP-FDD system. Then we examine the spreading and scrambling mechanisms of the 3GPP-TDD system, finally discussing the cell search scheme.

---

[1]TD-CDMA stands for *time division-code division multiple access.*
[2]In the future deployment, the data transmission bandwidth will be expanded up to 20MHz [53].

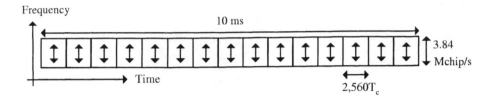

Figure 6.1.   TDD frame structure [122].

1.   TRANSPORT CHANNELS AND PHYSICAL CHANNELS

Transport channels of the 3GPP-TDD system include *dedicated channel* (DCH), *broadcast channel* (BCH), *paging channel* (PCH), *forward access channel* (FACH), *random access channel* (RACH), *downlink shared channel* (DSCH), and *uplink shared channel* (USCH). The USCH, which has not been used in the 3GPP-FDD system, is an uplink transport channel shared by several UEs carrying dedicated or traffic channels. [3]

Fig. 6.1 depicts the physical channel structure of the 3GPP-TDD system in time and frequency dimension (T_c denotes the chip interval). A physical channel in the 3GPP-TDD system is a burst which is transmitted in a particular *time slot* (TS) of length 2,560 chips within the allocated radio frames of length 38,400 chips (10ms). A burst is the combination of two data parts, a midamble, and a guard period. The duration of a burst is one time slot, and several bursts can be transmitted at the same time from one transmitter, in which case, the data part must use different OVSF channelization codes and the same scrambling code. For the midamble part, the same *basic midamble code* must be used within a cell but different users can use different midambles (refer to Section 1.2). A guard period is used for fear of the timing inaccuracies, power ramping, delay spread, and the propagation delay (in case no timing advance mechanism is used).

The channelization code is an OVSF code with the available spreading factors of 1, 2, 4, 8, and 16. By allowing only a small number of codes in the system, the 3GPP-TDD facilitates the implementation of the *joint detection* (JD) receiver for multiple user signals with moderate complexity. The midamble part of the burst can contain two different types of midambles - - a short midamble of length 256 chips and a long midamble of length 512 chips. The scrambling code and the basic midamble code are broadcast within a cell.

---

[3]The *common packet channel* (CPCH) of the 3GPP-FDD system is not defined in the 3GPP-TDD system.

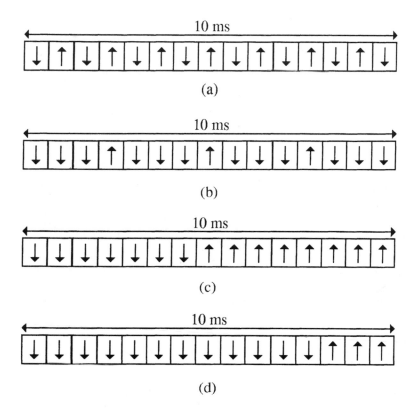

Figure 6.2. Examples of TDD frame structure: (a) multiple-switching-point configuration (symmetric DL/UL allocation), (b) multiple-switching-point configuration (asymmetric DL/UL allocation), (c) single-switching-point configuration (symmetric DL/UL allocation), (d) single-switching-point configuration (asymmetric DL/UL allocation) [122].

The 15 time slots in a frame can be allocated to either the uplink or the downlink. Thus a physical channel in the 3GPP-TDD system is defined by frequency, time slot, channelization code, burst type, and radio frame allocation, which is a mixture of the TDMA and the CDMA components. With such a flexibility, the 3GPP-TDD system can be adapted to different environments and deployment scenarios. Examples of multiple and single switching point configurations as well as of symmetric and asymmetric uplink/downlink allocations are given in Fig. 6.2. [4]

---

[4]In any configuration at least one time slot should be allocated for the downlink, and at least one time slot for the uplink.

1.1   DEDICATED PHYSICAL CHANNELS

The DCH is mapped onto the *dedicated physical channels* (DPCH).

Spreading Factor: Downlink DPCH uses the spreading factor of SF=16. Multiple parallel physical channels with different channelization codes can be used to support higher data rates (multi-code transmission). Operation with a single code with spreading factor 1 is also provided for the downlink DPCH in order to transmit high rate data when the inter-cell interference or background noise is small.

Uplink DPCH uses a spreading factor ranging from 16 down to 1 and prefers single channelization code to multi-code in order not to increase PAR of the UE transmitter. For multi-code transmission a UE can use a maximum of two different channelization codes per time slot simultaneously.

Burst Type: There are three types of bursts for the DPCH - - the *burst type-1* with a long midamble of 512 chips (see Fig. 6.3 (a)), the *burst type-2* with a short midamble of 256 chips (see Fig. 6.3 (b)), and the *burst type-3* with a long midamble of 512 chips and a long guard period (see Fig. 6.3 (c)). The data field of the burst type-1 is 976 chips long, whereas the data field of the burst type-2 is 1,104 chips long. The guard period for the burst type-1, as well as for the type-2, is 96 chip periods long. Due to the longer midamble, the burst type-1 is suitable for the uplink, where up to 16 different channel impulse responses can be jointly estimated. However, it may be also used for the downlink. The burst type-2 can be used for the downlink, or for the uplink when the bursts within a time slot are allocated to a small number of users. The burst type-3 has a slot structure similar to that of the burst type-1, but the second data symbol field is made shorter than the first symbol data field by 96 chips so as to provide additional guard time at the end of the time slot. It is used for uplink only. Due to the long guard period it is suitable for initial access. The number of symbols per data field, N_{symbol}, is related to the number of chips per data field, N_{chip}, by

$$N_{symbol} = N_{chips}/SF, \quad SF = 1, 2, 4, 8, 16. \tag{6.1}$$

The TPC and/or the TFCI may be transmitted over the DPCH, as shown in Fig. 6.4. The transmission is negotiated at call setup and can be re-negotiated during the call. If applied, they are transmitted using a part of the data fields in the burst, and their channelization codes are always the first allocated code of SF=16 according to the order in the higher layer allocation message. The TPC is transmitted (only in the uplink slot) directly after the midamble (if applied) and the TFCI is located on the positions adjacent to the midamble. This allocation allows for the best possible transmission of those important data, as interference

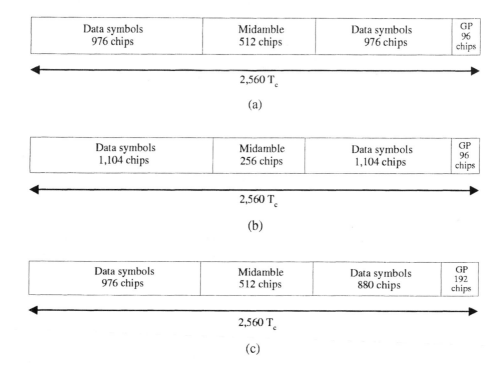

Figure 6.3. Structure of the bursts: (a) burst type-1, (b) burst type-2, (c) burst type-3 (GP: guard period) [122].

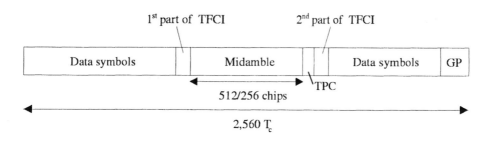

Figure 6.4. Position of TPC and TFCI information in the traffic burst [122].

from midamble can be cancelled in the receiving side and the channel estimation is more reliable for the bits adjacent to the midamble. When multiple DPCHs are transmitted for a connection in a concurrent or sequential way, only one of the DPCHs carries the TFCI information. For every user, the TPC information is transmitted at least once per transmitted frame.

1.2   MIDAMBLES FOR SPREAD BURSTS

The midambles (or, *training sequences*) of different users which are active in the same cell and in the same slot are cyclically shifted versions of one basic midamble code. Two sets of basic midamble codes are defined for application to a maximum of 128 cells in the 3GPP-TDD system - - $\{\mathbf{m}_{PL,m} : m = 0, 1, \cdots, 127\}$ for the burst type-1 or 3 and $\{\mathbf{m}_{PS,m} : m = 0, 1, \cdots, 127\}$ for the burst type-2. The length of the basic midamble codes, P, is 456 for the burst type-1 or 3 and 192 for the burst type-2. Different burst types must not be mixed in the same time slot of one cell.

We denote a particular basic midamble code by a binary (or, ± 1) vector of length P, $\mathbf{m}_P = (m_1, m_2, \cdots, m_P)$. Then, as QPSK modulation is used for the uplink and the downlink in the 3GPP-TDD system, the code is first transformed into a *complex basic midamble code* $\hat{\mathbf{m}}_P = (\hat{m}_1, \hat{m}_2, \cdots, \hat{m}_P)$, for $\hat{m}_i = (j)^i \cdot m_i, i = 1, 2, \cdots, P$. The vector $\hat{\mathbf{m}}_P$ is then extended to a *periodic basic midamble code* $\hat{\mathbf{m}} = (\hat{m}_1, \hat{m}_2, \cdots, \hat{m}_{I_{max}})$, where the vector length I_{max} is given by $I_{max} = L_m + (K' - 1)W + \lfloor P/K \rfloor$ for the midamble length L_m, the minimum time shift between two midambles in a time slot W, the basic midamble code length P, the maximum number of channel estimates per time slot for joint detection K, and $K' = K/2$. For the burst type-1 or 3, $L_m = 512$, $W = 57$, P=456, and $K' = 8$, while, for the burst type-2, $L_m = 256$, $W = 64$, P=192, and $K' = 3$. The first P elements of the periodic basic midamble code correspond to the P elements of the complex basic midamble code, and the rest $I_{max} - P$ elements are given by the periodic extension $\hat{m}_i = \hat{m}_{i-P}$, $i = P+1, P+2, \cdots, I_{max}$. When we denote the midamble code of the kth user $(k = 1, 2, \cdots, K)$ by $\hat{\mathbf{m}}^{(k)} = (\hat{m}_1^{(k)}, \hat{m}_2^{(k)}, \cdots, \hat{m}_{L_m}^{(k)})$, the midamble element $\hat{m}_i^{(k)}$ is generated by

$$\hat{m}_i^{(k)} = \hat{m}_{i+(K'-k)W}, \; i = 1, 2, \cdots, L_m, \; k = 1, 2, \cdots, K', \qquad (6.2)$$

for the first K' users, while it is generated by

$$\hat{m}_i^{(k)} = \begin{cases} \hat{m}_{i+(K-k-1)W+\lfloor P/K \rfloor}, \\ \qquad i = 1, 2, \cdots, L_m, \; k = K'+1, K'+2, \cdots, K-1, \\ \hat{m}_{i+(K'-1)W+\lfloor P/K \rfloor}, \\ \qquad i = 1, 2, \cdots, L_m, \; k = K, \end{cases}$$

$$(6.3)$$

for the second K' users, introducing intermediate shifts. Whether or not intermediate shifts are allowed in a cell is broadcast on the BCH. The midamble codes are transmitted without being spread or scrambled by any additional codes. A *midamble code set* or a *midamble code family* represents K specific midamble codes $\hat{\mathbf{m}}^{(k)}, k = 1, 2, \cdots, K$, which are based on a single basic midamble code $\mathbf{m}_P$.

If all users in one time slot have a common midamble in the downlink, the transmit power of this common midamble is set such that there is no power offset between the data part and the midamble part of the transmit signal within the time slot. In the case of user specific midambles, the transmit power of the user specific midamble is set such that there is no power offset between the data parts and the midamble part for this user within one slot.

When downlink beamforming is used, the user who has a dedicated channel and is involved in the beamforming gets one individual midamble even in the downlink.

1.3   COMMON PHYSICAL CHANNELS

Primary Common Control Physical Channel (P-CCPCH): The P-CCPCH is used for carrying the BCH. The time slot and code of the P-CCPCH in a frame is known from the *synchronization channel* (SCH). The P-CCPCH always uses the burst type-1 and the first channelization code of length 16, $C_{Q=16}^{(k=1)}$, for spreading. No TFCI is applied to the P-CCPCH. For the time slots where the P-CCPCH is transmitted, two midambles, $\hat{m}^{(1)}$ and $\hat{m}^{(2)}$ are exclusively reserved for P-CCPCH in order to support the STTD and the *beacon function*. When the STTD antenna diversity is applied to the P-CCPCH, $\hat{m}^{(1)}$ is used for the first antenna and $\hat{m}^{(2)}$ for the diversity antenna, while only $\hat{m}^{(1)}$ is used in no diversity antenna case.

Secondary Common Control Physical Channel (S-CCPCH):   The S-CCPCH is used for carrying the PCH and the FACH. The PCH and the FACH may be mapped onto one or several S-CCPCHs to adapt capacity to different requirements. The S-CCPCH uses the burst type-1 or 2 and a fixed spreading factor of 16, and may use the TFCI.

Physical Random Access Channel (PRACH): The uplink RACH is mapped onto one or several PRACHs. In this way the capacity of RACH can be flexibly scaled depending on the operator's need. The PRACH uses either spreading factor SF = 16 or SF = 8. For SF = 16, the number of QPSK symbols transmitted in the data field-1 and the data field-2 are respectively 61 and 55, while for SF = 8, the number doubles to 122 and 110, respectively. The set of spreading codes available for the PRACH as well as the associated spreading factors are broadcast on the BCH. The mobile stations transmit the uplink access bursts of type-3 randomly in the PRACH.

The basic midamble codes used for the PRACH bursts are the same as for the burst type-1 or 3, and the necessary time shifts are obtained by choosing either all indices $k = 1, 2, 3, \cdots, K'$ (for cells with small radius) or odd indices

$k = 1, 3, 5, \cdots, \leq K'$ (for cells with large radius). Different cells use different periodic basic midamble codes, i.e., different midamble code sets. For cells with large radius, additional midambles may be derived from the *time-inverted basic midamble code*. Thus, the *second basic midamble code* $\mathbf{m}_2$ is the time inverted version of basic midamble code $\mathbf{m}_1$. In this way, a joint channel estimation for the channel impulse responses of all active users within one time slot can be performed by a maximum of two cyclic correlators (in cells with small radius, a single cyclic correlator suffices). Different channel impulse response estimates for different users are obtained sequentially in time at the output of the cyclic correlators.

For the PRACH there exists a fixed association between the training sequence and the channelization code. The association rule is based on the order of the channelization codes $C_Q^{(k)}$ given by k and the order of the midambles $\mathbf{m}_j^{(k)}$ given by k and j, with the constraint that the midamble for a spreading factor Q is the same as that in the upper branch for the spreading factor $2Q$. The index j ($j = 1, 2$) indicates whether the original basic midamble code ($j = 1$) or the time-inverted basic midamble code is used ($j = 2$). In the case all k's are allowed and only one periodic basic midamble code $\mathbf{m}_1$ is available for the RACH, the association depicted in Fig. 6.5 (a) is applied, while in the case only odd k's are allowed, the association in Fig. 6.5 (b) is applied.

Synchronization Channel (SCH): In the 3GPP-TDD system, the code group of a cell can be derived from the synchronization channel. In order not to limit the uplink/downlink asymmetry, the SCH is mapped only on one or two downlink slots per frame. There are two cases of SCH (and P-CCPCH) allocation as follows:

(Case 1) The SCH and the P-CCPCH are allocated in one time slot, TS$\#k$, $k = 0, 1, \cdots, 14$.
(Case 2) The SCH is allocated in two time slots, TS$\#k$ and TS$\#k + 8$, while the P-CCPCH is allocated in one time slot TS$\#k$, $k = 0, 1, \cdots, 6$.

The position of the SCH (or the value k) in the frame can change on a long term basis. Due to this association between the SCH and the P-CCPCH, the position of the P-CCPCH is known from the SCH. Fig. 6.6 depicts a transmission example of the SCH for $k = 0$ in Case 2. As depicted in the figure, the SCH consists of a primary and three secondary synchronization codes of length 256. As public 3GPP-TDD systems keep synchronization between base stations, a capture effect for the SCH can occur.[5] The time offset t_{offset} enables the

---

[5]The capture effect generally means the phenomenon that a (strong) signal prevents the detection of the other (weak) signals when they reach the receiver at the same time.

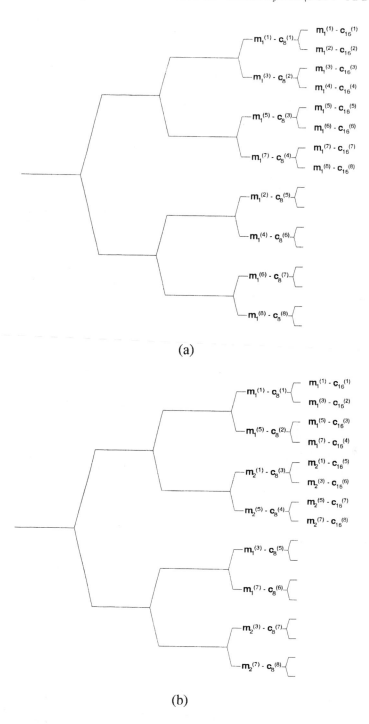

(a)

(b)

Figure 6.5.   Association of midambles to channelization codes in the OVSF tree: (a) for all *k*'s, (b) for odd *k*'s [122].

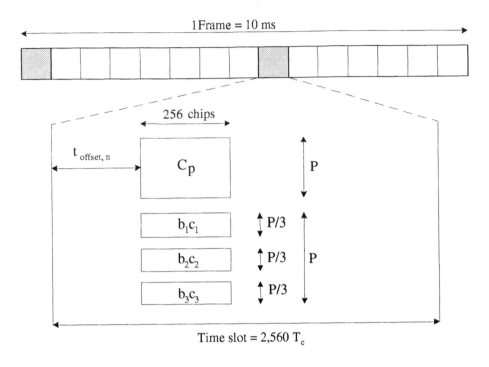

Figure 6.6. Synchronization channel composed of one primary synchronization code C_p and parallel secondary synchronization codes in slots k and $k + 8$ [122].

system to overcome the capture effect, which takes one of 32 values depending on the code group of the cell. The exact value for t_{offset} is given by

$$t_{offset,n} = n \cdot \lfloor \frac{2560 - 96 - 256}{31} \rfloor T_c = n \cdot 71T_c, \quad n = 0, 1, \cdots, 31, \quad (6.4)$$

for the code group index n.

Physical Uplink Shared Channel (PUSCH): The burst structure of the DPCH is used for the PUSCH. User specific physical layer parameters like power control, timing advance, or directive antenna settings are derived from the associated channel (FACH or DCH).

Physical Downlink Shared Channel (PDSCH): The burst structure of the DPCH is used for the PDSCH. User specific physical layer parameters are derived from the associated channel (FACH or DCH). There are three signalling methods available for indicating to the UE that there are data to decode on the DSCH: First, using the TFCI field of the associated channel or PDSCH; secondly, using the DSCH user specific midamble derived from the set of mi-

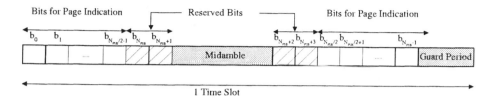

Figure 6.7.   Structure of the PICH burst [122].

dambles used for that cell; third, using higher layer signalling. When the midamble based method is used, the UE decodes the PDSCH if the PDSCH was transmitted with the midamble assigned to the UE by UTRAN. [6]

Page Indicator Channel (PICH): The PICH is a physical channel used to carry the *page indicators* (PI). Fig. 6.7 depicts the structure of a PICH burst and the numbering of the bits within the burst. The same burst type is used for the PICH in every cell. N_{PIB} bits in a normal burst of type-1 or type-2 are used to carry the PI, where $N_{PIB} = 240$ for the burst type-1 and $N_{PIB} = 272$ for the burst type-2. The bits $b_{N_{PIB}}, \cdots, b_{N_{PIB}+3}$, adjacent to the midamble, are reserved for possible future use. They are set to 0 and transmitted with the same power as the PI carrying bits. $N_{PI} (= N_{PIB}/2L_{PI})$ page indicators of length $L_{PI} (= 2, 4, 8)$ symbols are transmitted in one time slot. A *PICH block* is composed of the page indicators in the N_{PICH} consecutive frames, where N_{PICH} is configured by the higher layers. Thus, $N = N_{PICH} \times N_{PI}$ page indicators are transmitted in each PICH block. The PI calculated by the higher layers for a certain UE is mapped to the page indicator PI_p in the nth frame of a PICH block, where $p = PI \bmod N_{PI}$ and $n = \lfloor PI/N_{PI} \rfloor$. The page indicator PI_p in a time slot is mapped to the bits $\{b_{p \times L_{PI}}, \cdots, b_{p \times L_{PI}+L_{PI}-1}, b_{N_{PIB}/2+p \times L_{PI}}, \cdots, b_{N_{PIB}/2+p \times L_{PI}+L_{PI}-1}\}$ within the slot.

1.4   MIDAMBLE ALLOCATION FOR PHYSICAL CHANNELS

In general, midambles are part of the physical channel configuration which is performed by higher layers. Optionally, if no midamble is allocated nor the use of the common midamble is signalled by higher layers, a default midamble allocation is used. This default midamble allocation is given by a fixed associ-

---

[6]For this method no other physical channels may use the same time slot as the PDSCH and only one UE may share the PDSCH time slot at the same time.

ation between midambles and channelization codes, and is applied individually to all channelization codes within a time slot. Physical channels providing the beacon function always use the reserved midambles, while, for all other physical channels, the midamble allocation is signalled or given by default. (Refer to [122] for the default association between midamble codes and channelization codes.)

In the downlink, either a *common midamble* or *UE specific mediambles* may be assigned to the physical channels in a time slot, whose choice depends on the number of UEs in the slot, the beamforming/closed loop transmit diversity, and the PDSCH physical layer signaling methods. If no midamble is allocated by signalling, the UE derives the midamble from the associated channelization code and uses an individual midamble for each channelization code. For each association between midambles and channelization codes, there are one primary channelization code and a set of secondary channelization codes associated to each midamble. The higher layers allocate the channelization codes in a particular order. Primary channelization codes are allocated prior to associated secondary channelization codes. If midambles are reserved for the beacon function, all primary and secondary channelization codes that are associated with the reserved midambles are not used. Primary and its associated secondary channelization codes are not allocated to different UEs. In the case that secondary channelization codes are used, the secondary channelization codes of a set are allocated in ascending order with respect to their numbering.

In the uplink, if the midamble is explicitly allocated by the higher layers, the UE specific midambles are assigned to all UEs in a time slot. If no midamble is allocated by the higher layers, the UE derives the midamble from the assigned channelization code as for the downlink physical channels. If the UE changes the SF according to the data rate, it always varies the channelization code along the lower branch of the OVSF tree.

1.5   MAPPING OF TRANSPORT CHANNELS TO PHYSICAL CHANNELS

The following rule is applied for the mapping of the transport channels onto physical channels:

(a) The DCH is mapped onto one or several DPCHs.

(b) The BCH is mapped onto the P-CCPCH. The secondary SCH indicates in which time slot an MS can find the P-CCPCH. If the BCH information requires more resources than provided by the P-CCPCH, the BCH in the P-CCPCH will be composed of a pointer to additional S-CCPCH resources for the FACH where the additional BCH information is sent.

(c) The PCH is mapped onto one or several S-CCPCHs. The location of the PCH is indicated on the BCH. To allow an efficient *discontinuous reception*

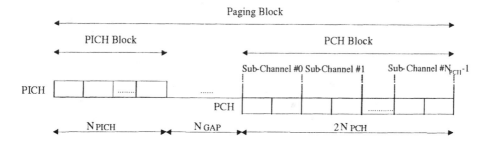

Figure 6.8. Paging sub-channels and the association of the PICH and PCH blocks [122]

(DRX) for battery saving, the PCH is divided into *PCH blocks*, each of which comprising N_{PCH} *paging sub-channels.* Each paging sub-channel is mapped onto two consecutive PCH frames within one PCH block. The assignment of UEs to paging sub-channels is independent of the assignment of UEs to page indicators. Fig. 6.8 depicts the structure of a *paging block* which consists of one *PICH block* and one *PCH block.* If a PI in a certain PICH block is set to 1 it is an indication that the UEs associated with this PI read their corresponding paging sub-channel within the same paging block. The value N_{GAP} ($N_{GAP} > 0$) denotes the number of frames between the end of the PICH block and the beginning of the PCH block, which is configured by the higher layers.

(d) The FACH is mapped onto one or several S-CCPCHs. The location of the FACH is indicated on the BCH, and both capacity and location can be changed, if required.

(e) The RACH is mapped onto one or several PRACHs. The same time slot may be used for the PRACH by more than one cell. Multiple transmissions using different spreading codes may be received in parallel. Furthermore, more than one slot per frame may be allocated for the PRACH. The location of the slots allocated to the PRACH is broadcast on the BCH.

(f) The USCH and the DSCH are respectively mapped onto one or several PUSCHs and PDSCHs.

2. TRANSMIT DIVERSITY AND BEACON FUNCTIONS

The transmit diversity schemes are applicable to the downlink physical channels with the following mapping conditions: TSTD for the SCH, STTD for the P-CCPCH, and *closed loop transmit diversity* for the DPCH. The closed loop transmit diversity either takes the form of *selective transmit diversity* (STD) or *transmit adaptive antenna* (TxAA): In the STD scheme, the transmit power is exclusively allocated to the antenna receiving the higher uplink power, while in the TxAA scheme, the power weights of both antennas are adjusted in order to maximize the receiving power after path combination in the uplink. As the propagation channel is reciprocal for both uplink and downlink in the TDD system, the STD and the TxAA based on the uplink channel estimation will provide a satisfactory downlink diversity performance. [7]

On the other hand, for the purpose of measurement, a *beacon function* is provided by particular physical channels - - either (Case 1) all physical channels using the channelization code $C_{Q=16}^{(k=1)}$ in TS#m, $m = 0, 1, \cdots, 14$ or (Case 2) all physical channels using the channelization code $C_{Q=16}^{(k=1)}$ in TS#m and TS#m+8, $m = 0, 1, \cdots, 6$, depending on the SCH allocation cases. Therefore, the P-CCPCH always provides the beacon function. The physical channels providing the beacon function are transmitted with reference power but without beam forming, and use the burst type-1. Furthermore, they use midambles $\hat{m}^{(1)}$ and $\hat{m}^{(2)}$ exclusively, and leave $\hat{m}^{(9)}$ and $\hat{m}^{(10)}$ unused in the corresponding time slot if 16 midambles are allowed in the cell. The reference power for any physical channel providing the beacon function is equally divided and allocated to $\hat{m}^{(1)}$ and $\hat{m}^{(2)}$ if STTD is applied, while it is exclusively allocated to $\hat{m}^{(1)}$, otherwise.

3.   TRANSMIT POWER CONTROL

Power control is applied to the 3GPP-TDD system to limit the interference level within the system, thus reducing the inter-cell interference level and to reduce the power consumption in the MS.

For the uplink DPCH and PUSCH, power control takes an open loop transmitter power control technique exploiting the uplink/downlink channel reciprocity in the TDD scheme. Measuring the received power of the physical channel providing the beacon function, the MS can calculate the path loss. Together with the information on the interference level in the uplink level and the target SIR, which is transmitted from the BS, the MS can set and adjust its transmit power in order to meet the required quality of services. Thus the power control update rate is typically once or twice per frame, which is equal to the P-CCPCH (providing the beacon function) transmission rate. The target SIR is also adjusted in the BS through the outer loop power control on a long-term basis. On the other hand, the uplink PRACH power is set by the higher layers.

---

[7]Conceptually, closed loop transmit diversity presumes that the estimation is made in the opposite side (or, in the MS) and the weight adjusting parameters are returned to the transmit side (or, to the BS), based on which the weights are adjusted. However, in the TDD system, the estimation is made directly in the BS side exploiting the channel reciprocity, in which context the STD and the TxAA may be regarded as open loop control schemes.

For the downlink DPCH and PDSCH, the power control rate may be different depending on the slot allocation for each user, and the SIR-based inner loop power control is used. The initial transmission power of the downlink DPCH and the PDSCH is set by the network. After the initial transmission, the BS transits into SIR-based inner loop power control. The MS generates TPC commands to control the network transmit power and sends them in the TPC field of the uplink DPCH and PUSCH. As a response to the received TPC command, the BS may adjust the transmit power of all downlink DPCHs and PDSCHs of the corresponding radio link (toward the same MS). The higher layer outer loop power control for adjusting the target SIR is also applied in order to meet the long-term quality requirements. On the other hand, the P-CCPCH transmit power is set by higher layer signaling and can be changed based on network determination on a slow basis. The reference transmit power of the P-CCPCH is signaled on the BCH. The relative transmit powers of the S-CCPCH and the PICH, compared to the P-CCPCH transmit power, are set by higher layer signaling.

4.   TIMING ADVANCE

A BS may adjust the MS transmission timing with *timing advance* technique. The initial value for timing advance is determined in the BS by measuring the timing of the uplink PRACH. When timing advance is used, the BS continuously measures the timing of a transmission from the MS and sends the necessary timing advance value. On receiving this value, the MS adjusts the timing of its transmissions accordingly in steps of 4 chips. The required timing advance is represented as a 6 bit number (0-63), being the multiple of 4 chips. Upon receiving the timing advance command, the MS adjusts its transmission timing according to the timing advance command at the frame number specified by a higher layer signaling. When a TDD to TDD handover takes place, the MS transmits in the new cell with timing advance adjusted by the relative timing difference between the new and the old cells.

If uplink synchronization is used, the timing advance is made sub-chip granular with high accuracy in order to enable synchronous CDMA in the uplink. The required timing advance is represented as a multiple of 1/4 chip. The BS continuously measures the timing of transmission from the MS and sends the necessary timing advance value. On receipt of this value the MS adjusts the timing of its transmission accordingly in steps of 1/4 chip. Support of uplink synchronization is optional for the MS.

5.   MULTIPLEXING, CHANNEL CODING, AND MEASUREMENTS

The multiplexing and channel coding procedure of the 3GPP-TDD system is nearly the same as that of the 3GPP-FDD uplink procedure (refer to Fig. 5.20) [123]. A remarkable feature in the 3GPP-TDD procedure is that two options are available for the second interleaving: The interleaving is applied either to the bits of all TrCHs of the corresponding CCTrCH or to the bits to be transmitted within one time slot.

The random access procedure and the measurement parameters are also similar to those of the 3GPP-FDD system. The required measurement parameters in the 3GPP-TDD system are as follows [124]: In the UE, P-CCPCH RSCP, FDD-cell CPICH RSCP, time slot ISCP, UTRA/GSM carrier RSSI, SIR, FDD-cell CPICH Ec/No, transport channel BLER, UE transmit power, SFN-SFN/CFN observed time difference, observed time difference to GSM cell are measured. In the network, RSCP, time slot ISCP, RSSI, SIR, transport channel BER, physical channel BER, transmitted carrier power, transmitted code power, and Rx timing deviation are measured. The timing deviation can be used for timing advance calculation or location services. (For detailed description of the other parameters, refer to Section 6 of Chapter 5.)

6.   DATA MODULATION

The basic chip rate of the 3GPP-TDD system is the same as that of the 3GPP-FDD system, i.e., 3.84 Mchip/s, but a lower chip rate service of 1.28 Mchip/s is also made possible. QPSK data modulation is used for both uplink and downlink, and the spreading factor Q for data symbols can take the values 1, 2, 4, 8, and 16.

Data modulation is performed by mapping two consecutive binary bits to a complex valued data symbol. For the burst type-1 and type-2, each user burst has two data carrying parts, termed *data blocks*

$$\mathbf{d}^{(k,i)} = (d_1^{(k,i)}, d_2^{(k,i)}, \cdots, d_{N_k}^{(k,i)})^T, \quad i = 1, 2, \quad k = 1, 2, \cdots, K, \quad (6.5)$$

where N_k is the number of symbols per data field for the user k. In the case of the burst type-3, the number of symbols in the second data block $\mathbf{d}^{(k,2)}$ decreases by $96/Q_k$ from the first data block $\mathbf{d}^{(k,1)}$, for the spreading factor of the kth user Q_k. Data block $\mathbf{d}^{(k,1)}$ is transmitted before the midamble and data block $\mathbf{d}^{(k,2)}$ after the midamble. Each of the N_k complex (QPSK) data symbols $d_n^{(k,i)}$, $i = 1, 2, k = 1, 2, \cdots, K, n = 1, 2, \cdots, N_k$ is generated from two consecutive binary data bits (provided from the output of the physical channel mapping procedure) $b_{1,n}^{(k,i)}, b_{2,n}^{(k,i)} \in \{0, 1\}, k = 1, 2, \cdots, K, n = 1, 2, \cdots, N_k, i = 1, 2,$

using the mapping relation given by

$$d_n^{(k,i)} = (-1)^{b_{1,n}^{(k,i)}} \cdot (j)^{b_{1,n}^{(k,i)} + b_{2,n}^{(k,i)} + 1}, \tag{6.6}$$

where the additions among exponents are the binary operations (i.e., exclusive-OR operations).

7. SPREADING AND SCRAMBLING

As the spreading or channelization operation, each complex valued data symbol $d_n^{(k,i)}$ is first multiplied by a real valued channelization code $C_{Q_k}^{(k)}$ of length $Q_k \in \{1, 2, 4, 8, 16\}$. The resulting complex valued sequence is then randomized by a complex valued scrambling sequence $\hat{v}$ of length 16.

7.1 CHIP MODULATION

The complex-valued chip signal generated by the spreading and scrambling process is QPSK-modulated as shown in Fig. 5.29. As in the 3GPP-FDD case, the real and the imaginary parts of the input complex chip signal S are split and independently pulse-shaped, up-converted by the cosine and the sine carriers, and then combined and amplified by the power amplifier for transmission. The employed pulse shaping filter is an RRC filter with the roll-off factor $\alpha = 0.22$.

7.2 CHANNELIZATION CODES

Each element $c_q^{(k)}$, $q = 1, 2, \cdots, Q_k$, of the real-valued channelization codes $C_{Q_k}^{(k)} = (c_1^{(k)}, c_2^{(k)}, \cdots, c_{Q_k}^{(k)})$, $k = 1, 2, \cdots, K$, takes the value $+1$ or -1. The $C_{Q_k}^{(k)}$ is one of the OVSF codes defined based on the code tree shown in Fig. 6.9, which allows to use plural codes with different spreading factors in the same time slot without violating the orthogonality. The spreading factor may go up to $Q_{MAX} = 16$.

7.3 SCRAMBLING CODES

The channelized chip sequence is then randomized by a cell specific scrambling code of length 16, $\hat{v} = (\hat{v}_1, \hat{v}_2, \cdots, \hat{v}_{16})$, where each element $\hat{v}_{16}$ takes one of the four values of $+1$, $+j$, -1, and $-j$. The complex scrambling code $\hat{v}$ is generated from the 128 binary cell-specific scrambling codes given in Table 6.1 [125], by using the transformation

$$\hat{v}_i = (j)^i \cdot v_i, \quad i = 1, 2, \cdots, 16. \tag{6.7}$$

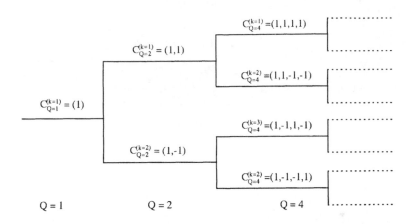

Figure 6.9.   Code-tree to generate the OVSF codes for channelization operation [125].

Table 6.1.   Cell specific binary scrambling codes [125].

Scrambling code	v_1	v_2	v_3	v_4	v_5	v_6	v_7	v_8	v_9	v_{10}	v_{11}	v_{12}	v_{13}	v_{14}	v_{15}	v_{16}
Code 0	-1	1	-1	-1	-1	1	-1	-1	1	-1	1	1	-1	1	-1	-1
Code 1	1	1	1	1	1	-1	1	-1	1	-1	-1	1	1	1	-1	-1
Code 2	1	-1	1	1	1	-1	1	1	-1	1	1	1	1	-1	-1	-1
Code 3	1	1	1	-1	-1	-1	-1	1	-1	-1	1	-1	-1	-1	1	-1
Code 4	1	1	1	-1	-1	-1	-1	1	1	1	1	-1	1	1	1	-1
Code 5	-1	1	1	-1	-1	-1	1	1	1	1	1	1	1	-1	1	-1
Code 6	-1	1	-1	-1	-1	1	-1	-1	-1	1	1	1	1	-1	-1	-1
Code 7	1	-1	1	-1	-1	-1	-1	-1	1	1	-1	-1	-1	1	1	-1
Code 8	1	1	1	-1	-1	-1	1	-1	1	1	-1	1	1	1	1	-1
Code 9	1	1	-1	1	1	1	1	-1	1	1	-1	-1	-1	1	1	-1
Code 10	1	-1	1	-1	1	1	1	1	-1	-1	1	1	-1	1	1	-1
...																
Code 124	-1	-1	1	1	1	1	1	1	1	-1	1	-1	-1	1	1	-1
Code 125	1	-1	-1	1	1	-1	1	-1	1	1	1	1	1	1	-1	-1
Code 126	1	1	1	1	-1	1	-1	1	-1	1	1	-1	1	1	-1	-1
Code 127	1	-1	1	-1	-1	-1	-1	-1	1	-1	-1	1	1	1	-1	-1

The combination of a user specific channelization code and a cell specific scrambling code can be seen as a user and cell specific *data block spreading code* $S^{(k)} = (s_1^{(k)}, s_2^{(k)}, \cdots, s_{N_k Q_k}^{(k)})$, whose element $s_p^{(k)}$ is given by

$$s_p^{(k)} = c_{1+\{(p-1) \bmod Q_k\}}^{(k)} \cdot \hat{v}_{1+\{(p-1) \bmod Q_{max}\}}, \qquad (6.8)$$

for $k = 1, 2, \cdots, K$, $p = 1, 2, \cdots, N_k Q_k$. Therefore, the spreading and scrambling processing can be seen as a multiplication process between the data blocks and the data block spreading codes.

Table 6.2. Code allocation for case 1 [125].

Code Group	Code Set	Frame with SFN mod 2 = 1			Frame with SFN mod 2 = 0			Associated t_{offset}
0	1	C_1	C_3	C_5	C_1	C_3	$-C_5$	t_0
1	1	C_1	$-C_3$	C_5	C_1	$-C_3$	$-C_5$	t_1
2	1	$-C_1$	C_3	C_5	$-C_1$	C_3	$-C_5$	t_2
3	1	$-C_1$	$-C_3$	C_5	$-C_1$	$-C_3$	$-C_5$	t_3
4	1	jC_1	jC_3	C_5	jC_1	jC_3	$-C_5$	t_4
5	1	jC_1	$-jC_3$	C_5	jC_1	$-jC_3$	$-C_5$	t_5
6	1	$-jC_1$	jC_3	C_5	$-jC_1$	jC_3	$-C_5$	t_6
7	1	$-jC_1$	$-jC_3$	C_5	$-jC_1$	$-jC_3$	$-C_5$	t_7
8	1	jC_1	jC_5	C_3	jC_1	jC_5	$-C_3$	t_8
9	1	jC_1	$-jC_5$	C_3	jC_1	$-jC_5$	$-C_3$	t_9
10	1	$-jC_1$	jC_5	C_3	$-jC_1$	jC_5	$-C_3$	t_{10}
11	1	$-jC_1$	$-jC_5$	C_3	$-jC_1$	$-jC_5$	$-C_3$	t_{11}
12	1	jC_3	jC_5	C_1	jC_3	jC_5	$-C_1$	t_{12}
13	1	jC_3	$-jC_5$	C_1	jC_3	$-jC_5$	$-C_1$	t_{13}
14	1	$-jC_3$	jC_5	C_1	$-jC_3$	jC_5	$-C_1$	t_{14}
15	1	$-jC_3$	$-jC_5$	C_1	$-jC_3$	$-jC_5$	$-C_1$	t_{15}
16	2	C_{10}	C_{13}	C_{14}	C_{10}	C_{13}	$-C_{14}$	t_{16}
17	2	C_{10}	$-C_{13}$	C_{14}	C_{10}	$-C_{13}$	$-C_{14}$	t_{17}
				...				
20	2	jC_{10}	jC_{13}	C_{14}	jC_{10}	jC_{13}	$-C_{14}$	t_{20}
				...				
24	2	jC_{10}	jC_{14}	C_{13}	jC_{10}	jC_{14}	$-C_{13}$	t_{24}
				...				
31	2	$-jC_{13}$	$-jC_{14}$	C_{10}	$-jC_{13}$	$-jC_{14}$	$-C_{10}$	t_{31}

8. SYNCHRONIZATION CODES

The *primary synchronization code* (PSC) and the *secondary synchronization codes* (SSC) in the 3GPP-TDD system are generated by exactly the same procedure that is taken for the 3GPP-FDD synchronization code generation (i.e., generalized hierarchical Golay sequence generation and Hadamard modulation methods). The difference is that the 3GPP-TDD system takes only the 12 codes $\{C_0, C_1, C_3, C_4, C_5, C_6, C_8, C_{10}, C_{12}, C_{13}, C_{14}, C_{15}\}$ selected from the 16 SSCs that are used in the 3GPP-FDD system.

8.1 CODE ALLOCATION

There are two types of code allocations depending on the number of slots assigned to the SCH in one frame - - one slot for the SCH (Case 1) and two slots for the SCH (Case-2).

For the frame synchronization and scrambling code identification, three SSCs are QPSK-modulated and transmitted in parallel with the PSC over the allocated

Table 6.3.   Code allocation for case 2 [125].

Code group	Code set	Frame with SFN mod 2 = 1						Frame with SFN mod 2 = 0						Associated t_{offset}
		Slot k			Slot k+8			Slot k			Slot k+8			
0	1	C_1	C_3	C_5	C_1	C_3	$-C_5$	$-C_1$	$-C_3$	C_5	$-C_1$	$-C_3$	$-C_5$	t_0
1	1	C_1	$-C_3$	C_5	C_1	$-C_3$	$-C_5$	$-C_1$	C_3	C_5	$-C_1$	C_3	$-C_5$	t_1
2	1	jC_1	jC_3	C_5	jC_1	jC_3	$-C_5$	$-jC_1$	$-jC_3$	C_5	$-jC_1$	$-jC_3$	$-C_5$	t_2
3	1	jC_1	$-jC_3$	C_5	jC_1	$-jC_3$	$-C_5$	$-jC_1$	jC_3	C_5	$-jC_1$	jC_3	$-C_5$	t_3
4	1	jC_1	jC_5	C_3	jC_1	jC_5	$-C_3$	$-jC_1$	$-jC_5$	C_3	$-jC_1$	$-jC_5$	$-C_3$	t_4
5	1	jC_1	$-jC_5$	C_3	jC_1	$-jC_5$	$-C_3$	$-jC_1$	jC_5	C_3	$-jC_1$	jC_5	$-C_3$	t_5
6	1	jC_3	jC_5	C_1	jC_3	jC_5	$-C_1$	$-jC_3$	$-jC_5$	C_1	$-jC_3$	$-jC_5$	$-C_1$	t_6
7	1	jC_3	$-jC_5$	C_1	jC_3	$-jC_5$	$-C_1$	$-jC_3$	jC_5	C_1	$-jC_3$	jC_5	$-C_1$	t_7
8	2	C_{10}	C_{13}	C_{14}	C_{10}	C_{13}	$-C_{14}$	$-C_{10}$	$-C_{13}$	C_{14}	$-C_{10}$	$-C_{13}$	$-C_{14}$	t_8
9	2	C_{10}	$-C_{13}$	C_{14}	C_{10}	$-C_{13}$	$-C_{14}$	$-C_{10}$	C_{13}	C_{14}	$-C_{10}$	C_{13}	$-C_{14}$	t_9
10	2	jC_{10}	jC_{13}	C_{14}	jC_{10}	jC_{13}	$-C_{14}$	$-jC_{10}$	$-jC_{13}$	C_{14}	$-jC_{10}$	$-jC_{13}$	$-C_{14}$	t_{10}
11	2	jC_{10}	$-jC_{13}$	C_{14}	jC_{10}	$-jC_{13}$	$-C_{14}$	$-jC_{10}$	jC_{13}	C_{14}	$-jC_{10}$	jC_{13}	$-C_{14}$	t_{11}
12	2	jC_{10}	jC_{14}	C_{13}	jC_{10}	jC_{14}	$-C_{13}$	$-jC_{10}$	$-jC_{14}$	C_{13}	$-jC_{10}$	$-jC_{14}$	$-C_{13}$	t_{12}
13	2	jC_{10}	$-jC_{14}$	C_{13}	jC_{10}	$-jC_{14}$	$-C_{13}$	$-jC_{10}$	jC_{14}	C_{13}	$-jC_{10}$	jC_{14}	$-C_{13}$	t_{13}
14	2	jC_{13}	jC_{14}	C_{10}	jC_{13}	jC_{14}	$-C_{10}$	$-jC_{13}$	$-jC_{14}$	C_{10}	$-jC_{13}$	$-jC_{14}$	$-C_{10}$	t_{14}
15	2	jC_{13}	$-jC_{14}$	C_{10}	jC_{13}	$-jC_{14}$	$-C_{10}$	$-jC_{13}$	jC_{14}	C_{10}	$-jC_{13}$	jC_{14}	$-C_{10}$	t_{15}
16	3	C_0	C_6	C_{12}	C_0	C_6	$-C_{12}$	$-C_0$	$-C_6$	C_{12}	$-C_0$	$-C_6$	$-C_{12}$	t_{16}
						...								
23	3	jC_6	$-jC_{12}$	C_0	jC_6	$-jC_{12}$	$-C_0$	$-jC_6$	jC_{12}	C_0	$-jC_6$	jC_{12}	$-C_0$	t_{20}
24	4	C_4	C_8	C_{15}	C_4	C_8	$-C_{15}$	$-C_4$	$-C_8$	C_{15}	$-C_4$	$-C_8$	$-C_{15}$	t_{24}
						...								
31	4	jC_8	$-jC_{15}$	C_4	jC_8	$-jC_{15}$	$-C_4$	$-jC_8$	jC_{15}	C_4	$-jC_8$	jC_{15}	$-C_4$	t_{31}

time slots. The QPSK modulation has informations in the code group to which the base station belongs (5 bits); the position of the frame within an interleaving period of 20 (ms) (1 bit); and the position of the slot within the frame (1 bit, Case 2 only).

The modulated S-SCH codes are also constructed such that their cyclic-shifts are unique according to the comma-free construction rule. That is, a non-zero cyclic shift less than 2 (Case 1) and 4 (Case 2) of any of the sequences is not equivalent to any cyclic shift of any other sequences. Also, a non-zero cyclic shift less than 2 (Case 1) and 4 (Case 2) of any of the sequences is not equivalent to itself with any other cyclic shift less than 8. The secondary synchronization codes are partitioned into two code sets for Case 1 and four code sets for Case 2. The *code set* is used to provide the following information: For Case 1, code set i ($i = 1, 2$) is associated with the 16 code groups from $16(i-1)$ to $16i-1$, and the exact code group and frame position information is provided by modulating the secondary codes in the code set. For Case 2, code set i ($i= 1, 2, 3, 4$) is associated with the 8 code groups from $8(i-1)$ to $8i-1$, and the slot boundary and frame position information is provided by the Comma-free property of the codeword while the code group information is provided by modulating the

Table 6.4.   Mapping scheme for cell parameters, code groups, scrambling codes, midambles and t_{offset} [125].

Cell parameter	Code group	Associated codes			Associated t_{offset}
		Scrambling code	Longbasic midamblecode	Short basic midamblecode	
0		Code 0	m_{PL0}	m_{SL0}	
1	Group 0	Code 1	m_{PL1}	m_{SL1}	t_0
2		Code 2	m_{PL2}	m_{SL2}	
3		Code 3	m_{PL3}	m_{SL3}	
4		Code 4	m_{PL4}	m_{SL4}	
5	Group 1	Code 5	m_{PL5}	m_{SL5}	t_1
6		Code 6	m_{PL6}	m_{SL6}	
7		Code 7	m_{PL7}	m_{SL7}	
		...			
124		Code 124	m_{PL124}	m_{SL124}	
125	Group 31	Code 125	m_{PL125}	m_{SL125}	t_{31}
126		Code 126	m_{PL126}	m_{SL126}	
127		Code 127	m_{PL127}	m_{SL127}	

secondary codes in the code set. The specific SSC allocation to each code set is as follows:

(Case 1)
Code set 1: C_1, C_3, C_5.
Code set 2: C_{10}, C_{13}, C_{14}.

(Case 2)
Code set 1: C_1, C_3, C_5.
Code set 2: C_{10}, C_{13}, C_{14}.
Code set 3: C_0, C_6, C_{12}.
Code set 4: C_4, C_8, C_{15}.

Table 6.2 and Table 6.3 list the association of code groups, code sets, modulated S-SCH code allocations, and associated timing offsets t_{offset} for Case 1 and Case 2, respectively (refer to Section 1.3).

8.2   EVALUATION OF SYNCHRONIZATION CODES

The evaluation of information transmitted in the SCH is done according to the association list shown in Table 6.4. Each of 32 code groups contains four specific scrambling codes, and each scrambling code is associated with a specific short and long basic midamble code. Each code group is additionally linked to a specific t_{offset}, thus to a specific frame timing. By using this scheme, the UE can derive the position of the frame boundary from the position

Table 6.5.   Alignment of cell parameter cycling and SFN [125].

Initial cell parameter assignment	Code group	Cell parameter used when SFN mod 2 = 0	Cell parameter used when SFN mod 2 = 1
0		0	1
1	Group 0	1	0
2		2	3
3		3	2
4		4	5
5	Group 1	5	4
6		6	7
7		7	6
...			
124		124	125
125	Group 31	125	124
126		126	127
127		127	126

of the SCH code and the knowledge of t_{offset}. (As to the long and short basic midamble codes m_{PL} and m_{PS}, refer to Section 1.2). On the other hand, in the actual cell parameter allocation, each cell cycles through two sets of cell parameters in a code group with the cell parameters changing each frame. This parameter cycling contributes to randomizing interference among base stations and facilitating the network planning through the averaging property [11]. Table 6.5 shows how the cell parameters are cycled according to the SFN.

9.   CELL SEARCH

The initial cell search of the 3GPP-TDD system is accomplished in three steps as in the 3GPP-FDD case (refer to Section 9 of Chapter 5) - - PSC acquisition, code group identification and slot synchronization, and scrambling code and basic midamble code identification and frame synchronization. However, the 3GPP-TDD channel structure is different from that of the 3GPP-FDD (refer to Section 1.3), so the specific cell search method takes a different form. In the following, we examine the three steps, assuming that there are 128 TDD cells in a system, which is the maximum number of cells that can be distinguished in the 3GPP-TDD system.

(Step 1) PSC Acquisition: In the first step of the cell search procedure, the MS carries out a matched filtering on the incoming PSC, C_p, which is common to all cells, in search for the strongest cell. For a cell with the SCH slot configuration of Case 1, the PSC arrives once in every frame time, while for a cell with the SCH slot configuration of Case 2, the PSC can be received twice in every frame time with the inter-arrival time of either 7 or 8 slots. (Refer to Section 1.3 for the SCH slot configurations.)

(Step 2) Code Group Identification and Slot Synchronization: In the second step, the MS uses the incoming three modulated SSCs to identify one out of 32 code groups. This is typically accomplished by correlating the received signal with the SSCs at the peak positions detected in the first step and then accumulating the resulting SSC correlation metrics to the code group metrics according to Table 6.2 (for Case 1 configuration) or Table 6.3 (for Case 2 configuration). The code group of the cell can then be uniquely identified by detecting the code group index that produces the maximum accumulated metric. The PSC provides the phase reference for coherent estimation of the incoming SSCs. As each code group is linked to a particular t_{offset} parameter (refer to Section 1.3) and contains only four cells, the MS can determine the slot timing without ambiguity in this step, reducing the cell identification uncertainty down to four. Furthermore, the modulation symbols of the SSCs enable the MS to determine the SFN (modulo 2) of the frames which contain the acquired SCH slots. For the configuration of Case 2, the incoming PSC can be exploited (possibly together with the incoming SSCs) to produce a reliable metric for discriminating the first and the last SCH slot positions within a frame.

(Step 3) Scrambling Code and Basic Midamble Code Identification, and Frame Synchronization: In the third step, the MS determines the downlink scrambling code, the basic midamble code, and the frame timing of the found cell. The long basic midamble code is identified by correlating the incoming P-CCPCH with the four possible long midamble codes of the code group identified in the second step and taking the maximally correlated one. As the P-CCPCH always uses the first midamble $\mathbf{m}^{(1)}$ (as well as the second midamble $\mathbf{m}^{(2)}$ in case of STTD) derived from the long basic midamble code $\mathbf{m}_{PL}$ and always uses the first channelization code $C_{Q=16}^{(h=1)}$, the P-CCPCH correlation can be immediately performed by using the long basic midamble code uniquely associated with each cell (refer to Table 6.4). When the long basic midamble code has been identified through the P-CCPCH correlation, all the cell parameters including the downlink scrambling code and the short basic midamble code are disclosed. The MS can read the system and cell specific BCH information and acquire the frame synchronization.

The PSC and the SSCs are constructed in the same way as in the 3GPP-FDD system, which facilitates inter-mode handover between the 3GPP-FDD and the 3GPP-TDD, and allows for an efficient hardware reuse in the MS. The handover or the neighboring cell search can be accomplished more easily in the 3GPP-TDD system than in the 3GPP-FDD system, as the frame boundaries of the neighboring cells are aligned to that of the camping cell in the 3GPP-TDD system. After reading the neighboring cell parameters conveyed from the camping cell, the MS can acquire the best handover cell by taking a simple correlation approach in a small uncertainty time window, as is typically done in the inter-cell synchronous systems.

Chapter 7

INTER-CELL SYNCHRONOUS IS-95 AND
CDMA2000 SYSTEMS (3GPP-2)

Inter-cell synchronous DS/CDMA cellular systems have been well known under the names IS-95 and cdma2000, and most of their core technologies are widely available in literature [7, 9, 10, 106, 126, 127]. So we omit its overview but, instead, directly discuss the spreading and scrambling issues briefly in comparison with those of the W-CDMA system. As the cdma2000 system has evolved from the IS-95 system, its spreading and scrambling techniques contain those of the IS-95 system as a subset, more specifically, as a part of the *spreading rate-1* techniques. [1]

1.   TIMING ALIGNMENT THROUGH EXTERNAL TIMING REFERENCE

One of the essential ingredients of the IS-95 and the cdma2000 systems, which has facilitated the system operation but has accompanied several defects in terms of system deployment as well, is the external timing reference source, or the *global positioning system* (GPS). All the base stations in the IS-95 or the cdma2000 systems align their reference timings to that of the GPS and all the mobile stations align their reference timing to that of the base stations with which they are currently associated. Therefore, all the base and mobile stations in operation are time-aligned in the inter-cell synchronous DS/CDMA systems, and the systems cannot operate in the event when the GPS timing services are not provided.

Fig. 7.1 depicts the system time alignment relations for signal transmission and reception at a BS and an MS in the cdma2000 system. The start of the

---

[1]The cdma2000 supports the multi-carrier mode of *spreading rate-3* (or, *3x mode* at 3.6864Mcps rate) as well as the IS-95 based single-carrier mode of *spreading rate-1* (or, *1x mode* at 1.2288Mcps rate).

system time is January 6, 1980, 00:00:00 UTC (*universal coordinated time*), to which the initial states of the long code of period $2^{42} - 1$ and the *zero offset* I- and Q- PN sequences of period 2^{15} are aligned. The initial state of the long code is the state in which the output of the long code SRG is the first 1 following 41 consecutive 0s, with the sequence generating vector (or, the binary mask) consisting of 1 (in the MSB) followed by 41 0s. [2] The initial state of each of the 2^{15} I- and Q- PN sequences, which are used for the downlink and the spreading rate-1 uplink, is the state in which the output of the corresponding PN sequence SRG is the first 1 following 15 consecutive 0s. For the spreading rate-3 uplink, another pair of I- and Q- PN sequences of period 3×2^{15} are used, which are derived by truncating an original m-sequence of period 2^{20}-1. The initial state of the I- PN sequence is the state in which the output of the corresponding SRG of length 20 is the first 1 following 19 consecutive 0s, and the Q- PN sequence is the I- PN sequence delayed by 2^{19} chips. Equivalently, the initial states of the I- and the Q- PN sequences are the states in which the first 20 outputs of the corresponding SRGs are 1000 0000 0001 0001 0100 and 1001 0000 0010 0100 0101, respectively. (Refer to the following sections for details of the long code and the I- and Q- PN sequences.)

As the reference timings of all the base and mobile stations are aligned to the external timing reference (with the deviation of signal propagation delay) and all the stations can keep track of the absolute time provided from the external timing source, the relative code phase differences among different stations can be kept constant within a small deviation range. Taking advantage of this fact, the IS-95 and the cdma2000 systems have alleviated the big burden of CDMA code allocation and timing acquisition in the following way:

All the base stations in the system employ a pair of common I- and Q- pilot PN sequences (or, a complex pilot PN sequence) of period 2^{15} for downlink chip scrambling, but different base stations keep different sequence phases, with the phase of the nth cell ($n = 0, 1, \cdots, 511$) being delayed by $64 \times n$ chips with respect to that of the zero offset pilot PN sequence. The pilot PN sequence period is 80/3(ms), whose start time corresponds to the *sync-channel frame* boundary. The *sync-channel superframe* is 80(ms) long and composed of three consecutive sync-channel frames. The start time of every 25th sync-channel superframe associated with a zero offset pilot PN sequence aligns with an *even second time mark*, which is the basic system reference epoch appearing at every 2 seconds. The long code state vector of length 42 coordinated with the GPS time is conveyed from a base station to mobile stations as an element of the *sync-channel message* over the sync-channel. Mobile stations first acquire the common pilot PN sequence of the cell where they are located.

---

[2]In the figure, $0^{(n)}$ denotes the sequence consisting of n consecutive zeros.

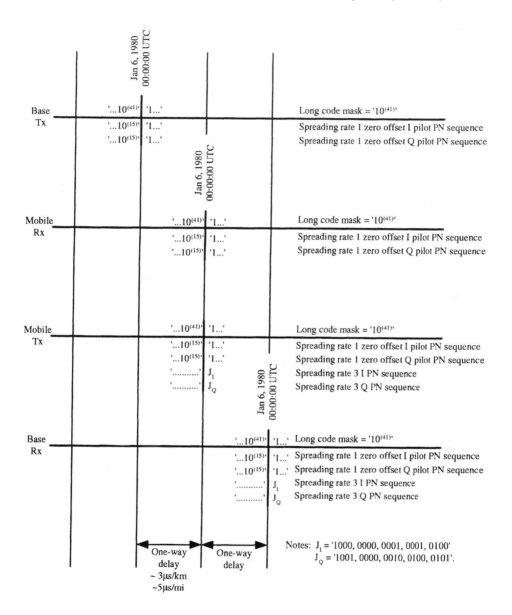

Figure 7.1. System time alignment relations [127].

The specific acquisition method is an implementation issue, but simple serial or parallel search methods are typically applied, as the uncertainty to be resolved is just the frame boundary timing of the common pilot sequence of period 2^{15} chips. After the acquisition of the common pilot sequence, each mobile station delineates and decodes the synch-channel message and loads the conveyed state vector as its long code SRG state at a pre-determined epoch. The state loading time is equal to 320 (ms) - $64 \times nT_c$ after the end of the last *sync-channel super-frame* containing the corresponding synchronization channel message, where T_c and n respectively denote the chip interval and the PN offset index of the current cell. Then, according to the timing relations depicted in Fig. 7.1, the synchronized mobile stations can despread/descramble the downlink channels, spread/scramble the uplink channels, and communicate with their base stations. The typical frame length of the traffic channels is 20 ms and the start time of every 100th frame aligns with an even second time mark regardless of the pilot PN offset index if the FRAME_OFFSET parameter is zero. A base station may provide the traffic channel frames with the non-zero FRAME_OFFSET parameter, which begin $1.25 \times$ FRAME_OFFSET(ms) later than the frames with zero frame offset. [3]

2.   UPLINK SPREADING AND SCRAMBLING

The cdma2000 uplink spreading and scrambling process takes three different forms depending on the radio configurations - - orthogonal modulation based original IS-95 configuration, BPSK modulation based cdma2000 1x configuration, and BPSK modulation based cdma2000 3x configuration. [4] Typical spreading/scrambling schemes for the three configurations are respectively depicted in Figs. 7.2 to 7.4.

2.1   CHIP MODULATION

In the IS-95 configuration, the chip rate is 1.2288Mcps and the complex-valued chip signal generated by the spreading and scrambling process is offset-QPSK modulated. The imaginary part of the input complex chip signal is delayed by a half chip with respect to the real part and each of them is independently pulse-shaped, up-converted by the cosine and the sine carriers, and then combined and amplified by the power amplifier for transmission. The employed baseband filter is a 48-tap symmetric lowpass filter whose roll-off factor α is about 0.11. The ratio of the baseband filtering rate to the chip rate (or, oversampling ratio) is 4. The offset QPSK modulation scheme contributes to lowering the PAR of the uplink channel. When multiple uplink channel signals are transmitted in parallel, additional channel signals take the carrier phases different from those readily chosen by the existing channel signals, with

---

[3]In the cdma2000 system, a short frame of length 5 ms is also employed [127].

[4]In fact, the cdma2000 specification defines six different uplink radio configurations depending on the modulation scheme, chip rate, and the data rate [127].

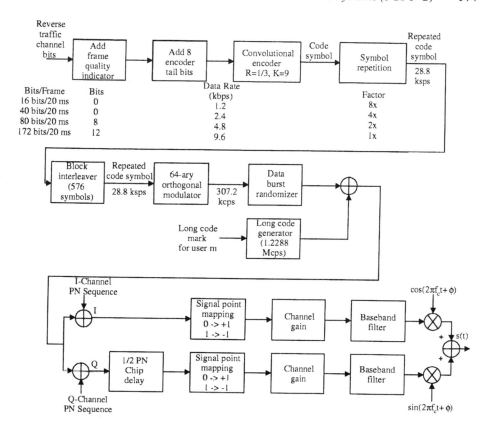

Figure 7.2.   Uplink spreading and scrambling for IS-95 configuration [127].

the relative phase offsets being 0, $\pi/4$, $\pi/2$, or $3\pi/4$, whichever contributes to lowering the PAR in the multi-code transmission. [5]

In the cdma2000 1x configuration, the chip rate is 1.2288Mcps and the complex-valued chip signal is QPSK modulated. The PAR is lowered in advance by applying the HPSK spreading/scrambling technique. The same baseband filter and oversampling ratio as in the IS-95 configuration are employed.

In the cdma2000 3x configuration, the chip rate is 3.6864Mcps and the complex-valued chip signal is QPSK modulated. The employed baseband filter is a 108-tap symmetric lowpass filter whose roll-off factor α is about 0.069. The oversampling ratio is 4.

---

[5]The maximum number of parallel channels in IS-95B system is eight. Thus, some parallel channels may take the same carrier phase when more than four channels are used simultaneously.

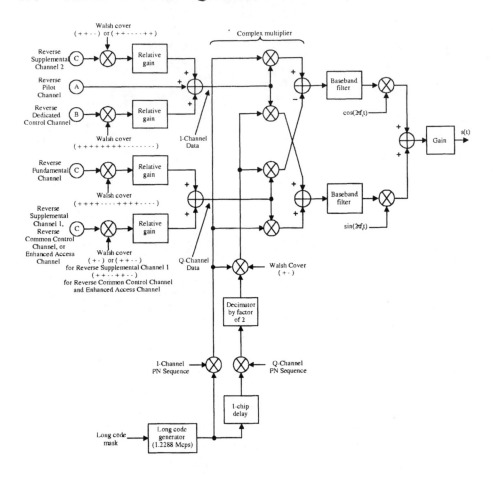

Figure 7.3.   Uplink spreading and scrambling for cdma2000 1x configuration [127].

2.2   CHANNELIZATION CODES

In the IS-95 configuration, there is no channelization code to separate different channel signals transmitted from a mobile station. The channel separation is made by applying different long code masks to different channel signals. The 64-ary Walsh code in Fig. 7.2 is not used for channelization but for orthogonal modulation. However, as six input bits to the orthogonal modulator are mapped to one of the 64 Walsh codewords of length 64, the effective signal bandwidth is spread by 64/6 during the orthogonal modulation processing. For reference, the generation procedures of the modulator input bits (i.e., frame quality indicator attachment, tail bit attachment, channel encoding, symbol repetition, and block interleaving) are also shown in Fig. 7.2.

In the cdma2000 1x and 3x configurations, the OVSF code (or, a union of Walsh codes with various lengths) is used for channelization of the parallel

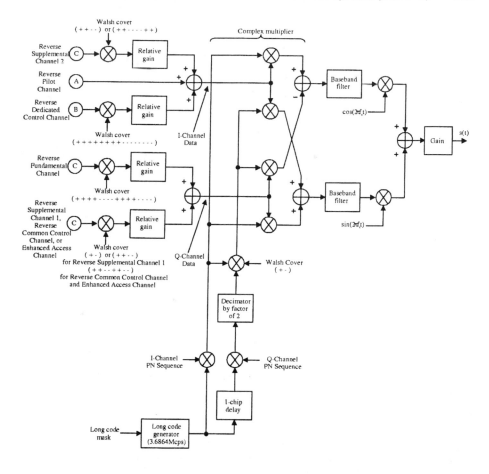

Figure 7.4.   Uplink spreading and scrambling for cdma2000 3x configuration [127].

channel signals transmitted from a mobile station. The BPSK modulated control and traffic channels are channelized by different OVSF codewords, and then some channels are assigned to the the I-phase while the others are to the Q-phase according to the data assignment rule shown in Fig. 7.3 and Fig. 7.4.

2.3   SCRAMBLING CODES

In the IS-95 configuration, the Walsh chips of rate 307.2kcps are spread by the long code of rate 1.2288Mcps, which results in an additional bandwidth expansion by the factor of 4. Before the long code spreading, the Walsh symbol bursts are selectively punctured by the factor equal to the symbol repetition factor shown in Fig. 7.2. The symbols to be punctured are determined by the pseudorandom pattern derived from the previously generated long code sequence values. Fig. 7.5 depicts an exemplary implementation of the employed

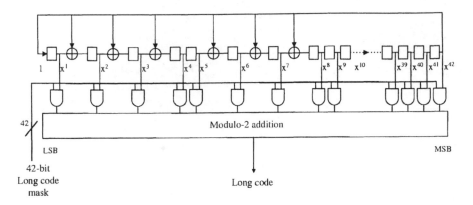

Figure 7.5.   Long code generator [127].

long code generator. The long code generator is characterized by the recursion polynomial of $p(x) = x^{42} + x^{35} + x^{33} + x^{31} + x^{27} + x^{26} + x^{25} + x^{22} + x^{21} + x^{19} + x^{18} + x^{17} + x^{16} + x^{10} + x^7 + x^6 + x^5 + x^3 + x^2 + x^1 + 1$ and generates any shifts of an m-sequence of period 2^{42}-1. Each common or user-specific long code is generated by the modulo-2 inner product of a specific 42-bit mask (or, generating vector) and the 42-bit state vector of the long code generator. After the burst randomization by puncturing and the long code spreading, the chip sequences are scrambled by the cell-specific complex PN sequence of length 2^{15} and then modulated by the offset QPSK processing. The I-phase (or, real part) and Q-phase (or, imaginary part) PN sequences are the *extended m-sequences* whose characteristic polynomials are $p_I(x) = x^{15} + x^{13} + x^9 + x^8 + x^7 + x^5 + 1$ and $p_Q(x) = x^{15} + x^{12} + x^{11} + x^{10} + x^6 + x^5 + x^4 + x^3 + 1$, respectively. [6] The extension is made by inserting a 0 after 14 consecutive 0s for the m-sequences of period 2^{15}-1.

In the cdma2000 1x configuration, the channelized and summed chip signals of rate 1.2288Mcps of the I- and Q- phases are regarded as the real and imaginary parts of a complex signal. The complex chip signal is scrambled by the complex scrambling sequence that is constructed by a user-specific long code of period 2^{42}-1 (specified by the assigned mask) and a cell-specific complex PN sequence of period 2^{15} through the HPSK construction method (refer to Section 7.4 of Chapter 5). Note that the two independent input sequences needed for the HPSK construction are respectively prepared by multiplying the long code and the delayed long code (by 1 chip) to the I-phase and the Q-phase PN sequences.

---

[6]The use of complex PN sequence contributes to randomizing the multiple access interference regardless of the relative carrier phases among multiple users [7, 106].

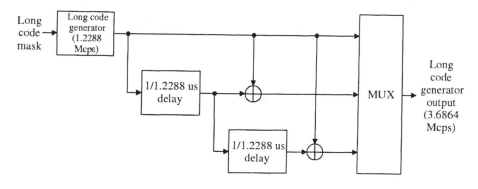

Figure 7.6.   Long code generator for spreading rate-3 [127].

The real and imaginary parts of the scrambled complex signal are fed to the baseband filters for chip pulse shaping.

The scrambling method of the cdma2000 3x configuration is exactly the same as that of the 1x configuration. However, the chip rate of the scrambling sequence as well as the OVSF sequences is three times faster than that of the 1x configuration (i.e., 3.6864Mcps). As the current state vector of the long code generator has a one-to-one correspondence with the current GPS time, the long code for the 3x configuration is generated by multiplexing three delayed versions of the original long code operating at 1.2288Mcps. The specific generation method of the long code for the 3x configuration is depicted in Fig. 7.6. The resulting period of the code becomes $3 \times (2^{42} - 1)$. Furthermore, in order to maintain the frame timing relations of the various channels with the even second time mark at the chip rate of 3.6864Mcps, the I-phase and the Q-phase PN sequences of period 3×2^{15} for the 3x configuration are formed from a new m-sequence of period $2^{20}-1$ using different starting positions and truncating the sequences after 3×2^{15} chips. The starting state of the I-phase PN sequence is the state at which the output of the corresponding sequence generator is the first 1 following 19 consecutive 0s. The starting state of the Q-phase PN sequence is that of the I-phase PN sequence delayed by 2^{19} chips in the untruncated m-sequence. The characteristic polynomial of the untruncated m-sequence is $p(x) = x^{20} + x^9 + x^5 + x^3 + 1$.

3.   DOWNLINK SPREADING AND SCRAMBLING

The cdma2000 downlink spreading and scrambling process takes two different forms depending on the number of carriers that are simultaneously transmitted - - the 1x single-carrier configuration and the 3x multi-carrier con-

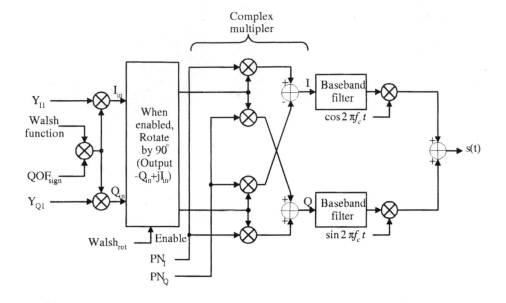

Figure 7.7.   Downlink spreading and scrambling for single-carrier configuration [127].

figuration. [7] Typical spreading/scrambling schemes for the two configurations are respectively depicted in Fig. 7.7 and Fig. 7.8.

As is shown in the figures, the QPSK data stream is regarded as the basic input data format to the spreading/scrambling circuit. However, the BPSK data stream for the pilot channel, the sync-channel, and the paging/traffic channels of the original IS-95 based radio configurations is also supported by inputting zero values to the Q-phase branches. On the other hand, in the case of 1x single-carrier configuration, *orthogonal transmit diversity* (OTD) or *space time spreading* (STS, or STTD) may be applied to obtain diversity gain. Here, dual downlink antennas transmit the properly repeated copies of the data symbols in parallel after applying to each copy the same spreading and scrambling operation shown in Fig. 7.7.

3.1   CHIP MODULATION

In the downlink, the chip rate is 1.2288Mcps for both 1x and 3x configurations and the complex-valued chip signal generated by the spreading and scrambling process is QPSK modulated. The I-phase and Q-phase chip signals are independently pulse-shaped, up-converted by the cosine and the sine

---

[7]In fact, the cdma2000 specification defines nine different downlink radio configurations [127]. However, the difference is small as far as the spreading and scrambling process is concerned.

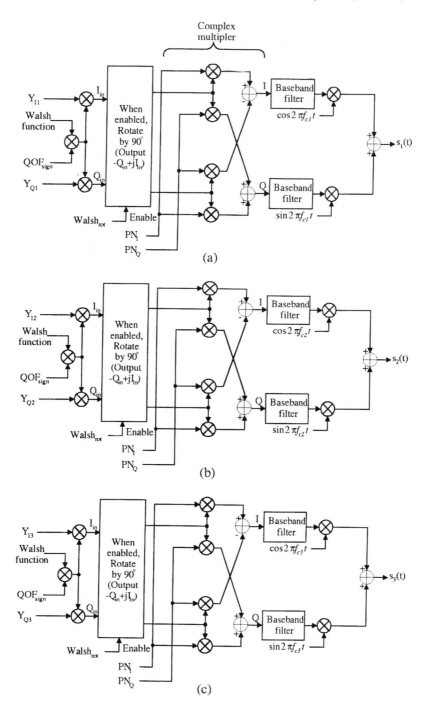

Figure 7.8.   Downlink spreading and scrambling for multi-carrier configuration [127]: (a) for carrier 1, (b) for carrier 2, (c) for carrier 3.

184 SCRAMBLING TECHNIQUES FOR CDMA COMMUNICATIONS

carriers, and then combined and amplified by the power amplifier for transmission. The employed baseband filter is a 48-tap symmetric lowpass filter whose roll-off factor α is about 0.11, and the oversampling ratio is 4. In the 3x configuration, a three times higher rate data stream is first demultiplexed to three low rate parallel streams, each of which is spread and scrambled by the same channelization code and the cell-specific complex PN sequence of length 2^{15}, and then independently pulse-shaped and up-converted onto three different carrier frequency bands for transmission.

3.2   CHANNELIZATION CODES

The basic channelization code is the OVSF code composed of variable length Walsh functions. The minimum Walsh length is 4 and the maximum Walsh length is 128 for the 1x configuration, 256 for the 3x configuration, and 512 when the *auxiliary pilot channels* are employed. Among the nth Walsh functions of length N, W_n^N, W_0^{64} is used for the downlink pilot channel, W_{32}^{64} for the sync-channel, and W_1^{64} to W_7^{64} for the paging channels if they exist. If the *transmit diversity pilot channel* is present, W_{16}^{128} is assigned to it. If *quick paging channels* are present for the 1x configuration, W_{80}^{128}, W_{48}^{128}, and W_{112}^{128} are assigned to them. The function W_{64}^{128} is not used in the 1x configuration, and the functions W_{64}^{128}, W_{64}^{256}, W_{128}^{256}, and W_{192}^{256} are not used in the 3x configuration. Other Walsh functions are usable for other channels provided that they are chosen to be orthogonal or quasi-orthogonal to all other code channels in use.

The *quasi-orthogonal functions* (QOF) have been introduced to cope with the case when the available OVSF codewords are exhausted. For the creation of a QOF, the repeated sequence of an appropriate Walsh function is first multiplied by the repeated sequence with symbol +1 or -1 which correspond to the *sign multiplier QOF mask* (QOF_{sign}) values of 0 or 1. Then the sequence is multiplied by the repeated sequence with symbols 1 (or, 0 rotation) or j (or, $\pi/2$ rotation) which correspond to the *rotate enable Walsh function* ($\text{Walsh}_{\text{rot}}$) values of 0 or 1. The resulting quasi-orthogonal functions are not orthogonal to the Walsh functions, but careful choices of the QOF_{sign} and $\text{Walsh}_{\text{rot}}$ sequences can make the QOFs nearly orthogonal to the Walsh functions (and among themselves). The QOF_{sign} and $\text{Walsh}_{\text{rot}}$ sequences that are available for the cdma2000 downlink are as listed in Table 7.1. The corresponding channelization process with QOF is illustrated in Figs. 7.7 and 7.8. If the QOF is not employed, QOF_{sign} and $\text{Walsh}_{\text{rot}}$ always take bit 0.

3.3   SCRAMBLING CODES

The long code generator and the I-phase and Q-phase PN sequence generators that are employed in the uplink are also employed in the downlink. A difference is that the long code is not used for chip scrambling in the downlink

Table 7.1.   Masking function for quas-orthogonal functions with length 256 [127].

Function	Masking function	
	Hexadecimal representation of QOF_{sign}	$Walsh_{tot}$
0	0000000000000000000000000000000 0000000000000000000000000000000	W_0^{256}
1	7228d7724eebebb1eb4eb1ebd78d8d28 278282d81b41be1b411b1bbe7dd8277d	W_{130}^{256}
2	114b1e4444e14beeee4be144bbe1b4ee dd872d77882d78dd2287d277772d87dd	W_{173}^{256}
3	1724bd71b28118d48ebddb172b187eb2 e7d4b27ebd8ee82481b22be7dbe871bd	W_{47}^{256}

but for data scrambling and randomization of power control bit positions after being decimated to a lower rate sequence. Thus the long code operation is not shown in the chip spreading/scrambling blocks of Figs. 7.7 and 7.8. The MS receiver employs the same decimated long code to descramble the data symbols and extract the power control bits. The channelized complex chip signal of rate 1.2288Mcps is scrambled by the cell-specific complex PN sequence of length 2^{15} at the same rate. In the case of multi-carrier transmission (or, 3x configuration) or transmit diversity, the same copies of the cell-specific complex PN sequence are used for scrambling of all parallel carriers.

III

DSA-BASED SCRAMBLING CODE ACQUISITION

Chapter 8

DISTRIBUTED SAMPLE ACQUISITION (DSA) TECHNIQUES

Among the various rapid acquisition schemes introduced in the previous chapter, the *distributed sample acquisition* (DSA) distinguishes itself from others by conveying the state information of the transmitter SRG to the receiver reliably and noncoherently. In support of this, the DSA employs the novel concept of *igniter sequence* and applies the state synchronization method of the *distributed sample scrambling* (DSS) techniques [69, 70, 77, 78, 128], thereby achieving the code acquisition in very short time even in the DS/CDMA environment with a long-period PN code (or *m*-sequence). This new approach resolves the coherent acquisition problem which used to be unavoidable in the conventional sequential estimation trials.

In this chapter, we describe the DSA scheme in detail, introducing other members of the DSA family as well, namely, *parallel DSA* (PDSA), *batch DSA* (BDSA), and *differential DSA* (D^2SA). As the DSA technique is relatively new and not yet widely referenced, we provide a thorough description on its theoretical background and performance analyses.

1. PRINCIPLES OF THE DSA

A pair of SRGs of identical structure, in principle, can be synchronized by loading the same state values (i.e., the L values stored in the SRG of length L) to each SRG at the same time. This implies that, if some conveyance means are available it is much faster to acquire synchronization of the receiver SRG by conveying the transmitter SRG state values to the receiver SRG than by searching for the phase of peak correlation value. In order to achieve such SRG state based synchronization in the CDMA environment, however, it is critical to resolve the following two problems: *How to convey the state samples reliably in practical low-SNR CDMA channels*, and *how to manipulate the conveyed state samples to acquire synchronization.*

The DSA technique is designed to resolve those above two problems by employing the *igniter sequence* concept and the *DSS* technique respectively to handle the sample conveyance problem and the SRG synchronization problem. The igniter sequence refers to an auxiliary sequence which is newly introduced in the DSA scheme to aid the synchronization of the main sequence. Its major function is to reliably convey the state values of the main SRG, but it also provides a timing reference for sampling the SRG state values and correcting the receiver SRG state. On the other hand, the DSS is a scrambling technique which was recently introduced for use in the *asynchronous transfer mode* (ATM) transmission of digital signals [69]. It has the distinctive feature that the transmitter main SRG state samples are conveyed to the receiver SRG in a *distributed* form, based on which the synchronization can be done in a progressive manner. This technique is conceptually similar to the sequential estimation techniques discussed in the previous chapter in that it acquires synchronization by directly controlling the receiver SRG using the conveyed state information, but is different in that it works on distributed state samples, not consecutive samples (chip values). Each distributed state sample contributes to the synchronization by triggering a correction process that corrects the receiver SRG state. [1]

1.1   DSA SYSTEM ORGANIZATION AND OPERATION

Fig. 8.1 depicts the functional block diagram of the acquisition-related part of the DS/CDMA system that employs the DSA scheme. The transmitter part consists of a DSA-spreader and a sample-spreader, and the receiver part contains their despreading counterparts, that is, DSA-despreader and sample-despreader. The DSA-spreader/despreader pair take the synchronization function while the sample-spreader/despreader pair take the sample conveyance function. Those two functions are supported by two different SRGs - - the *main SRG* residing in the DSA-spreader and the *igniter SRG* residing in the sample-spreader.

The main SRG generates the *main sequence*, that is, the long-period PN sequence employed for the data spreading, whose fast synchronization is our ultimate goal. Each user data is separated by the channelization sequence (i.e., one of the orthogonal Walsh sequences) and then spread by the common main sequence. Note that once the main sequence is acquired the channelization sequence boundary (i.e., the data symbol boundary) can be determined immediately [10].

---

[1]For the details of the DSS technique, refer to [69, 70, 77, 78, 128]. Reference [69] provides the most comprehensive description of the DSS.

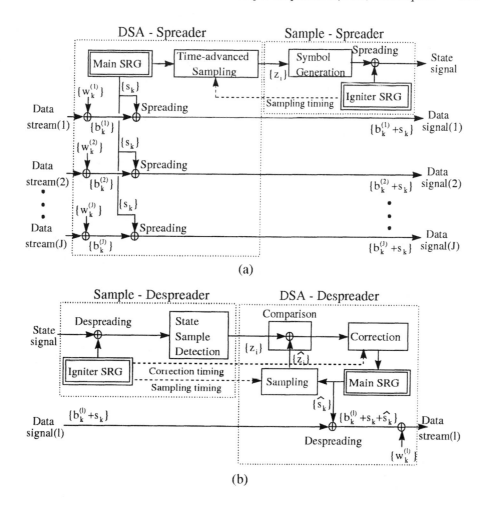

Figure 8.1. Functional block diagram of the acquisition-related part of the DS/CDMA system employing the DSA scheme: (a) Base-station (transmitter), (b) mobile-station (receiver).

The igniter SRG generates the *igniter sequence* which conveys the state samples of the main SRG for acquisition of the main sequence. The period of the igniter sequence, N_I, is designed to be much shorter than the main sequence period, $N_M (= 2^L - 1)$. The time-advanced sampling block takes the state sample z_i of the main SRG in advance : More specifically, it takes the sample z_i at time $(r + i - 1)N_I$ which is yet to be generated at time $(r + i)N_I$, for a reference value r. Then the symbol generation block maps the sample z_i to the corresponding binary orthogonal symbol and spreads it using the igniter SRG.

In the receiver, the sample-despreader despreads the received state signal and matches it with each binary orthogonal symbol to detect the conveyed sample

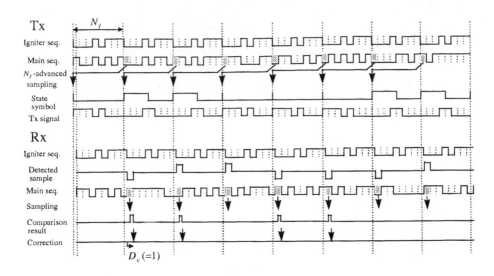

Figure 8.2.   Timing diagram of the DSA system.

z_i, passing it to the DSA-despreader. The DSA-despreader generates its own SRG state sample $\hat{z}_i$ and compares it with the conveyed sample z_i and initiates a correction process as a normal DSS descrambler does. The state sampling and correction processes are triggered by the sampling and correction pulses issued by the igniter SRG.

1.1.1   DSA-BASED ACQUISITION PROCEDURE

Now we consider the DSA-based acquisition procedure in the broadcasting DS/CDMA communication system like the cellular system employing the pilot channel for synchronization and channel estimation [10]. Differently from the conventional serial-search acquisition scheme employing the unmodulated long-peiod main sequence as the pilot channel sequence, the DSA base station transmits the modulated igniter sequence instead of the main sequence itself, while the traffic channel data are still spread by the channelizing orthogonal sequence and the main sequence. So, in the receiver, or a *mobile station* (MS), the acquisition is done in two stages - - the igniter sequence acquisition stage and the main sequence acquisition stage.

In the first stage, the MS acquires the igniter sequence through serial search, [2] which is composed of igniter sequence acquisition and verification process. It

---

[2]The serial search scheme employed for the igniter sequence acquisition is a modified version of the conventional one in that it utilizes two sufficient statistics. Refer to the next subsection of this section for more details.

can be done very fast because the period of the igniter sequence is made much shorter than that of the main sequence. Once the igniter sequence acquisition is completed, the second stage begins: The sample-despreader restores the conveyed state symbols by despreading the state signal for each igniter sequence period N_I, and detects the conveyed state sample z_i by applying a conventional noncoherent detection method. Then the DSA-despreader compares the conveyed state sample z_i, which is determined at the end of each igniter-sequence period N_I, with the receiver-generated state sample $\hat{z}_i$, which is generated by the receiver SRG at the beginning of the next igniter sequence period. Since the DSA-spreader is designed to take the state sample z_i of the transmitter SRG one igniter-sequence period earlier, whereas it takes one igniter-sequence period to detect the conveyed state sample, the sampling time of $\hat{z}_i$ coincides with that of z_i. If the two state samples differ, a correction process is triggered to correct the main SRG state at a certain time within one igniter-sequence period, while no action is taken if they coincide. The principles how and when to correct the SRG state will be detailed in the following sections. The sampling and correction timing pulses are provided by the igniter SRG. The related timing diagrams is shown in Fig. 8.2. [3]

Once L comparison-correction operations are made, the main SRG of length L is supposed to reach the synchronization state. [4] In case any of those L conveyed samples are corrupted by error during transmission, however, it leads to false-synchronization. Therefore, we need to carry out a verification process after the synchronization process to check if the synchronization is correctly done. While there could be various sophisticated ways to verify the synchronization, we take a simple verification process in this chapter that checks whether or not the conveyed and receiver-generated samples coincide V more times after the L comparison-correction operations. If all the V sample-sets coincide, then the mobile station declares completion of synchronization of the main sequence, beginning to track and estimate the channel characteristic and carrier phase.[5] Note that once the main sequence as well as the igniter sequence is acquired, the modulated igniter sequence can be independently generated in the MS. This implies that once the synchronization process is completed the incoming modulated igniter sequence is eligible for the estimation of the channel characteristic

---

[3]In fact, the receiver SRG state should be synchronized to the delayed version of the transmitter SRG state for the propagation delay between the two systems, but, since the receiver SRG synchronization is done based only on the waveform arriving at the front end of the receiver, the propagation delay does not affect the synchronization process. So we do not set forth the propagation delay factor explicitly in the figure.

[4]The comparison-correction operations may be completed in less than L state symbol transmissions by conveying several state samples per symbol and applying the BDSA or the PDSA technique to be introduced in Sections 3 and 4.

[5]Once the igniter sequence is acquired the chip-timing alignment and clock frequency tracking can be done even before the main sequence acquisition, thus providing more reliable state symbol stream for the state sample detection block.

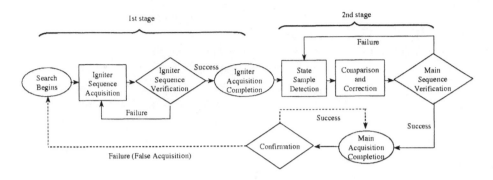

Figure 8.3.   Flow diagram outlining the DSA acquisition process.

and carrier phase. After the synchronization process, the main data is decoded by multiplying the main sequence and the channelizing orthogonal sequence to the incoming data signal. If there happens any discrepancy in V sample-sets, the MS repeats the sample detection and main SRG correction procedure. (Optionally, the MS can be made to resume the initial igniter sequence search mode and repeats the acquisition procedure.) Fig. 8.3 depicts the overall acquisition process (the solid line part).

Despite the above verification process, there still exists room for *false synchronization* (or *false alarm*), that is, it is still probable, no matter how small the probability may be, that the V sample sets coincide without having true SRG synchronization. So we need to confirm, while decoding the main data signal, whether the acquired synchronization state is true or false. Noting that the data decoding performance such as the BER performance drops below the desired level in the case of false synchronization, we monitor the data decoding performance for certain period of time, for example, about K times ($K \gg 1$) the searching stage dwell time τ_D. In case the acquisition is determined to be true we continue decoding the main data without interruption, but, otherwise, we recall the completion declaration and resume the initial igniter sequence searching stage at the cost of *false alarm penalty time* of $K\tau_D$. The dotted part of Fig. 8.3 depicts this synch-confirmation process.

1.1.2   IGNITER SEQUENCE ACQUISITION AND STATE SAMPLE DETECTION

Now we detail the operation of the sample-despreader which performs two different processes - - igniter sequence acquisition and state sample detection - - to supply reliable state samples to the DSA-despreader. In support of this discussion we redraw the sample-despreader in Fig. 8.1 (b) in the form of Fig. 8.4 (a), focusing on the physical layer operations. The received state signal $r(t)$ is

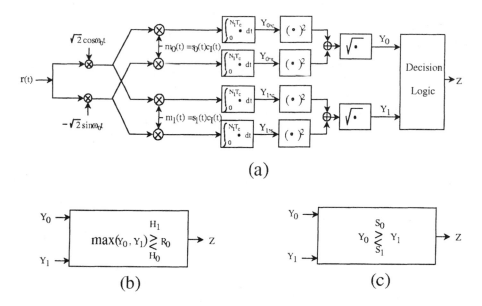

Figure 8.4. Detailed operational structure of the sample despreader: (a) Operational structure; (b) decision logic in the igniter sequence acquisition mode; (c) decision logic in the sample detection mode.

first down-converted by the in-phase and quadrature carriers generated in the receiver, and the resulting in-phase and quadrature components are multiplied by the local igniter sequence (i.e., $c_I(t)$) and matched to each orthogonal symbol (i.e., $s_0(t)$, $s_1(t)$). The matched outputs (i.e., $Y_{0,c}$, $Y_{0,s}$, $Y_{1,c}$, $Y_{1,s}$) are then squared, summed, and square-rooted to generate the matched-signal energy (i.e., Y_0, Y_1), which are used for deciding the igniter sequence and the conveyed state sample. In the igniter sequence acquisition mode, the phase of each modulated igniter sequence (i.e., $m_0(t)$, $m_1(t)$) advances serially by chip unit until the igniter sequence in-phase state is acquired. Once the igniter sequence acquisition is declared, the sample-despreader shifts to the sample detection mode.

The decision logics for the final process in the DSA sample despreading are as given in Fig. 8.4 (b) and (c), respectively for igniter sequence acquisition and the sample detection. In igniter sequence acquisition, the matched-signal energies Y_0 and Y_1 are both small if the phase of the local sequence does not coincide with that of the incoming sequence, while one of them becomes large if the phase coincides. So we arrange the decision logic such that it declares acquisition state (H_1) if the larger of Y_0 and Y_1 exceeds a properly set threshold (R_0), and declares out-of-phase state (H_0) otherwise. In sample detection, we arrange the decision logic such that it declares detection of symbol s_0 if Y_0

is larger than Y_1, and symbol s_1 otherwise. Note that we need two sufficient statistics Y_0 and Y_1 even in the igniter sequence acquisition mode as the igniter sequence is modulated by one of the two orthogonal symbols. This contrasts to the conventional serial search scheme in which only one sufficient statistic (Y_0) suffices because the pilot sequence is unmodulated. The resulting *false alarm probability per cell* of this modified (i.e., two-branched) serial search acquisition scheme could be somewhat higher than that of the conventional serial search scheme for a given threshold, as the "in-phase" could be falsely declared even in the "out-of-phase" state if either of the two statistics happens to exceed the threshold due to the channel noise.

1.2   DSA SYNCHRONIZATION PARAMETER DESIGN

Now that the sample spreader/despreader pair, both supported by the igniter sequence, provides a reliable means for sample conveyance, we concentrate on the synchronization problem of the main SRGs residing in the DSA spreader/despreader pair. For this purpose, we reorganize Fig. 8.1 into the simplified form in Fig. 8.5 (a) and (b), in which the sample spreader/despreader pair are put together in a sample conveyance block. This dotted block has the overall processing delay of N_I, while, in contrast, the time-advanced sampling block has the same amount of *negative* delay. Therefore if we combine those two blocks we obtain a pure sampling block which has the effect of sampling z_i at the *virtual sampling time* $(r + i)N_I$. Fig. 8.5 (a) depicts the resulting equivalent block diagram of the DSA spreader, and Fig. 8.5 (b) remains to be its counterpart despreader.

For the lth user despreader symbol stream $\{ b_k^{(l)} + s_k + \hat{s}_k \}$ in the DSA despreader to be identical to the original symbol stream $\{ b_k^{(l)} \}$, the despreader main sequence $\{ \hat{s}_k \}$ should be identical to the spreader main sequence $\{ s_k \}$ at all time, and this becomes possible only when the despreader SRG is synchronized to the spreader SRG. In order to achieve this synchronization, we compare the sample z_i, which is taken out of the main SRG sequence $\{ s_k \}$ at the virtual sampling time $(r + i)N_I$, with the sample $\hat{z}_i$ taken out of $\{ \hat{s}_k \}$ at the same sampling time, and reflect their discrepancy to correcting the state of the despreader SRG. This synchronization mechanism is known as the *DSS synchronization* technique, and according to the DSS theories fully developed in [69], a sum of L comparison-correction processes are required to synchronize an SRG of length L.

In the following, we will investigate when to sample, how to do time-advanced sampling, and how to correct the SRG state, based on a rigorous mathematical modeling of the DSA spreader/despreader pair.

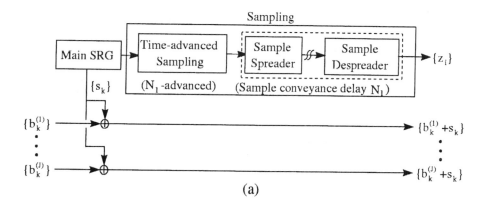

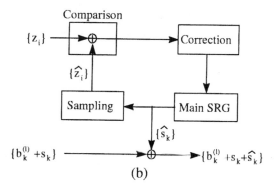

Figure 8.5. DSA spreader and despreader : (a) Spreader, (b) despreader.

1.3 MATHEMATICAL MODELING FOR DSA SYNCHRONIZATION

Let $\mathbf{d}_k$ and $\hat{\mathbf{d}}_k$ denote the *state vectors* of the spreader and despreader SRGs at time k, respectively, and let $\mathbf{T}$ denote the *state transition matrix* that relates two successive vectors such that [6]

$$\mathbf{d}_{k+1} = \mathbf{T} \cdot \mathbf{d}_k, \tag{8.1a}$$

$$\hat{\mathbf{d}}_{k+1} = \mathbf{T} \cdot \hat{\mathbf{d}}_k. \tag{8.1b}$$

In addition, let $\mathbf{h}$ denote the *generating vector* that generates the sequence value s_k (or $\hat{s}_k$) out of the state vector $\mathbf{d}_k$ (or $\hat{\mathbf{d}}_k$) through the relation

$$s_k = \mathbf{h}^t \cdot \mathbf{d}_k, \tag{8.2a}$$

---

[6]The relation $\hat{\mathbf{d}}_{k+1} = \mathbf{T} \cdot \hat{\mathbf{d}}_k$ holds for the despreader SRG when the correction process is not applied. With the correction applied, this relation changes to the form in Eq.(8.5).

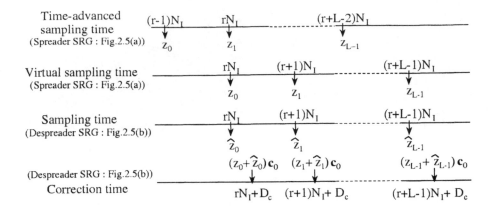

Figure 8.6.   Sampling and correction timing diagram for DSA.

$$\hat{s}_k = \mathbf{h}^t \cdot \hat{\mathbf{d}}_k, \tag{8.2b}$$

and let $\mathbf{c}_0$ denote the *correction vector* that corrects the old state vector $\hat{\mathbf{d}}_{old}$ to a new state vector $\hat{\mathbf{d}}_{new}$ through the relation

$$\hat{\mathbf{d}}_{new} = \hat{\mathbf{d}}_{old} + (z_i + \hat{z}_i)\mathbf{c}_0. \tag{8.3}$$

According to the discussions in the previous section, samples z_i and $\hat{z}_i$, $i = 0$, $1, \cdots, L - 1$, are both taken at the same sampling time $(r + i)N_I$, respectively from the (virtual) spreader in Fig. 8.5 (a) and the despreader in Fig. 8.5 (b), and the related correction is made, as necessary, at the correction time $(r+i)N_I + D_c$ for a delay D_c in $0 < D_c \le N_I$. [7] Fig. 8.6 depicts this timing relationship. So, the samples z_i and $\hat{z}_i$ are related to the state vectors by the relations

$$z_i = s_{(r+i)N_I} = \mathbf{h}^t \cdot \mathbf{d}_{(r+i)N_I}, \tag{8.4a}$$

$$\hat{z}_i = \hat{s}_{(r+i)N_I} = \mathbf{h}^t \cdot \hat{\mathbf{d}}_{(r+i)N_I}, \; i = 0, 1, \cdots, L - 1, \tag{8.4b}$$

and the state vectors at the correction time takes the expressions

$$\mathbf{d}_{(r+i)N_I + D_c} = \mathbf{T} \cdot \mathbf{d}_{(r+i)N_I + D_c - 1}, \tag{8.5a}$$

$$\hat{\mathbf{d}}_{(r+i)N_I + D_c} = \mathbf{T} \cdot \hat{\mathbf{d}}_{(r+i)N_I + D_c - 1} + (z_i + \hat{z}_i)\mathbf{c}_0, \tag{8.5b}$$

$$i = 0, 1, \cdots, L - 1.$$

Now, noting that the spreader and despreader SRGs get synchronized if the state vectors $\mathbf{d}_k$ and $\hat{\mathbf{d}}_k$ become identical, we define the *state distance vector*

---

[7]The reference time rN_I denotes the time when the igniter sequence acquisition is completed and the sample detection is initiated.

δ_k to be the L-vector representing the difference of them, that is,

$$\delta_k \equiv \mathbf{d}_k + \hat{\mathbf{d}}_k. \tag{8.6}$$

If we combine Eq.(8.1), Eq.(8.4) and Eq.(8.5) at the correction time and insert the result to Eq.(8.6), then we can easily get the recursive relation

$$\delta_{(r+i)N_I+D_c} = \begin{cases} (\mathbf{T}^{D_c} + \mathbf{c}_0 \cdot \mathbf{h}^t) \cdot \delta_{rN_I}, \\ \qquad\qquad i = 0, \\ (\mathbf{T}^{N_I} + \mathbf{c}_0 \cdot \mathbf{h}^t \cdot \mathbf{T}^{N_I-D_c}) \cdot \delta_{(r+i-1)N_I+D_c}, \\ \qquad\qquad i = 1, 2, \cdots, L-1. \end{cases} \tag{8.7}$$

By applying this relation repeatedly for L times of corrections, we obtain the finally corrected state distance vector

$$\delta_{(r+L-1)N_I+D_c} = \Lambda \cdot \delta_{rN_I} \tag{8.8}$$

for the $L \times L$ *correction matrix*

$$\Lambda \equiv (\mathbf{T}^{N_I} + \mathbf{c}_0 \cdot \mathbf{h}^t \cdot \mathbf{T}^{N_I-D_c})^{L-1} \cdot (\mathbf{T}^{D_c} + \mathbf{c}_0 \cdot \mathbf{h}^t). \tag{8.9}$$

In order to achieve the synchronization by L corrections, it is necessary to make the final state distance vector $\delta_{(r+L-1)N_I+D_c}$ a zero vector regardless of the initial state distance vector δ_{rN_I}, which can be done only by making the correction matrix Λ a zero matrix. Therefore, for a given DSA spreader/despreader pair whose constituent SRGs are structured by $\mathbf{T}$ and $\mathbf{h}$, [8] the synchronization problem turns to a problem of determining appropriate values of N_I, D_c and $\mathbf{c}_0$ that make the correction matrix null.

1.4   SAMPLING AND CORRECTION CONDITIONS

For an SRG of length L structured by the state transition matrix $\mathbf{T}$ and the generating vector $\mathbf{h}$, we define the *discrimination matrix* $\Delta_{\mathbf{T},\mathbf{h}}$ to be the $L \times L$ matrix

$$\Delta_{\mathbf{T},\mathbf{h}} \equiv \begin{bmatrix} \mathbf{h} & (\mathbf{T}^{N_I})^t \cdot \mathbf{h} & (\mathbf{T}^{2N_I})^t \cdot \mathbf{h} & \cdots & (\mathbf{T}^{(L-1)N_I})^t \cdot \mathbf{h} \end{bmatrix}^t, \tag{8.10}$$

where N_I denotes the sampling interval [9]. Then we can determine N_I based on the following theorem.

---

[8]The structure of an SRG is determined by the state transition matrix $\mathbf{T}$, and the sampling structure of an SRG is determined by the generating vector $\mathbf{h}$. Therefore the overall SRG structure is fully determined by $\mathbf{T}$ and $\mathbf{h}$, and hence we use the wording "an SRG *structured* by $\mathbf{T}$ and $\mathbf{h}$".
[9]Note that this sampling interval corresponds to the igniter sequence period.

THEOREM 8.1 (SAMPLING TIME CONDITION) *For a nonsingular state transition matrix* $\mathbf{T}$, *the correction matrix* Λ *in Eq.(8.9) can be zero only if the igniter sequence period* N_I *is chosen such that the the the discrimination matrix* $\Delta_{\mathbf{T},h}$ *in Eq.(8.10) becomes nonsingular.*

Proof: We prove that if N_I is chosen such that $\Delta_{\mathbf{T},h}$ in Eq.(8.10) becomes singular, the correction matrix Λ cannot be zero. To prove this by contradiction, we suppose Λ becomes zero for some choice of D_c and $\mathbf{c}_0$, and choose a nonzero vector $\hat{\delta}$ such that $\Delta_{\mathbf{T},h} \cdot \hat{\delta} = 0$. Then by Eq.(8.10) we have $\mathbf{h}^t \cdot \mathbf{T}^{iN_I} \cdot \hat{\delta} = 0$ for $i = 0, 1, \cdots, L - 1$. Applying this relation repeatedly to $\Lambda \cdot \hat{\delta}$ for the correction matrix Λ in Eq.(8.9), we obtain the equality $\Lambda \cdot \hat{\delta} = \mathbf{T}^{(L-1)N_I + D_c} \cdot \hat{\delta}$ or, equivalently, $\hat{\delta} = \mathbf{T}^{-(L-1)N_I - D_c} \cdot \Lambda \cdot \hat{\delta}$. This implies that $\hat{\delta} = 0$ since Λ is a zero matrix, which contradicts the assumption $\hat{\delta} \neq 0$.   ∎

This theorem provides a necessary condition on choosing N_I necessitated to make the despreader SRG synchronizable to the spreader SRG. According to the theorem, the period of the igniter sequence should be chosen such that the relevant discrimination matrix becomes *nonsingular*.

Once the period of the igniter sequence is chosen such that the discrimination matrix is nonsingular, we can determine the correction delay D_c and the correction vector $\mathbf{c}_0$ as follows :

THEOREM 8.2 (CORRECTION TIME AND VECTOR CONDITION) *For a non-singular state transition matrix* $\mathbf{T}$, *let the sampling interval* N_I *be chosen such that the discrimination matrix* $\Delta_{\mathbf{T},h}$ *in Eq.(8.10) is nonsingular. Then, the correction matrix* Λ *in Eq.(8.9) becomes zero if and only if the correction vector* $\mathbf{c}_0$ *takes, for an arbitrary correction delay* D_c *in* $0 < D_c \leq N_I$, *the expression*

$$\mathbf{c}_0 = \mathbf{T}^{(L-1)N_I + D_c} \cdot \Delta_{\mathbf{T},h}^{-1} \cdot \mathbf{e}_{L-1}, \qquad (8.11)$$

where the L-vector $\mathbf{e}_i$, $i = 0, 1, \cdots, L - 1$, *denotes the ith standard basis vector whose ith element is 1 and the others are 0.* [10]

Proof: To begin with, we expand the equation $\Delta_{\mathbf{T},h} \cdot \Delta_{\mathbf{T},h}^{-1} = \mathbf{I}$ for $\Delta_{\mathbf{T},h}$ in Eq.(8.10), to get the relations

$$\mathbf{h}^t \cdot \mathbf{T}^{iN_I} \cdot \Delta_{\mathbf{T},h}^{-1} = \mathbf{e}_i^t, i = 0, 1, \cdots, L - 1. \qquad (8.12)$$

We first prove the "if" part of the theorem. Since $\mathbf{T}$ is nonsingular, we can rewrite Eq.(8.9) as $\Lambda = \Lambda_*^L \cdot \mathbf{T}^{D_c - N_I}$ for $\Lambda_* = \mathbf{T}^{N_I} + \mathbf{c}_0 \cdot \mathbf{h}^t \cdot \mathbf{T}^{N_I - D_c}$. Due to

---

[10]In this section we use indices 0 through $L - 1$, instead of 1 through L, to indicate the entry positions in L-vectors or $L \times L$ matrices.

Eq.(8.11), this Λ_* can be rewritten as $\Lambda_* = \mathbf{T}^{(L-1)N_I+D_c} \cdot \Delta_{\mathbf{T},h}^{-1} \cdot \mathbf{A}_e \cdot \Delta_{\mathbf{T},h} \cdot \mathbf{T}^{-(L-1)N_I-D_c}$ for $\mathbf{A}_e = \Delta_{\mathbf{T},h} \cdot \mathbf{T}^{N_I} \cdot \Delta_{\mathbf{T},h}^{-1} + \mathbf{e}_{L-1} \cdot \mathbf{h}^t \cdot \mathbf{T}^{LN_I} \cdot \Delta_{\mathbf{T},h}^{-1}$. If the relations in Eq.(8.12) are applied to this, it reduces to $\mathbf{A}_e = [0 \ \mathbf{e}_0 \ \mathbf{e}_1 \ \cdots \ \mathbf{e}_{L-2}]$, which is a nilpotent matrix of nilpotency L. Therefore, $\mathbf{A}_e^L = 0$ and $\Lambda_*^L = 0$, and hence $\Lambda = 0$.

For the proof of the "only if" part, we evaluate $\Lambda \cdot \Delta_{\mathbf{T},h}^{-1} \cdot \mathbf{e}_{L-1}$ by applying Eq.(8.9) and Eq.(8.12) repeatedly. Then, we finally obtain the relation $\Lambda \cdot \Delta_{\mathbf{T},h}^{-1} \cdot \mathbf{e}_{L-1} = \mathbf{T}^{(L-1)N_I+D_c} \cdot \Delta_{\mathbf{T},h}^{-1} \cdot \mathbf{e}_{L-1} + \mathbf{c}_0$. Therefore, if $\Lambda = 0$, then $\mathbf{c}_0 = \mathbf{T}^{(L-1)N_I+D_c} \cdot \Delta_{\mathbf{T},h}^{-1} \cdot \mathbf{e}_{L-1}$. ∎

Note that the theorem does not impose any restrictions on the choice of correction time, which implies that the correction time may be arbitrarily chosen.

The above two theorems enable to formulate the following corollary :

COROLLARY 8.1 *For an m-sequence(PRBS) of period $2^L - 1$, $\{s_k\}$, generated by the spreader SRG having the transition matrix $\mathbf{T}$ and the generating vector $\mathbf{h}$, if we take an arbitrary sequence whose period, N_I, is relatively prime to the period of the m-sequence as the igniter sequence, then we can synchronize the despreader SRG using the DSA scheme equipped with the single correction vector $\mathbf{c}_0$ in Eq.(8.11).*

Proof: Let $\{t_k\}$ be an igniter sequence of period N_I and let N_I be relatively prime to $2^L - 1$. Then the sequence generated by sampling the given m-sequence $\{s_k\}$ at each start of the igniter sequence period is an N_I-decimated sequence of the original m-sequence, and thus becomes an m-sequence of period $2^L - 1$ (refer to Theorems 7.2 and 7.29 in [69]). If we denote this new sequence by $\{\tilde{s}_k\}$, then $\{\tilde{s}_k\}$ has the transition matrix $\tilde{\mathbf{T}} = \mathbf{T}^{N_I}$ and the generating vector $\mathbf{h}$.

Suppose that $\Delta_{\mathbf{T},h}$ is singular. Then, there exists a lowest degree nonconstant polynomial $\Psi_{\tilde{\mathbf{T}},h}(x)$ whose degree is lower than L, and $\mathbf{h}^t \cdot \Psi_{\tilde{\mathbf{T}},h}(\tilde{\mathbf{T}}) = 0$. Since $\{\tilde{s}_k\}$ is a binary m-sequence of period $2^L - 1$, the characteristic polynomial of the $L \times L$ matrix $\tilde{\mathbf{T}}$, $C_{\tilde{\mathbf{T}}}(x)$, is a primitive polynomial of degree L over GF(2) [4, 69]. Dividing $C_{\tilde{\mathbf{T}}}(x)$ by $\Psi_{\tilde{\mathbf{T}},h}(x)$, we get the relation $C_{\tilde{\mathbf{T}}}(x) = \Psi_{\tilde{\mathbf{T}},h}(x) \cdot Q(x) + R(x)$, where the degree of $R(x)$ is lower than that of $\Psi_{\tilde{\mathbf{T}},h}(x)$. Inserting $x = \tilde{\mathbf{T}}$, then multiplying $\mathbf{h}^t$ to the left of each sides of the relation, and finally applying the relation $C_{\tilde{\mathbf{T}}}(\tilde{\mathbf{T}}) = 0$, we obtain $\mathbf{h}^t \cdot R(\tilde{\mathbf{T}}) = 0$. Then, since the degree of $R(x)$ is lower than that of $\Psi_{\tilde{\mathbf{T}},h}(x)$, $R(x)$ should be zero by the definition of $\Psi_{\tilde{\mathbf{T}},h}(x)$, so $\Psi_{\tilde{\mathbf{T}},h}(x)$ divides $C_{\tilde{\mathbf{T}}}(x)$. But it is a contradiction because no non-constant polynomial whose degree is lower than L can divide a primitive polynomial of degree L. This proves that $\Delta_{\mathbf{T},h}$ is nonsingular.

Therefore, by Theorem 8.1, the despreader SRG is synchronizable for the igniter sequence $\{t_k\}$ of period N_I, and, by Theorem 8.2, the synchronization can be done using the single correction vector $\mathbf{c}_0$ specified in Eq.(8.11). ∎

According to the corollary, we may use any of the extended m-sequences as the igniter sequence, since the period of each extended m-sequence is relatively prime to those of the m-sequences. The resulting DSA despreader has simple circuitry as the single correction vector c_0 can work on it.

We finally consider how to realize the time-advanced sampling, which can be done by employing a new sampling vector as specified in the following theorem:

THEOREM 8.3 (TIME-ADVANCED SAMPLING VECTOR) *Let* $(r+i)N_I$ *be the sampling time when a sample* z_i *is taken from an SRG structured by the state transition matrix* $\mathbf{T}$ *and the generating vector* $\mathbf{h}$. *Then, the sample* z_i *is identical to the sample taken at time* $(r+i-1)N_I$ *using the sampling vector*

$$\mathbf{v}_0 = (\mathbf{T}^{N_I})^t \cdot \mathbf{h}. \qquad (8.13)$$

Proof : By Eq.(8.1) and Eq.(8.2), the sequence data generated at time $(r+i)N_I$ can be represented by $s_{(r+i)N_I} = \mathbf{h}^t \cdot \mathbf{T}^{(r+i)N_I} \cdot \mathbf{d}_0 = ((\mathbf{T}^{N_I})^t \cdot \mathbf{h})^t \cdot \mathbf{T}^{(r+i-1)N_I} \cdot \mathbf{d}_0 = ((\mathbf{T}^{N_I})^t \cdot \mathbf{h})^t \cdot \mathbf{d}_{(r+i-1)N_I}$. This implies that $s_{(r+i)N_I} = \mathbf{v}_0^t \cdot \mathbf{d}_{(r+i-1)N_I}$ for $\mathbf{v}_0 = (\mathbf{T}^{N_I})^t \cdot \mathbf{h}$, that is, if we take the sample at time $(r+i-1)N_I$ using the sampling vector $(\mathbf{T}^{N_I})^t \cdot \mathbf{h}$, then the sampled data is identical to the sequence data generated at time $(r+i)N_I$. ∎

Based on the theories and corollary we have discussed so far, we can design the DSA synchronization parameters in the following procedure: Given an SRG structured by the state transition matrix $\mathbf{T}$ and the generating vector $\mathbf{h}$, we first take the igniter sequence period N_I to be an integer relatively prime to the period $2^L - 1$ of the main SRG sequence. Then, we take an arbitrary value of correction delay D_c within $0 < D_c \leq N_I$. [11] Finally, we take the correction vector c_0 and the time-advanced sampling vector $\mathbf{v}_0$ as specified in Eq.(8.11) and Eq.(8.13), respectively.

EXAMPLE 8.1 We assume that the main SRG sequence is an m-sequence whose characteristic polynomial is $\Psi(x){=}x^{15}{+}x^{13}{+}x^9{+}x^8 {+}x^7{+}x^5{+}1$, with the transition matrix $\mathbf{T}$ and the sequence generating vector $\mathbf{h}$ of the main SRG given by

$$\mathbf{T} = \begin{bmatrix} \mathbf{0} & \mathbf{I}_{14\times14} \\ 1 & \mathbf{t} \end{bmatrix}, \quad \mathbf{t} = [0\,0\,0\,0\,1\,0\,1\,1\,1\,0\,0\,0\,1\,0], \quad (8.14a)$$

$$\mathbf{h} = [1\,0\,0\,0\,0\,0\,0\,0\,0\,0\,0\,0\,0\,0\,0]^t. \qquad (8.14b)$$

---

[11] While any arbitrary value may be chosen for D_c, it is desirable to take $D_c{=}1$ as it enables us to employ the same timing pulse for the sampling and correction process, and enables to achieve the fastest possible synchronization. The correction employing the correction delay $D_c{=}1$ is called *immediate correction* (refer to Chapter 12 of [69]).

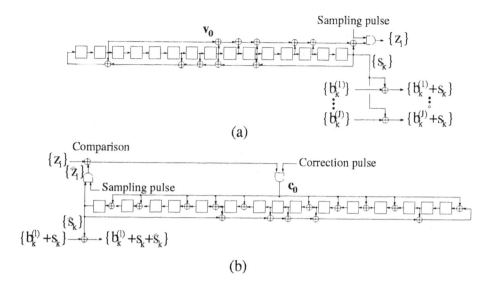

(a)

(b)

Figure 8.7.   DSA spreader(a) and despreader(b) circuits designed in Example 8.1.

Note that the SRG length L is 15. We take the igniter sequence to be an extended m-sequence of period $128(=2^7)$, whose start is marked up, for example, by the symbol "1" in the 8-bit string "00000001". Then the sampling interval N_I becomes 128 and the sampling matrix $\Delta_{T,h}$ in Eq.(8.10) becomes nonsingular. We take the correction delay D_c of 1 . Then, by Eq.(8.10), Eq.(8.11) and Eq.(8.13), we get the following correction and time-advanced sampling vectors:

$$\mathbf{c_0} = [\,1\,1\,0\,1\,1\,1\,0\,0\,1\,1\,1\,1\,0\,0\,1\,]^t, \qquad (8.15a)$$

$$\mathbf{v_0} = [\,1\,0\,1\,1\,0\,1\,1\,1\,0\,0\,0\,0\,0\,1\,0\,]^t. \qquad (8.15b)$$

Fig. 8.7 (a) and (b) depict the resulting DSA spreader and despreader circuits that incorporate the designed sampling and correction functions. The two circuits exactly match with the DSA spreader and despreader blocks in Fig. 8.1 (a) and (b).

2.   PERFORMANCE ANALYSIS OF THE DSA

It is a complex and challenging work to analyze the performance of the DSA scheme rigorously. It is because the synchronization mechanism of the DSA scheme is composed of several different types of processes - - namely, the igniter sequence acquisition employing the serial search and the threshold detection, the main SRG correction employing the orthogonal binary symbol detection, the synch-verification employing the counter detection, and the resetting of the whole synchronization processes that can occasionally happen. Throughout these series of processes there are involved numerous unexpected chances of false synchronization and resettings that influence the performance of the overall system.

In this section, we carry out a rigorous performance analysis of the DSA scheme in support of its operation introduced in the previous sections.

2.1   STATE TRANSITION DIAGRAMS FOR DSA ACQUISITION

In support of a complete analysis of the DSA performances, we employ state transition diagrams and take the transform domain approach (or, the moment generating function approach) [14].

2.1.1   ACQUISITION RELATED PARAMETERS

We define by *acquisition time* the total time spent in the synchronization process, from the instant when the igniter sequence conveying the main SRG state symbol begins to the instant when the transmission of the true synchronization is confirmed. It includes the time for igniter sequence acquisition, main SRG correction, and synch-verification, additionally including the false alarm penalty time in the case of false acquisition (see Fig. 8.3).

If we denote by $p(n)$ the probability to reach the main sequence acquisition state in n time units (i.e., $n \times \tau_D$), its moment generating function $P_{ACQ}(z)$ is given by $P_{ACQ}(z) = \sum_{n=0}^{\infty} p(n)z^n$. Once $P_{ACQ}(z)$ is given, the mean the acquisition time, T_{acq}, is determined by the relation $E\{T_{acq}\} = \frac{dP_{ACQ}(z)}{dz}|_{z=1} \times \tau_D$.

For the performance analysis, we assume that it takes the same fixed time τ_D for each of the following processes: One phase comparison in the igniter sequence searching stage, one state sample detection in the main SRG correction stage, and one verification in the main sequence synch-verification stage. [12]

Concerning the igniter sequence synchronization, we define by *detection probability per run*, $P_{d,r}$, the probability that the detector declares a true "in-phase" state per run (i.e., one round of shift of the igniter sequence), and by *false acquisition probability per cell*, $P_{fa,c}$, the probability that the detector declares a false "in-phase" state per phase (or "cell"). After the igniter sequence acquisition, the conveyed state sample is detected out of the acquired igniter sequence regardless of the trueness of the acquisition. Noting that the detected state samples may be probably corrupted by channel noise, we define *in-phase*

---

[12] We assume that the local phase advances by the step size of 1-chip. Under this assumption, in fact, there exist two states in the in-phase region, which can be approximated by the single in-phase state model with properly modified probability parameters. Refer to [15].

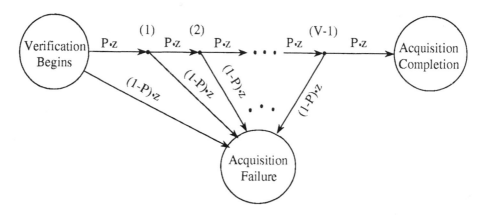

Figure 8.8.   Flowgraph for the verification process in Fig. 8.3.

coincidence probability, P_{it}, and *out-of-phase coincidence probability, P_{ot},* such that the former refers to the probability that the SRG sample detected in the state sample detector coincides with the *transmitted* main SRG sample when the igniter sequence is *truly* acquired, while the latter refers to the same probability when the igniter sequence is *falsely* acquired. As to the synch-verification process that follows the L comparison-correction steps, we define *in-phase false verification probability, P_{ir},* and *out-of-phase false verification probability, P_{or},* such that the former refers to the probability that the SRG sample determined in the detector coincides with the *receiver* main SRG sample when the receiver main SRG is not synchronized to the transmitter main SRG and the igniter sequence acquisition is *truly* acquired, while the latter refers to the same probability except that the igniter sequence is *falsely* acquired. If we denote by P_e the average decision error probability caused by the channel noise when the igniter sequence is synchronized, then the above probabilities are interrelated by $P_{it} = 1 - P_e \equiv P_c$ and $P_{ot} = P_{ir} = P_{or} = \frac{1}{2} \equiv P_f$.

Fig. 8.8 depicts the flow of the verification logic in the DSA system operation, which indicates that the synch-verification becomes successful only when V consecutive coincidence tests pass. If we denote by $P_{VS}(P, n)$ the probability to reach the acquisition success state (i.e., the acquisition completion state) in n steps, and by $P_{VF}(P, n)$ the probability to reach the acquisition-failure state in n steps, then the moment generating functions of $P_{VS}(P, n)$ and $P_{VF}(P, n)$ respectively become $H_{VS}(P, z) = P^V z^V$ and $H_{VF}(P, z) = \frac{(1-P)(1-P^V z^V)z}{1-Pz}$, where P denotes the *one-step success probability* (i.e., the probability that one coincidence test passes).

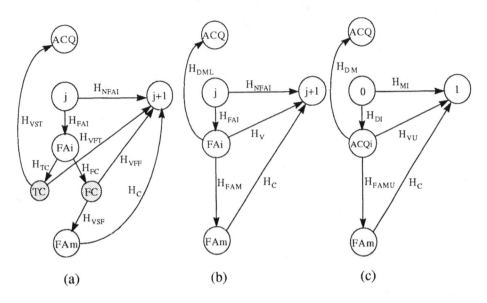

Figure 8.9. State transition diagrams for (a) state $i=1, 2, \cdots, N_I-1$ and (c) state $i=0$ (true "in-phase" state) in the DSA acquisition. The state diagram in (b) is a simplified version of that in (a).

2.1.2   FORMULATION OF STATE TRANSITION DIAGRAMS

In order to analyze the acquisition process of the DSA scheme, we need to determine the state transition diagram of the DSA-based acquisition process. For this we define *state j* to be the state of the receiver-generated igniter sequence whose phase advances the received igniter sequence by $j \bmod N_I$ chips. Among the N_I possible states, 0 through N_I-1, state 0 corresponds to the true "in-phase" state. We define four more states FA_i, ACQ_i, FA_m, and ACQ to indicate the *igniter sequence false alarm, igniter sequence acquisition, main sequence false alarm*, and *main sequence acquisition* states, respectively.

Fig. 8.9 depicts the state transitions among two adjacent states j and $j+1$ respectively for $j=1, 2, \cdots, N_I-1$ (a and b) and for $j=0$ (c), which is the true "in-phase" state. The weight for each transition in the diagram denotes the *transfer function* of the particular transition, which is defined to be the moment-generating function of the corresponding n-step state transition probability.

In Fig. 8.9 (a), which is for the case when the phase difference between the local igniter sequence and the incoming sequence is $j \bmod N_I$ chips, there are two different possible transitions: One is the transition to state $j+1$, the next igniter sequence phase comparison state (i.e., igniter sequence non-false alarm), and the other is the transition to the state FA_i, the igniter sequence acquisition declaration state (i.e., igniter sequence false alarm). The transfer functions of the igniter sequence non-false alarm, $H_{NFAI}(z)$, and of the igniter sequence false alarm, $H_{FAI}(z)$, are respectively given by $H_{NFAI}(z) = (1 - P_{fa,c})z$ and $H_{FAI}(z) = P_{fa,c}z$.

Once the state transition is made to state FA_i, the L comparison-correction process follows and then the synch-verification process begins. If any of the L detected state samples is different from the corresponding transmitter main SRG sample, [13] (which is highly probable as the detected samples are yet not reliable), the receiver main SRG will not get synchronized to the transmitter one, so the transition is made to the *false correction* state FC. In contrast, if all the L detected samples "luckily" coincide with the corresponding transmitter main SRG samples, transition is made to the *true correction* state TC.

In the FC state, state transition is made to state $j+1$ if the verification process fails (highly probable), and to state FA_m, otherwise. On the other hand, in the TC state, state transition is made to state $j+1$ if the verification process fails (highly probable), and to state ACQ, otherwise. Note that the probability to reach the state ACQ is negligibly small, as it can happen only when two successive unprobable events (that is, L correct comparison-corrections and V consecutive coincidence in the verification stage) occur.

In state FA_m, state transition is made to state $j+1$ after the recall process of the synch-confirmation logic, which is modeled by the penalty time of $K\tau_D$ as in [14].

Fig. 8.9 (b) is a simplified version of Fig. 8.9 (a), in which the intermediate states FC and TC are eliminated. The transfer functions defined on this simplified state transition diagram can be easily determined to be [14] $H_V(z) = P_{ot}^L z^L \cdot H_{VF}(P_{ot}, z) + (1 - P_{ot}^L)z^L \cdot H_{VF}(P_{or}, z)$, $H_{FAM}(z) = (1 - P_{ot}^L)z^L \cdot H_{VS}(P_{or}, z)$, $H_{DML}(z) = P_{ot}^L z^L \cdot H_{VS}(P_{ot}, z)$, and $H_C(z) = z^K$.

Now we consider Fig. 8.9 (c), which depicts the state transition for the case when the local igniter sequence is initially "in-phase" to the incoming igniter sequence. It is a simplified diagram obtained through a simplification process similar to that applied in obtaining Fig. 8.9 (b). Note that state FA_i in Fig. 8.9 (b) is replaced with ACQ_i, the true igniter sequence acquisition state. Differently from the "out-of-phase" case, the transition to the ACQ_i state in this case is highly probable. The transfer functions from state 0 to state ACQ_i (i.e., igniter sequence detection) and to state 1 (i.e., igniter sequence miss) are respectively given by $H_{DI}(z) = P_{d,r}z$ and $H_{MI}(z) = (1 - P_{d,r})z$. In state ACQ_i, there

---

[13] Note that, in this case, the corresponding transmitter main SRG sample indicates the transmitter main sequence value which was generated j chip units earlier than the conveyed state sample.

[14] In determining these transfer functions, the out-of-phase coincidence probability $P_{ot}^{(j)}$ should be employed for $j=1,2,\cdots,N_I-1$, which refers to the probability that the SRG sample detected in the state sample detector coincides with the transmitter main SRG sample that was generated j chip units earlier than the conveyed sample. In reality, however, $P_{ot}^{(j)}$ is the same for all j, so we drop off the superscript j, which yields the common transfer functions for all $N_I - 1$ states.

are three possible transitions as for the "out-of-phase" case - - to state 1, to state FA_m, and to state ACQ - -, which respectively result from verification failure, false correction (FC) followed by verification success, and true correction (TC) followed by verification success. Their relevant transfer functions are [15] $H_{VU}(z) = P_{it}^L z^L \cdot H_{VF}(P_{it}, z) + (1 - P_{it}^L)z^L \cdot H_{VF}(P_{ir}, z)$, $H_{FAMU}(z) = (1 - P_c^L)z^L \cdot H_{VS}(P_{ir}, z)$, and $H_{DM}(z) = P_{it}^L z^L \cdot H_{VS}(P_{it}, z)$. The transfer function from FA_m to state 1 is given by $H_C(z)$ as before.

If we construct the state transition diagram of the overall acquisition process based on Fig. 8.9, we obtain the circular diagram shown in Fig. 8.10. The prior probabilities are not exhibited in the figure, for which we will consider the uniform distribution and the worst-case distribution as in [14]. Notice that states ACQ_i and FA_i are put twice for each segment to make the graph planar.

2.2   MEAN ACQUISITION TIME ANALYSIS

Now we carry out the performance analysis based on the state transition diagrams established in the previous section. We focus on the mean acquisition time in the performance analysis, briefly considering the implementation issue in the next section.

2.2.1   ANALYSIS BASED ON STATE TRANSITION DIAGRAM

Referring to the state transition diagram in Fig. 8.10, we define four basic transfer functions $H_D(z)$, $H_M(z)$, $H_0(z)$, and $H_{FD}(z)$ as follows:

$$H_D(z) = H_{DI}(z)H_{DM}(z), \tag{8.16a}$$

$$H_M(z) = H_{MI}(z) + H_{DI}(z)(H_{VU}(z) + H_{FAMU}(z)H_C(z)), \tag{8.16b}$$

$$H_0(z) = H_{NFAI}(z) + H_{FAI}(z)(H_V(z) + H_{FAM}(z)H_C(z)), \tag{8.16c}$$

$$H_{FD}(z) = H_{FAI}(z)H_{DML}(z). \tag{8.16d}$$

Then, by applying the loop-reduction method, we can obtain the overall transfer functions of the following forms

$$P_{ACQ}^W(z) = \frac{H_D(z)H_0^{N_I-1}(z)}{1 - H_M(z)H_0^{N_I-1}(z)}$$
$$+ \frac{H_{FD}(z)(1 - H_0^{N_I-1}(z))}{(1 - H_M(z)H_0^{N_I-1}(z))(1 - H_0(z))}, \tag{8.17a}$$

---

[15] These relations are derived based on the assumption that the PN sequence is perfectly random, that is, each sequence value is independent. In case V is larger than L, the latter V-L sequence values are uniquely determined by the preceding L values, so the independence assumption may be violated. However, even in this case, the corresponding complex transfer functions can be approximated by the same relations.

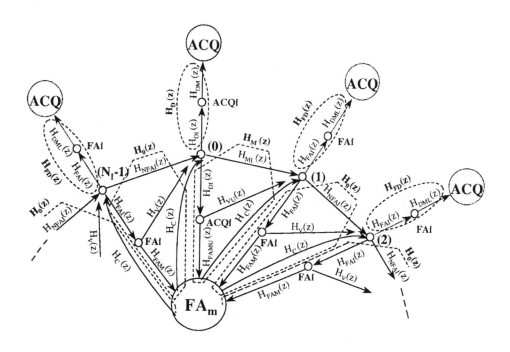

Figure 8.10. State transition diagram for the overall DSA acquisition process (reconfigured in planar form).

$$P_{ACQ}^{U}(z) = \frac{1}{N_I} \frac{H_D(z)(1 - H_0^{N_I}(z))}{(1 - H_M(z)H_0^{N_I-1}(z))(1 - H_0(z))}$$

$$+ \frac{H_{FD}(z)}{N_I} \left\{ \frac{N_I - 1}{1 - H_0(z)} \right.$$

$$\left. + \frac{(H_M(z) - H_0(z))(1 - H_0^{N_I-1}(z))}{(1 - H_M(z)H_0^{N_I-1}(z))(1 - H_0(z))^2} \right\}, \qquad (8.17b)$$

where 'W' and 'U' denote the worst-case and uniform prior probability distributions for the initial state.

While it is straightforward to evaluate the mean acquisition time using the state transition diagram in Fig. 8.10, the resulting expressions are exceedingly bulky. In order to generate meaningful practical expressions, therefore, we modify the moment generating functions by applying the fact that L and V are set, in practice, large enough to make $P_f^{L+V} \approx 0$. Then, from Eq.(8.16), we obtain $H_0(z) \approx (1 - P_{fa,c})z + P_{fa,c}z[(1 - P_f)\frac{1-P_f^V z^V}{1-P_f z}z^{L+1} + P_f^V z^{L+V+K}]$ and $H_{FD}(z) \approx 0$. [16] This modification then simplifies the expressions in Eq.(8.17a) and Eq.(8.17b) as follows:

$$P_{ACQ}^W(z) = \frac{H_D(z)H_0^{N_I-1}(z)}{1 - H_M(z)H_0^{N_I-1}(z)}, \tag{8.18a}$$

$$P_{ACQ}^U(z) = \frac{1}{N_I}\frac{H_D(z)(1 - H_0^{N_I}(z))}{(1 - H_M(z)H_0^{N_I-1}(z))(1 - H_0(z))}. \tag{8.18b}$$

Since these simplified equations coincide with those in [14, Eq.(5) and (6)], we can apply equations [14, Eq.(8) and (9)] in evaluating the mean acquisition time of the DSA scheme. Consequently, putting $P_f = 1/2$, we can get the following mean acquisition times of the DSA scheme

$$E\{T_{acq}^{W,DSA}\} = \frac{1}{P_{d,r}P_c^{L+V}}\left[1 + P_{d,r}\{L + 3 - P_c^L + \frac{P_c - P_c^V}{1 - P_c}P_c^L\right.$$
$$+ \frac{(K-2)(1-P_c^L)}{2^V}\}$$
$$+ (N_I - 1)\{1 + P_{fa,c}(L + 2 + \frac{K-2}{2^V})\}\right] \cdot \tau_D, \tag{8.19a}$$

$$E\{T_{acq}^{U,DSA}\} = \frac{1}{P_{d,r}P_c^{L+V}}\left[1 + P_{d,r}\{L + 3 - P_c^L + \frac{P_c - P_c^V}{1 - P_c}P_c^L\right.$$
$$+ \frac{(K-2)(1-P_c^L)}{2^V}\}$$
$$+(N_I - 1)\{1 + P_{fa,c}(L + 2 + \frac{K-2}{2^V})\} (1 - \frac{P_{d,r}P_c^{L+V}}{2})\right] \cdot \tau_D, \tag{8.19b}$$

for the *worst-case* (W) and *uniform* (U) prior probability distributions, respectively.

---

[16]This modification has the effect of interpreting the "lucky" main sequence acquisition which jumps into the ACQ state from the igniter sequence false alarm state (i.e., state $1,2,\cdots,N_I-1$) as the main sequence false acquisition, even though it is a true acquisition. Since this modification ignores some of those true acquisition paths really existent, the evaluated DSA acquisition time will become longer than the actual one, but the probability of the "lucky" main sequence acquisition is small enough to make the difference negligible.

On the other hand, the mean acquisition time of the conventional serial search acquisition [17] (SSA, in short) scheme employing the unmodulated main sequence itself as the pilot channel sequence can be easily determined by applying [14, Eq.(8), (9), and Fig.3], and the results are

$$
E\{T_{acq}^{W,SSA}\} = \frac{1}{\hat{P}_{d,r}^{V+1}}[1 + \hat{P}_{d,r}\frac{1 - \hat{P}_{d,r}^V}{1 - \hat{P}_{d,r}}
$$

$$
+ (N_M - 1)\{1 + \hat{P}_{fa,c}(\frac{1 - \hat{P}_{fa,c}^V}{1 - \hat{P}_{fa,c}} + K\hat{P}_{fa,c}^V)\}] \cdot \tau_D, \tag{8.20a}
$$

$$
E\{T_{acq}^{U,SSA}\} = \frac{1}{\hat{P}_{d,r}^{V+1}}[1 + \hat{P}_{d,r}\frac{1 - \hat{P}_{d,r}^V}{1 - \hat{P}_{d,r}}
$$

$$
+ (N_M - 1)\{1 + \hat{P}_{fa,c}(\frac{1 - \hat{P}_{fa,c}^V}{1 - \hat{P}_{fa,c}} + K\hat{P}_{fa,c}^V)\}(1 - \frac{\hat{P}_{d,r}^{V+1}}{2})] \cdot \tau_D,
$$

$$\tag{8.20b}$$

where $\hat{P}_{d,r}$ and $\hat{P}_{fa,c}$ respectively denotes the *detection probability per run* and the *false acquisition probability per cell* when the unmodulated pilot sequence is employed. This reduces to equation Eq.(8.20) in [14] when $V=0$. Note that the number of candidate cells in the searching mode, N_M, is $2^L - 1$ when the SRG length is L.

2.2.2   DETERMINATION OF PROBABILITY PARAMETERS

In the following, we elaborate how to determine the acquisition-related probability parameters. Let m_j(t), j = 0,1, denote the normalized state symbol s_j (t) spread by the igniter sequence $c_I(t)$, i.e., $m_j(t) = s_j(t)c_I(t)$, $\int_0^{N_I T_c} |m_j(t)|^2 dt = 1$. Then $\int_0^{N_I T_c} m_0(t)m_1(t)dt = 0$, since the symbols $s_0(t)$ and $s_1(t)$ are orthogonal, where T_c denotes the chip interval. We assume, without loss of generality, that the symbol s_0 (i.e.,state sample "0") is transmitted at time 0. Then the received signal takes the expression $r(t) = \sqrt{2P}m(t + lT_c + \eta T_c) \cos(\omega_0 t + \theta) + n(t)$, where $m(t) = \sum_{k=-\infty}^{\infty} m_{j(k)}(t - kN_I T_c)$, $j(k) \in \{0, 1\}$, $j(0) = 0$, and P denotes the state signal power and $n(t)$ the white gaussian noise with the two-side spectral density $N_0/2$. [18] In addition, l

---

[17]We assume for the serial search verification logic that, after declaring the "in-phase" state, the acquisition completion is declared if V consecutive threshold overtakings happen, and the search mode is resumed otherwise. Also we assume, for simplicity, that the dwell time in the verification mode is the same as that in the search mode, which is equal to the dwell time of the DSA, $\tau_D(=N_I T_c)$.

[18]This noise term $n(t)$ includes both the channel noise and the *multiple access interference* (MAI). When the long-period PN sequence is employed for the multiple access, the MAI may be approximated by the white gaussian noise [7].

and η respectively denote the integer and fractional part of the phase difference between the incoming and receiver-generated sequences normalized by T_c, and θ is an unknown carrier phase.

Referring to Fig. 8.4, we can represent each of matched components $Y_{j,c}$ and $Y_{j,s}$ (j=0,1) by $Y_{j,c} = X_j \cos\theta + N_{j,c} = \bar{X}_j \cos\theta + (X_j^r \cos\theta + N_{j,c})$ and $Y_{j,s} = X_j \sin\theta + N_{j,s} = \bar{X}_j \sin\theta + (X_j^r \sin\theta + N_{j,s})$, where $X_j = \sqrt{P} \int_0^{N_I T_c} m(t + lT_c + \eta T_c) m_j(t) dt$, $N_{j,c} = \sqrt{2} \int_0^{N_I T_c} n(t) m_j(t) \cos\omega_0 t dt$, and $N_{j,s} = \sqrt{2} \int_0^{N_I T_c} n(t) m_j(t) \sin\omega_0 t dt$, and $\bar{X}_j$ and X_j^r (j=0,1) respectively represent the deterministic and random parts of the correlation value X_j. Since X_j^r, $N_{j,c}$, and $N_{j,s}$ are all random variables, we integrate these terms into two composite noise terms $\hat{N}_{j,c}$ and $\hat{N}_{j,s}$, such that $\hat{N}_{j,c} = X_j^r \cos\theta + N_{j,c}$ and $\hat{N}_{j,s} = X_j^r \sin\theta + N_{j,s}$. Then, according to [15], we can approximate them very closely to zero-mean independent gaussian random variables, getting the deterministic parts

$$\bar{X}_0 = \begin{cases} \sqrt{P} N_I T_c (1 - |\eta|), & \text{if } H_1 \text{(in-phase)}, \\ 0, & \text{if } H_0 \text{(out-of-phase)}, \end{cases}$$

$$\bar{X}_1 = 0,$$

and the sums of the composite noise variances

$$Var\{\hat{N}_{0,c}\} + Var\{\hat{N}_{0,s}\}$$
$$= Var\{\hat{N}_{1,c}\} + Var\{\hat{N}_{1,s}\}$$
$$= \begin{cases} N_0 N_I T_c [1 + \gamma_c |\eta|^2], & \text{if } H_1 \text{(in-phase)}, \\ N_0 N_I T_c [1 + \gamma_c (1 - 2|\eta| + 2|\eta|^2)], & \text{if } H_0 \text{(out-of-phase)}, \end{cases}$$

where $\gamma_c \equiv PT_c/N_0$ denotes the chip-SNR.

Now from the relation $Y_j = \sqrt{Y_{j,c}^2 + Y_{j,s}^2}$, as well as the gaussian approximation of the composite random variables, we can determine the probabilities that Y_j exceeds the given threshold R_0, $P_{jk} \equiv Pr\{Y_j > R_0 | H_k\} = \int_{R_0}^{\infty} p_{Y_j|H_k}(y_j|H_k) dy_j$, $j, k = 0, 1$, getting the results

$$P_{00}(|\eta|) = \exp\{-\frac{c}{2[1 + \gamma_c (1 - 2|\eta| + 2|\eta|^2)]}\}, \qquad (8.21a)$$

$$P_{10}(|\eta|) = \exp\{-\frac{c}{2[1 + \gamma_c (1 - 2|\eta| + 2|\eta|^2)]}\}, \qquad (8.21b)$$

$$P_{01}(|\eta|) = Q(\sqrt{\frac{2N_I \gamma_c (1 - |\eta|)^2}{1 + \gamma_c |\eta|^2}}, \sqrt{\frac{c}{1 + \gamma_c |\eta|^2}}), \qquad (8.21c)$$

$$P_{11}(|\eta|) = \exp\{-\frac{c}{2[1 + \gamma_c |\eta|^2]}\}, \qquad (8.21d)$$

where c and $Q(\alpha, \beta)$ respectively denote the *normalized threshold* and the *Marcum's Q-function* [23] defined by $c \equiv \frac{2R_0^2}{N_0 N_I T_c}$ and $Q(\alpha, \beta) \equiv \int_\beta^\infty x \exp\{-\frac{1}{2}(x^2 + \alpha^2)\} I_0(\alpha x) dx$ for the 0th-order modified Bessel function $I_0(x)$.

Since the igniter sequence false acquisition happens when either Y_0 or Y_1 exceeds R_0 under H_0, we can determine, by applying Eq.(8.21a) and Eq.(8.21b), the *false acquisition probability per cell*

$$P_{fa,c}(|\eta|) = 1 - (1 - P_{00}(|\eta|))(1 - P_{10}(|\eta|)). \tag{8.22}$$

Thus, when the desired false acquisition probability is given (as in the Neyman-Pearson Criterion [23]), the threshold c is determined by

$$c = -2[1 + \gamma_c(1 - 2|\eta| + 2|\eta|^2)] \log(1 - \sqrt{1 - P_{fa,c}(|\eta|)}). \tag{8.23}$$

On the other hand, applying Eq.(8.21c) and Eq.(8.21d), we get the detection probability for a cell with the phase difference $|\eta|$, $Q_d(|\eta|) = 1 - (1 - P_{01}(|\eta|))(1 - P_{11}(|\eta|))$. Since we assume that the local phase advances by the step size T_c, there exist two cells in the in-phase region whose phase differences are $|\eta|$ and $1 - |\eta|$, respectively. Therefore, the *detection probability per run* is determined by

$$P_{d,r}(|\eta|) = Q_d(|\eta|) + (1 - Q_d(|\eta|))Q_d(1 - |\eta|). \tag{8.24}$$

Finally, assuming that the symbol decision is made with the smaller phase difference out of $|\eta|$ and $1 - |\eta|$ if both cells in the in-phase region are detected [19], we can determine the *probability of correct sample decision (under H_1)*

$$\begin{aligned}
P_c(|\eta|) &\approx \frac{Q_d(|\eta|)(1 - Q_d(1 - |\eta|))(1 - \hat{P}_e(|\eta|))}{Q_d(|\eta|) + Q_d(1 - |\eta|) - Q_d(|\eta|)Q_d(1 - |\eta|)} \\
&+ \frac{(1 - Q_d(|\eta|))Q_d(1 - |\eta|)(1 - \hat{P}_e(1 - |\eta|))}{Q_d(|\eta|) + Q_d(1 - |\eta|) - Q_d(|\eta|)Q_d(1 - |\eta|)} \\
&+ \frac{Q_d(|\eta|)Q_d(1 - |\eta|)(1 - \min\{\hat{P}_e(|\eta|), \hat{P}_e(1 - |\eta|)\})}{Q_d(|\eta|) + Q_d(1 - |\eta|) - Q_d(|\eta|)Q_d(1 - |\eta|)},
\end{aligned} \tag{8.25}$$

where $\hat{P}_e(|\eta|)$ denotes the decision error probability that the symbol s_1 is selected when the symbol s_0 is transmitted, which is easily determined to be [20]

---

[19]This assumption implies that we need to compare the correlation value of the current cell with that of the next cell for one or a few more symbol intervals whenever the igniter sequence acquisition is declared. This results in the increase of the mean acquisition times of Eq.(8.19), which, however, is negligible in the practical CDMA environment where $N_I + L + V$ is large enough.

[20]$\hat{P}_e$ is computed by the relation $\hat{P}_e = \int_0^\infty \int_{y_0}^\infty p_{Y_1|H_1}(y_1|H_1) p_{Y_0|H_1}(y_0|H_1) dy_1 dy_0$.

$\hat{P}_e(|\eta|) = \frac{1}{2}\exp\{-\frac{N_I\gamma_c(1-|\eta|)^2}{2(1+\gamma_c|\eta|^2)}\}$. On the other hand, for the SSA scheme employing the unmodulated main sequence itself as the pilot channel sequence, the false acquisition probability per cell $\hat{P}_{fa,c}$ and the detection probability per run $\hat{P}_{d,r}$ are respectively given by

$$\hat{P}_{fa,c}(|\eta|) = P_{00}(|\eta|), \tag{8.26}$$

$$\hat{P}_{d,r}(|\eta|) = P_{01}(|\eta|) + (1 - P_{01}(|\eta|))P_{01}(1 - |\eta|). \tag{8.27}$$

Thus, the threshold c for a given false alarm probability is determined by

$$c = -2[1 + \gamma_c(1 - 2|\eta| + 2|\eta|^2)]\log(\hat{P}_{fa,c}(|\eta|)). \tag{8.28}$$

2.3   PERFORMACE EVALUATIONS

Now we evaluate the performance of the proposed DSA scheme by applying the two most important performance factors - - mean acquisition time and implementation complexity - - that have been widely adopted in most related literatures [4, 14, 15, 27].

2.3.1   MEAN ACQUISITION TIME EVALUATION

For a numerical evaluation of the mean acquisition time, we take the communication environment previously set in Example 8.1.

EXAMPLE 8.2 *(Example 8.1 continued)* : Based on Eq.(8.19) and Eq.(8.20), along with the the probability parameter analysis results, we compare the acquisition performances of the proposed DSA scheme and the conventional SSA scheme. We continue *Example 8.1* in Section 1, under the assumption that the false alarm penalty factor is 1000 and the fractional chip misalignment $\eta = 0.25$, applying the gaussian approximation for the multiple access interference and the channel noise. For the results of Fig. 8.11 and Fig. 8.12, we assume that the chip-SNR is -10dB [21], and the threshold c is set such that the false acquisition probability $P_{fa,c}$ becomes 0.01 for DSA and $\hat{P}_{fa,c}$ becomes 0.25 for SSA, respectively. [22] The resulting $P_{d,r}$, P_c, and $\hat{P}_{d,r}$ are 0.732, 0.981, and 0.994.

Fig. 8.11 compares the two schemes in terms of mean acquisition times and their ratio, for varing verification step sizes of 0 through $L-1$. We can observe dramatic improvements in acquisition time for the proposed DSA scheme. That is, the mean acquisition time of the DSA scheme becomes about 100 times shorter than that of the conventional SSA scheme.

---

[21] Though we operate the system in a lower SNR environment, for example, in -13dB chip-SNR environment, we can still maintain the fast acquisition performance of the DSA system by increasing N_I to 256 or by transmitting each state symbol twice.

[22] These false acquisition probabilities produce the mean acquisition performance close to optimal over wide SNR ranges. See Fig. 8.14 (a).

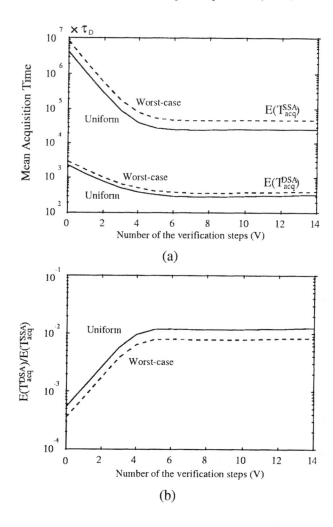

Figure 8.11.   Comparison of mean acquisition times of the DSA and conventional serial search acquisition (SSA) schemes for the worst case and uniform prior probability distributions (γ_c=-10dB, $\eta = 0.25$, $P_{fa,c} = 0.01$ for DSA, $\hat{P}_{fa,c} = 0.25$ for SSA, $L = 15$, $N_I = 128$, $K = 1000$): (a) Mean acquisition times, (b) mean acquisition time ratio.

Fig. 8.12 (a) and (b) plot the acquisition time ratio of the DSA scheme over the conventional SSA scheme in terms of false alarm penalty factor K (for L=15) and main sequence SRG length L (for K=1,000) respectively: Fig. 8.12 (a) shows that the ratio increases as the penalty time increases, but within the same order of magnitude. Fig. 8.12 (b) shows that the acquisition time ratio decreases

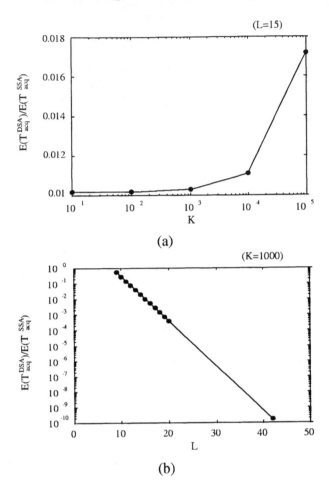

Figure 8.12.   Mean acquisition time ratio vs. false alarm penalty factor K (a), and vs. main SRG length L (b) (γ_c=-10dB, $\eta = 0.25$, $P_{fa,c} = 0.01$ for DSA, $\hat{P}_{fa,c} = 0.25$ for SSA, $N_I = 128$, uniform prior probability).

exponentially as the SRG length increases. This indicates that the advantage of the proposed DSA scheme becomes substantial for long PN sequences. [23]

Fig. 8.13 compares the acquisition times for varying threshold values when the chip-SNR is -10dB. The result shows that the acquisition time performance of the DSA scheme is less sensitive to the threshold value than that of the SSA scheme. In the acquisition stage, we can't estimate the system SNR that is needed to determine the optimal threshold, thus it is desirable to make the acquisition operation robust to poor threshold settings.

---

[23] In plotting Figs. 8.12 and 8.13, optimal verification step sizes are assumed.

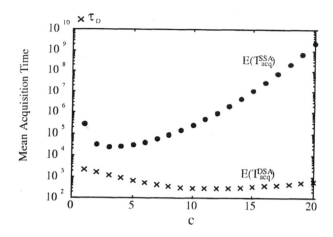

Figure 8.13.   Mean acquisition time vs. the normalized threshold c (γ_c=-10dB, $\eta = 0.25$, L=15, $N_I = 128$, K=1000, uniform prior probability).

Fig. 8.14 plots the acquisition times and their ratios for varying chip-SNR: Fig. 8.14 (a) and (b) plot the mean acquisition times for the chip misalignment $\eta = 0.25$ and $\eta = 0$ respectively, with the threshold set such that the false acquisition probability becomes 0.5, 0.25, 0.1, and 0.01. Fig. 8.14 (c) plots ratio of mean acquisition times for $\eta = 0.25$ and $\eta = 0$ when $\hat{P}_{fa,c} = 0.25$ and $P_{fa,c} = 0.01$ (i.e., when the best threshold is set for each scheme.).

We see in the figures that the acquisition time reduction ratio is maintained to be larger than 100 when the chip-SNR is larger than -10dB for $\eta = 0.25$. But the DSA acquisition time increases rapidly as the chip-SNR becomes very low, which happens because P_c becomes too small, that is, the state symbol SNR becomes too low to transfer the transmitter SRG information reliably. However, we can maintain the small acquisition time even in the very low SNR ranges by executing the chip timing alignment processing (i.e., by making η approach 0) before the main SRG correction. (See Fig. 8.14 (c).)

2.3.2   IMPLEMENTATION COMPLEXITY OF DSA SCHEME

Finally we consider the implementation complexity aspect of the proposed DSA scheme. In implementing the proposed DSA scheme in the transmitter and receiver blocks, we additionally need an igniter sequence generation block including a short length SRG, state symbol generation and detection block, a time-advanced sampling circuit in the transmitter, and a correction circuit and a verification circuit in the receiver (see Fig. 8.1). However, with all these functional increase, the required hardware increase is very little, as is well demonstrated in Fig. 8.7. In contrast, the parallel search scheme, which is the only

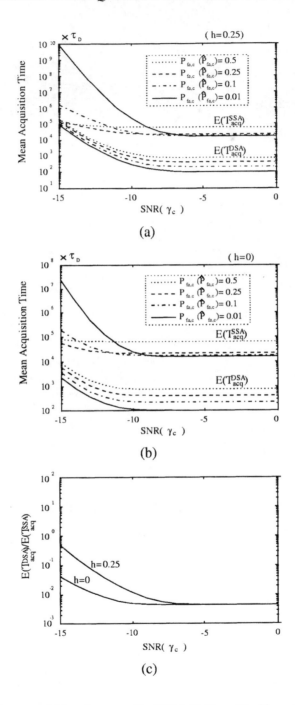

Figure 8.14.   Mean acquisition time vs. chip-SNR (γ_c) ($L = 15$, $N_I = 128$, $K = 1000$, uniform prior probability): (a) Mean acquisition times ($\eta = 0.25$), (b) mean acquisition times ($\eta = 0$), (c) mean acquisition time ratio. ($P_{fa,c} = 0.01$ for DSA, and $\hat{P}_{fa,c} = 0.25$ for SSA.)

Table 8.1. Performance comparison among SSA, PSA, and DSA schemes. (*M* denotes the degree of parallelism. All values in the table are referenced to the case of the SSA. α is a small value less than 1.)

	SSA	PSA	DSA
A. Acquisition time	1	$1/M$	$(L+N_f)/(2^L-1)$
B. System complexity	1	M	$1+\alpha$
C. Product of A and B	1	1	$(1+\alpha)(L+N_f)/(2^L-1)$

scheme comparable to the DSA scheme in terms of acquisition time, requires as much hardware penalty as the acquisition time gain, because the reduction in acquisition time is directly proportional to the number of duplications of parallel branches [27]. Consequently the required hardware complexity is much higher for the parallel search scheme than that of the DSA scheme. This may be well illustrated by Example 8.1: In order to achieve the acquisition time ratio of 0.01, the parallel search scheme requires about 100 parallel branch circuits each of which contains an SRG having its own initial state values and detection circuits. Table. 8.1 summarizes the performance distinctions among the three acquisition schemes, i.e., serial search acquisition (SSA), parallel search acquisition (PSA), and distributed sample acquisition (DSA), in more detail.

3. BATCH DSA (BDSA)

There are many DS/CDMA systems that adopt multi-bit transmission such as *M*-ary orthogonal signaling, QPSK signaling, multi-carrier signaling, and so on [9, 10, 53, 129, 130, 131]. In these cases the optimal number of synchronization information bits, i.e., *state samples*, to be simultaneously transmitted may be larger than one depending on the allocated power budget, number of parallel channels, and igniter sequence period. Therefore, it is desirable to devise a more generalized DSA scheme that can handle the SRG synchronization process for multi-bit transmission. Such a generalized signaling DSA scheme may also enable us to use error control coding techniques of proper code rates to convey the state samples more efficiently. In order to synchronize the SRG by employing a multi-bit based DSA scheme, we basically need to extend the underlying single sampling - single correction technique to a *multiple sampling - multiple correction* based one. The resulting variation of the DSA are the *batch*

DSA (BDSA) and the *parallel DSA* (PDSA). [24] In this section, we discuss the resulting BDSA, with the discussion of the PDSA deferred to the next section.

3.1   BDSA SYSTEM OPERATION

Fig. 8.15 depicts the functional block diagram of the acquisition-related part of the DS/CDMA system that employs the BDSA scheme. The transmitter part consists of a BDSA-spreader, a *serial input parallel output* (SIPO) connector and a *parallel sample*(PS)-spreader, and the receiver part contains their despreading counterparts, that is, a BDSA-despreader, a *parallel input serial output* (PISO) connector and a PS-despreader. The BDSA-spreader/despreader pair take the synchronization function while the PS-spreader/despreader pair take the parallel-sample conveyance function.

The time-advanced sampling block in the BDSA-spreader takes the state sample z_{ij}, $j = 0, 1, \cdots, b-1$, of the main SRG in advance : More specifically, it takes the sample z_{ij} at time $(R + i - 2)N_I + \alpha_j$, $0 \leq \alpha_0 < \alpha_1 < \cdots < \alpha_{b-1} < N_I$, for a reference value R, which is the main SRG sequence to be generated at time $(R + i)N_I + \alpha_j$. The b samples taken in the interval $[(R + i - 2)N_I, (R + i - 1)N_I)$ are stored in the SIPO connector block. Then the symbol generation block maps those b samples to the corresponding M-ary orthogonal symbol $\mathbf{x}_i$ at time $(R + i - 1)N_I$ and spreads it using the igniter SRG. We assume $M = 2^b$, that is, the symbol generation block in the PS-spreader generates one symbol out of 2^b cadidates in every N_I interval, from the b binary samples $z_{i0}, z_{i1}, \cdots, z_{i(b-1)}$ provided by the SIPO connector. In the case of multi-carrier DS/CDMA systems [9], the b samples are converted to b binary symbols instead of being mapped to a 2^b-ary symbol, spread by the igniter sequence, and then transmitted in parallel over each sub-carrier.

The PS-despreader despreads the received state signal and matches it with each M-ary orthogonal symbol (or each binary symbol in the case of multi-carrier systems) to detect the conveyed samples z_{ij}'s, passing them to the PISO block at time $(R + i)N_I$. Then the PISO connector conveys the samples to the BDSA-despreader in a sequential manner such that the sample z_{ij} is provided at time $(R + i)N_I + \alpha_j$, $j = 0, 1, \cdots, b - 1$. On the other hand, the BDSA-despreader generates its own SRG state sample $\hat{z}_{ij}$ at time $(R+i)N_I + \alpha_j$, $j = 0, 1, \cdots, b-1$, which corresponds to the main SRG sequence value $\hat{s}_{(R+i)N_I + \alpha_j}$, and compares it with the conveyed sample z_{ij}. If $\hat{z}_{ij}$ coincides with z_{ij}, no action takes place, but otherwise the jth correction circuit is triggered to correct the main SRG state at time $(R+i)N_I + \beta_j$, with β_j chosen such that the correction can be made before the next sample comparison (i.e., $\alpha_j < \beta_j \leq \alpha_{j+1}$ for $j = 0, 1, \cdots, b - 2$, and $\alpha_{b-1} < \beta_{b-1} \leq N_I + \alpha_0$).

---

[24]In view of the BDSA and the PDSA, the original DSA corresponds to a special case with $M = 2$.

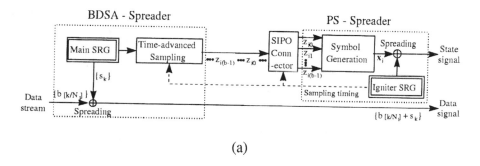

(a)

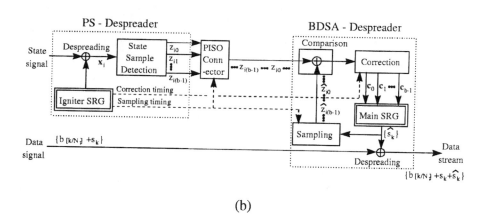

(b)

Figure 8.15. Functional block diagram of the acquisition-related part of the DS/CDMA system employing the BDSA scheme : (a) Transmitter, (b) receiver.

The state sampling and correction processes are timed by the sampling and correction pulses issued by the igniter SRG. The related timing diagrams are depicted in Fig. 8.16.

The acquisition and verification procedure of the BDSA is the same as that of the DSA of Section 1, except that the b state samples are manipulated simultaneously.

3.2   BDSA SYNCHRONIZATION PARAMETER DESIGN

Now that the PS spreader/despreader pair, both supported by the igniter sequence, provides a reliable means for sample conveyance, we concentrate on the synchronization problem of the main SRGs residing in the BDSA spreader/despreader pair.

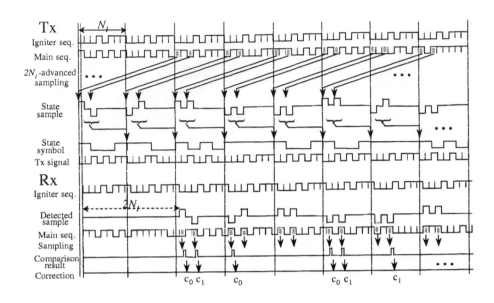

Figure 8.16.   System timing diagram (illustrating the case $b=2$, $N_I=8$).

In order to achieve the synchronization, we compare the sample z_{ij}, which is taken out of the main SRG sequence $\{s_k\}$ at the (virtual) sampling time $(R+i)N_I + \alpha_j$, with the sample $\hat{z}_{ij}$ taken out of $\{\hat{s}_k\}$ at the same sampling time, and reflect their discrepancy to correcting the state of the despreader SRG. In the following, we will investigate when to sample, how to do time-advanced sampling, and how to correct the SRG state, based on a rigorous mathematical modeling of the BDSA spreader/despreader pair.

3.2.1   MATHEMATICAL MODELING FOR SYNCHRONIZATION

Let $\mathbf{d}_k$ and $\hat{\mathbf{d}}_k$ denote the *state vectors* of the spreader and despreader SRGs at time k, respectively, and let $\mathbf{T}$ denote the *state transition matrix* that relates two successive state vectors such that $\mathbf{d}_{k+1} = \mathbf{T} \cdot \mathbf{d}_k$, $\hat{\mathbf{d}}_{k+1} = \mathbf{T} \cdot \hat{\mathbf{d}}_k$. [25] In addition, let $\mathbf{h}$ denote the *generating vector* that generates the sequence value s_k out of the state vector $\mathbf{d}_k$ through the relation $s_k = \mathbf{h}^t \cdot \mathbf{d}_k$, $\hat{s}_k = \mathbf{h}^t \cdot \hat{\mathbf{d}}_k$ and let $\mathbf{c}_j$, $j = 0, 1, \cdots, b-1$, denote the *jth correction vector* that corrects the old state vector $\mathbf{d}_{old}$ to a new state vector $\mathbf{d}_{new}$ through the relation $\hat{\mathbf{d}}_{new} = \hat{\mathbf{d}}_{old} + (z_{ij} + \hat{z}_{ij})\mathbf{c}_j$.

---

[25]The relation $\hat{\mathbf{d}}_{k+1} = \mathbf{T} \cdot \hat{\mathbf{d}}_k$ holds for the despreader SRG without correction process. If correction process is added, this relation changes to the form in Eq.(8.30b).

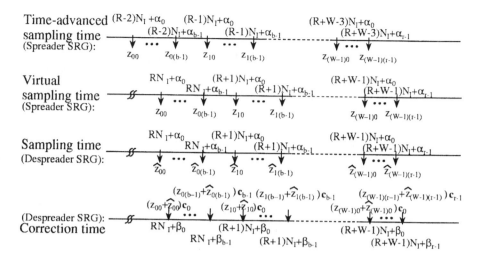

Figure 8.17.   Sampling and correction timing diagram for BDSA.

In compliance with this batch-type correction processing, we rearrange the L state samples in W batches of b samples. Then we get the relation $L - 1 = b(W - 1) + (r - 1)$, $1 \leq r \leq b$. Here we consider the case $L > b$, or equivalently, $W \geq 2$, as otherwise batch processing is meaningless and the synchronization process can be well handled by the existing DSA scheme. The SIPO and PISO connectors in Fig. 8.15 convert the Wb samples z_i, $i = 0, 1, \cdots, Wb - 1$ to z_{ij}, $i = 0, 1, \cdots, W - 1$, $j = 0, 1, \cdots, b - 1$ in support of this batch processing. In case $L < Wb$ (or equivalently $r \neq b$), only the first L out of Wb batch-processed samples are used for correcting the receiver SRG state, with the remaining $Wb - L$ samples being used, in conjunction with the subsequent batch-samples, for verification. The samples z_{ij} and $\hat{z}_{ij}$ are taken *virtually* at the same sampling time $(R + i)N_I + \alpha_j$, and the related correction is made, as necessary, at the correction time $(R + i)N_I + \beta_j$. Fig. 8.17 depicts this timing relationship. So, the samples z_{ij} and $\hat{z}_{ij}$ are related to the state vectors by the relations

$$z_{ij} = s_{(R+i)N_I + \alpha_j} = \mathbf{h}^t \cdot \mathbf{d}_{(R+i)N_I + \alpha_j}, \tag{8.29a}$$

$$\hat{z}_{ij} = \hat{s}_{(R+i)N_I + \alpha_j} = \mathbf{h}^t \cdot \hat{\mathbf{d}}_{(R+i)N_I + \alpha_j}, \tag{8.29b}$$

$$i = 0, 1, \cdots, W - 1, \quad j = 0, 1, \cdots, b - 1,$$

and the state vectors at the correction time takes the expressions

$$\mathbf{d}_{(R+i)N_I + \beta_j} = \mathbf{T} \cdot \mathbf{d}_{(R+i)N_I + \beta_j - 1}, \tag{8.30a}$$

$$\hat{\mathbf{d}}_{(R+i)N_I + \beta_j} = \mathbf{T} \cdot \hat{\mathbf{d}}_{(R+i)N_I + \beta_j - 1} + (z_{ij} + \hat{z}_{ij})\mathbf{c}_j, \tag{8.30b}$$

$$\begin{cases} i = 0, 1, \cdots, W - 2, \; j = 0, 1, \cdots, b - 1, \\ i = W - 1, \; j = 0, 1, \cdots, r - 1. \end{cases}$$

We define the *state distance vector* δ_k to be $\delta_k \stackrel{\triangle}{=} \mathbf{d}_k + \hat{\mathbf{d}}_k$. Then by combining Eq.(8.29) and Eq.(8.30) we can easily get the recursive relation

$$\delta_{(R+i)N_I + \beta_j} =$$

$$\begin{cases} (\mathbf{T}^{\beta_0} + \mathbf{c}_0 \cdot \mathbf{h}^t \cdot \mathbf{T}^{\alpha_0}) \cdot \delta_{RN_I}, \\ \quad i = 0, \; j = 0, \\ (\mathbf{T}^{N_I + \beta_0 - \beta_{b-1}} + \mathbf{c}_0 \cdot \mathbf{h}^t \cdot \mathbf{T}^{N_I + \alpha_0 - \beta_{b-1}}) \cdot \delta_{(R+i-1)N_I + \beta_{b-1}}, \\ \quad i = 1, 2, \cdots, W - 1, \; j = 0, \\ (\mathbf{T}^{\beta_j - \beta_{j-1}} + \mathbf{c}_j \cdot \mathbf{h}^t \cdot \mathbf{T}^{\alpha_j - \beta_{j-1}}) \cdot \delta_{(R+i)N_I + \beta_{j-1}}, \\ \quad i = 0, 1, \cdots, W - 2, \; j = 1, 2, \cdots b - 1, \; b \geq 2, \\ \quad i = W - 1, \; j = 1, 2, \cdots, r - 1, \; r \geq 2. \end{cases} \quad (8.31)$$

By applying this relation repeatedly for $L (= b(W - 1) + r)$ times of corrections, we obtain the finally corrected state distance vector $\delta_{(R+W-1)N_I + \beta_{r-1}}$ $= \Lambda \cdot \delta_{RN_I}$ for the $L \times L$ *correction matrix*

$$\Lambda \stackrel{\triangle}{=} \Lambda_r \cdot \Lambda_b^{W-2} \cdot \Lambda_o, \quad (8.32a)$$

where

$$\Lambda_r \stackrel{\triangle}{=} [(\mathbf{T}^{\beta_{r-1} - \beta_{r-2}} + \mathbf{c}_{r-1} \cdot \mathbf{h}^t \cdot \mathbf{T}^{\alpha_{r-1} - \beta_{r-2}}) \cdots .$$
$$(\mathbf{T}^{\beta_1 - \beta_0} + \mathbf{c}_1 \cdot \mathbf{h}^t \cdot \mathbf{T}^{\alpha_1 - \beta_0}) \times (\mathbf{T}^{N_I + \beta_0 - \beta_{b-1}} + \mathbf{c}_0 \cdot \mathbf{h}^t \cdot \mathbf{T}^{N_I + \alpha_0 - \beta_{b-1}})], \quad (8.32b)$$

$$\Lambda_b \stackrel{\triangle}{=} [(\mathbf{T}^{\beta_{b-1} - \beta_{b-2}} + \mathbf{c}_{b-1} \cdot \mathbf{h}^t \cdot \mathbf{T}^{\alpha_{b-1} - \beta_{b-2}}) \cdots .$$
$$(\mathbf{T}^{\beta_1 - \beta_0} + \mathbf{c}_1 \cdot \mathbf{h}^t \cdot \mathbf{T}^{\alpha_1 - \beta_0}) \times (\mathbf{T}^{N_I + \beta_0 - \beta_{b-1}} + \mathbf{c}_0 \cdot \mathbf{h}^t \cdot \mathbf{T}^{N_I + \alpha_0 - \beta_{b-1}})], \quad (8.32c)$$

$$\Lambda_o \stackrel{\triangle}{=} [(\mathbf{T}^{\beta_{b-1} - \beta_{b-2}} + \mathbf{c}_{b-1} \cdot \mathbf{h}^t \cdot \mathbf{T}^{\alpha_{b-1} - \beta_{b-2}}) \cdots .$$
$$(\mathbf{T}^{\beta_1 - \beta_0} + \mathbf{c}_1 \cdot \mathbf{h}^t \cdot \mathbf{T}^{\alpha_1 - \beta_0}) \times (\mathbf{T}^{\beta_0} + \mathbf{c}_0 \cdot \mathbf{h}^t \cdot \mathbf{T}^{\alpha_0})]. \quad (8.32d)$$

In order to achieve the synchronization by applying L corrections, it is necessary to make the final state distance vector $\delta_{(R+W-1)N_I + \beta_{r-1}}$ a zero vector regardless of the initial state distance vector δ_{RN_I}, which can be done only by making the correction matrix Λ a zero matrix. Therefore, for a given BDSA spreader/despreader pair whose constituent SRGs are structured by $\mathbf{T}$ and $\mathbf{h}$, the synchronization problem turns to a problem of determining appropriate N_I, α_j, β_j, and $\mathbf{c}_j$ $(j = 0, 1, \cdots, b - 1)$ that make the correction matrix null.

3.2.2   SAMPLING AND CORRECTION CONDITIONS

For an SRG of length L structured by the state transition matrix $\mathbf{T}$ and the generating vector $\mathbf{h}$, we define the *discrimination matrix* Δ to be the $L \times L$ matrix

$$\Delta \stackrel{\triangle}{=} [\, (\mathbf{T}^{\alpha_0})^t \cdot \mathbf{h} \, \cdots \, (\mathbf{T}^{\alpha_{b-1}})^t \cdot \mathbf{h} \;\; (\mathbf{T}^{N_I+\alpha_0})^t \cdot \mathbf{h} \, \cdots$$
$$(\mathbf{T}^{N_I+\alpha_{b-1}})^t \cdot \mathbf{h} \cdots (\mathbf{T}^{(W-1)N_I+\alpha_0})^t \cdot \mathbf{h} \cdots$$
$$(\mathbf{T}^{(W-1)N_I+\alpha_{r-1}})^t \cdot \mathbf{h} \,]^t. \tag{8.33}$$

Then we can determine the igniter sequence period N_I and sampling times, α_0, $\alpha_1, \cdots, \alpha_{b-1}$, based on the following theorem.

THEOREM 8.4 (BDSA SAMPLING TIME CONDITION) *For a nonsingular state transition matrix* $\mathbf{T}$, *if the igniter sequence period,* N_I, *and the sampling times,* α_j, $j = 0, 1, \cdots, b-1$, *are chosen such that the the discrimination matrix* Δ *in Eq.(8.33) becomes singular, then for any choice of correction times,* β_j, $j = 0, 1, \cdots, b-1$, *and correction vectors,* $\mathbf{c}_j$, $j = 0, 1, \cdots, b-1$, *the correction matrix* Λ *in Eq.(8.32a) cannot be zero.*

Proof : We prove this by contradiction. Let α_j, $j = 0, 1, \cdots, b-1$, and N_I be chosen such that Δ in Eq.(8.33) becomes singular. We suppose that the correction matrix Λ becomes a zero matrix for some choice of β_j and $\mathbf{c}_j$, $j = 0, 1, \cdots, b-1$. We choose a nonzero vector $\hat{\delta}$ such that $\Delta \cdot \hat{\delta} = 0$. Then by Eq.(8.33) we have $\mathbf{h}^t \cdot \mathbf{T}^{iN_I+\alpha_j} \cdot \hat{\delta} = 0$ for $i = 0, 1, \cdots, W-2$, $j = 0, 1, \cdots, b-1$, and $i = W-1, j = 0, 1, \cdots, r-1$. Applying this relation repeatedly to $\Lambda \cdot \hat{\delta}$ for the correction matrix Λ in Eq.(8.32a), we obtain the equality $\Lambda \cdot \hat{\delta} = \Lambda_r \cdot \Lambda_b^{W-2} \cdot \Lambda_o \cdot \hat{\delta} = \Lambda_r \cdot \Lambda_b^{W-2} \cdot \mathbf{T}^{\beta_{b-1}} \cdot \hat{\delta} = \Lambda_r \cdot \mathbf{T}^{(W-2)N_I+\beta_{b-1}} \cdot \hat{\delta} = \mathbf{T}^{(W-1)N_I+\beta_{r-1}} \cdot \hat{\delta}$ or, equivalently, $\hat{\delta} = \mathbf{T}^{-(W-1)N_I-\beta_{r-1}} \cdot \Lambda \cdot \hat{\delta}$. This implies that $\hat{\delta} = 0$ since Λ is a zero matrix, which contradicts the assumption $\hat{\delta} \neq 0$. ∎

Once the period of the igniter sequence and sampling times are chosen such that the discrimination matrix is nonsingular, we can determine the correction times, β_j, and the correction vector, $\mathbf{c}_j$, $j = 0, 1, \cdots, b-1$, as follows:

THEOREM 8.5 (BDSA CORRECTION TIME & VECTOR CONDITION) *For a non-singular state transition matrix* $\mathbf{T}$, *let the igniter sequence period* N_I *and the sampling times,* α_j, $j = 0, 1, \cdots, b-1$, *be chosen such that the discrimination matrix* Δ *in Eq.(8.33) is nonsingular. Then, the correction matrix* Λ *in Eq.(8.32a) becomes a zero matrix if and only if the correction vectors,* $\mathbf{c}_j$, $j = 0, 1, \cdots, b-1$, *take, for an arbitrary set of correction times* β_j's *that satisfy* $\alpha_j < \beta_j \le \alpha_{j+1}$ $(0 \le j \le b-2)$ *and* $\alpha_{b-1} < \beta_{b-1} \le N_I + \alpha_0$, *the expressions*

$$\mathbf{c}_j = $$
$$\begin{cases} \mathbf{T}^{(W-1)N_I+\beta_j} \cdot \Delta^{-1} \cdot (\mathbf{e}_{(W-1)b+j} + \sum_{k=(W-1)b+j+1}^{(W-1)b+(r-1)} u_{j,k}\mathbf{e}_k), \\ \qquad j = 0, 1, \cdots, r-2, \\ \mathbf{T}^{(W-1)N_I+\beta_{r-1}} \cdot \Delta^{-1} \cdot \mathbf{e}_{(W-1)b+(r-1)}, \\ \qquad j = r-1, \\ \mathbf{T}^{(W-2)N_I+\beta_j} \cdot \Delta^{-1} \cdot (\mathbf{e}_{(W-2)b+j} + \sum_{k=(W-2)b+j+1}^{(W-1)b+(r-1)} u_{j,k}\mathbf{e}_k), \\ \qquad j = r, r+1, \cdots, b-1, \end{cases}$$

$$(8.34)$$

where the L-vector $\mathbf{e}_i$, $i = 0, 1, \cdots, L-1$, *denotes the ith standard basis vector whose ith element is 1 and the others are 0, and* $u_{j,k}$ *is a binary number that can be arbitrarily set to either 0 or 1 for* $j = 0, 1, \cdots, b-1$ *and* $k = 0, 1, \cdots, L-1$.
[26]

Proof : First, we prove the case $b = 1$. In this case, we have $r = 1$, $W = L$, so the definitions in Eq.(8.32b–d) yield the relations $\Lambda_r = \Lambda_b = \mathbf{T}^{N_I+\beta_0-\beta_{b-1}} + \mathbf{c}_0 \cdot \mathbf{h}^t \cdot \mathbf{T}^{N_I+\alpha_0-\beta_{b-1}} = \mathbf{T}^{N_I} + \mathbf{c}_0 \cdot \mathbf{h}^t \cdot \mathbf{T}^{N_I-D_c}$, $\Lambda_o = \mathbf{T}^{\beta_0}+\mathbf{c}_0\cdot\mathbf{h}^t\cdot\mathbf{T}^{\alpha_0} = (\mathbf{T}^{D_c}+\mathbf{c}_0\cdot\mathbf{h}^t)\cdot\mathbf{T}^{\alpha_0}$ for the correction delay $D_c, 0 < D_c \equiv \beta_0 - \alpha_0 \le N_I$. Therefore, the correction matrix Λ in Eq.(8.32a) reduces to $\Lambda = \Lambda_{DSA}\cdot\mathbf{T}^{\alpha_0}$ for $\Lambda_{DSA}=(\mathbf{T}^{N_I}+\mathbf{c}_0\cdot\mathbf{h}^t\cdot\mathbf{T}^{N_I-D_c})^{L-1}\cdot(\mathbf{T}^{D_c}+\mathbf{c}_0\cdot\mathbf{h}^t)$, which is the correction matrix of the original DSA system (see Eq.(8.9)). Furthermore, the discrimination matrix Δ of Eq.(8.33) also becomes $\Delta_{DSA} \cdot \mathbf{T}^{\alpha_0}$ for $\Delta_{DSA} = [\mathbf{h} \ (\mathbf{T}^{N_I})^t\cdot\mathbf{h} \ (\mathbf{T}^{2N_I})^t\cdot\mathbf{h} \ \cdots \ (\mathbf{T}^{(L-1)N_I})^t \cdot \mathbf{h}]^t$, the discrimination matrix of the DSA system (see Eq.(8.10)). Since Δ and $\mathbf{T}$ are nonsingular matrices, Δ_{DSA} is also nonsingular, and Λ becomes zero if and only if Λ_{DSA} becomes zero. By Theorem 8.2, this is equivalent to that the correction vector of the system is $\hat{\mathbf{c}} = \mathbf{T}^{(L-1)N_I+D_c} \cdot \Delta_{DSA}^{-1} \cdot \mathbf{e}_{L-1}$. Therefore, it suffices to show that the correction vector $\mathbf{c}_0$ defined in Eq.(8.34) corresponds to this $\hat{\mathbf{c}}$. But from the relation $L - 1 = b(W - 1) + (r - 1), 1 \le r \le b$, and Eq.(8.34), we have the relation $\mathbf{c}_0 = \mathbf{T}^{(L-1)N_I+\beta_0} \cdot \Delta^{-1} \cdot \mathbf{e}_{L-1}$. So if we insert the relations $D_c = \beta_0 - \alpha_0$ and $\Delta_{DSA} = \Delta\cdot\mathbf{T}^{-\alpha_0}$ to this, we finally get the desired relation $\mathbf{c}_0 = \mathbf{T}^{(L-1)N_I+D_c} \cdot (\Delta \cdot \mathbf{T}^{-\alpha_0})^{-1}\cdot\mathbf{e}_{L-1} = \mathbf{T}^{(L-1)N_I+D_c} \cdot \Delta_{DSA}^{-1}\cdot\mathbf{e}_{L-1} = \hat{\mathbf{c}}$.

Then, we prove the case $b \ge 2$. We first prove the "if" part of the theorem. For this we show by induction that $\Lambda \cdot \Delta^{-1} \cdot \mathbf{e}_k = 0$, $k = 0, 1, \cdots, L-1$ for the $\mathbf{c}_j$'s in Eq.(8.34). The induction process is given in Appendix A.1.

---

[26]Note that we use indices 0 through $L-1$, instead of 1 through L, to indicate the entry positions in L-vectors or $L \times L$ matrices.

Next we prove the "only if" part of the theorem. Since $\mathbf{T}$ and Δ are both nonsingular, a vector $\mathbf{c}_j$ can be uniquely expressed as

$$\mathbf{c}_j = \mathbf{T}^{(W-1)N_I + \beta_j} \cdot \Delta^{-1} \cdot \sum_{k=0}^{(W-1)b+r-1} \mu_{j,k} \mathbf{e}_k,$$
$$j = 0, 1, \cdots, (r-1), \tag{8.35a}$$

$$\mathbf{c}_j = \mathbf{T}^{(W-2)N_I + \beta_j} \cdot \Delta^{-1} \cdot \sum_{k=0}^{(W-1)b+r-1} \nu_{j,k} \mathbf{e}_k,$$
$$j = r, r+1, \cdots, (b-1), \tag{8.35b}$$

for the basis vector $\mathbf{e}_k$ and the binary numbers $\mu_{j,k}$ and $\nu_{j,k}$. Note that Eq.(8.35b) is needed only when $r \le b-1$. To prove the theorem, it suffices to show, under the assumption that Λ is a zero matrix, that $\mu_{j,(W-1)b+j} = 1$, $\mu_{j,k} = 0$ for $k = 0, 1, \cdots, (W-1)b+j-1$, and, additionally, in case $r \le b-1$, $\nu_{j,(W-2)b+j} = 1$, and $\nu_{j,k} = 0$ for $k = 0, 1, \cdots, (W-2)b+j-1$. The detailed proof is given in Appendix A.2.   ∎

Note that the theorem does not impose any restrictions on the choice of the correction time, which implies that the correction time may be arbitrarily chosen as long as it appears before the next sampling time.

COROLLARY 8.2 *If b divides N_I, and if Δ is nonsingular for the choice of uniform sampling time $\alpha_j = jN_I/b$, $j = 0, 1, \cdots, b-1$, then for a constant correction delay $D_c (0 < D_c \le N_I/b)$ the b correction vectors can be chosen to be identical, and*

$$\mathbf{c}_0 = \mathbf{c}_1 = \cdots = \mathbf{c}_{b-1} = \mathbf{T}^{(L-1)N_I/b+D_c} \cdot \Delta^{-1} \cdot \mathbf{e}_{L-1} \overset{\triangle}{=} \mathbf{c}_{DSA}. \tag{8.36}$$

This uniform sampling case corresponds to the DSA in Section 1 with a single correction made in the contracted interval N_I/b. We can easily show, by applying the relations $L - 1 = b(W-1) + (r-1)$ $(1 \le r \le b)$ and $\alpha_j = jN_I/b$, $\beta_j = \alpha_j + D_c$, $j = 0, 1, \cdots, b-1$ for a correction delay D_c in $0 < D_c \le N_I/b$, that the above single correction vector $\mathbf{c}_{DSA}$ is identical to $\mathbf{c}_{r-1}$ in Eq.(8.34).

For example, we consider the case of employing the *extended m-sequence* as the igniter sequence. As well known [132], each extended m-sequence has the period of $N_I = 2^s$, for an integer s, which is always relatively prime to those of the PN-sequences (i.e., m-sequences). So if we take $b = 2^k$, for $k \le s$, then we can realize a single correction vector-based BDSA system. Note, in this case, that the sampling interval $N_I/b = 2^{s-k}$ is relatively prime to the period of the PN-sequence, and, according to Corollary 8.1, this relative-prime relation guarantees the non-singularity of Δ.

Finally, the time-advanced sampling can be accomplished by employing a new sampling vector as specified in the following theorem:

THEOREM 8.6 (BDSA TIME-ADVANCED SAMPLING VECTOR)   *Let* $(R + i)N_I + \alpha_j$ *be the sampling time when a sample* $z_{ij}(i = 0, 1, \cdots, W - 1,$ $j = 0, 1, \cdots, b - 1)$ *is taken from an SRG structured by the state transition matrix* $\mathbf{T}$ *and the generating vector* $\mathbf{h}$. *Then, the sample* z_{ij} *is identical to the sample taken at time* $(R + i - 2)N_I + \alpha_j$ *using the sampling vector*

$$\mathbf{v} = (\mathbf{T}^{2N_I})^t \cdot \mathbf{h}. \tag{8.37}$$

Proof : The theorem can be proved in a similar way to that of Theorem 8.3.   ∎

Based on the three theorems we have discussed so far, we can design the BDSA synchronization parameters in the following procedure: *Given an SRG structured by the state transition matrix* $\mathbf{T}$ *and the generating vector* $\mathbf{h}$, *we first take the igniter sequence period* N_I *and the sampling times* α_j, $j = 0, 1, \cdots, b-1$, *such that the discrimination matrix of Eq.(8.33) becomes nonsingular. Then, we take arbitrary correction times* β_j, $j = 0, 1, \cdots, b - 1$, *that satisfy* $\alpha_j < \beta_j \leq \alpha_{j+1}$ *for* $0 \leq j \leq b-2$ *and* $\alpha_{b-1} < \beta_{b-1} \leq N_I + \alpha_0$. [27] *Finally, we take the correction vectors* $\mathbf{c}_j$, $j = 0, 1, \cdots, b-1$, *and the time-advanced sampling vector* $\mathbf{v}$ *as specified in Eq.(8.34) and Eq.(8.37), respectively.*

3.3   BDSA SPREADER DESIGN EXAMPLE

We assume that the main SRG sequence is an m-sequence whose characteristic polynomial is $\Psi(x) = x^{15} + x^{13} + x^9 + x^8 + x^7 + x^5 + 1$, with the transition matrix $\mathbf{T}$ and the sequence generating vector $\mathbf{h}$ of the main SRG given by [28]

$$\mathbf{T} = \begin{bmatrix} \mathbf{0} & \mathbf{I}_{14 \times 14} \\ 1 & \mathbf{t} \end{bmatrix}, \quad \mathbf{t} = [0\,0\,0\,0\,1\,0\,1\,1\,1\,0\,0\,0\,1\,0], \tag{8.38a}$$

$$\mathbf{h} = [1\,0\,0\,0\,0\,0\,0\,0\,0\,0\,0\,0\,0\,0\,0]^t. \tag{8.38b}$$

---

[27]While any arbitrary value may be chosen for β_j, it is desirable to take $\beta_j = \alpha_j + 1, j = 0, 1, \cdots, b-1$, as it enables us to employ the same timing pulse for the sampling and correction process, and enables to achieve the fastest possible synchronization. This type of correction is called *immediate correction* [69, 128].

[28]Note that this SRG is the same one as was introduced in Section 1. This corresponds to the SRG for generating the in-phase pilot PN sequence of IS-95 [10].

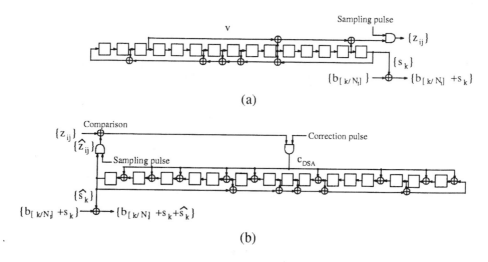

Figure 8.18. BDSA spreader (a) and despreader (b) circuits designed in Example 8.3.

We take an extended m-sequence of period $128(=2^7)$ whose start is marked up, for example, by the symbol "1" in the 8-bit string "00000001", as the igniter sequence.

EXAMPLE 8.3 We consider the case when *two* bits are conveyed per symbol (i.e., the 4-ary signaling system). Since $b = 2$ divides the igniter sequence period $N_I = 128$, we try to realize the BDSA system based on a single correction vector. If we insert $\alpha_0 = 0$, $\alpha_1 = N_I/b = 64$ and $N_I = 128$, along with those **T** and **h** in Eq.(8.38), into Eq.(8.33), the resulting sampling matrix Δ becomes nonsingular. Hence, we can employ a single correction vector $\mathbf{c}_{DSA}$ in Eq.(8.36) for the synchronization. In addition, we take $\beta_j = \alpha_j + 1, j = 0, 1$, for immediate correction [69], and insert $N_I = 128$, $L = 15$, $b = 2$ and $D_c = 1$ to Eq.(8.36). Then we obtain the single correction vector

$$\mathbf{c}_{DSA} = [1\,1\,1\,0\,1\,1\,0\,0\,1\,0\,0\,1\,0\,1\,1]^t. \qquad (8.39)$$

On the other hand, by Eq.(8.37), we get the time-advanced sampling vector

$$\mathbf{v} = [0\,1\,0\,0\,0\,1\,0\,0\,0\,0\,0\,0\,1\,0\,0]^t. \qquad (8.40)$$

Fig. 8.18 (a) and (b) depict the resulting BDSA spreader and despreader circuits that incorporate the designed sampling and correction functions.

EXAMPLE 8.4 *(Example 8.3 continued)* For the same SRG, we consider another signaling system in which *three* bits are conveyed per symbol (i.e., the 8-ary

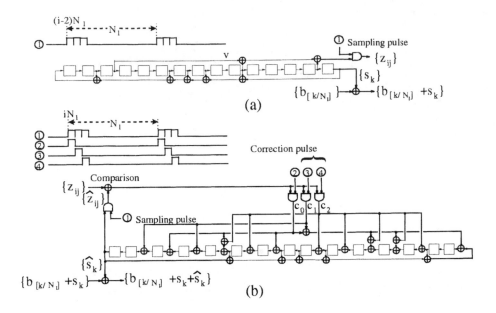

Figure 8.19.   BDSA spreader(a) and despreader(b) circuits designed in Example 8.4.

signaling system). Apparently this system cannot be realized in a single correction vector-based BDSA system, because $b = 3$ does not divide $N_I = 128$. From the relation $L = (W - 1)b + r \, (1 \leq r \leq b)$, we have $W = 5$ and $r = 3$. If we take $\alpha_j = j$ and $\beta_j = \alpha_j + 1$ for $j = 0, 1, 2$, the resulting sampling matrix Δ becomes nonsingular. Then, according to Theorem 8.5, we can obtain a maximum of eight correction vector sets by assigning different binary numbers for $u_{0,13}, u_{0,14}$, and $u_{1,14}$. When we set $u_{0,13} = u_{0,14} = u_{1,14} = 0$, we obtain the set

$$c_0 = [\,0\,0\,1\,0\,1\,0\,1\,0\,0\,0\,1\,1\,0\,0\,1\,]^t, \tag{8.41a}$$

$$c_1 = [\,0\,1\,0\,1\,0\,0\,0\,0\,0\,0\,1\,1\,0\,0\,1\,]^t, \tag{8.41b}$$

$$c_2 = [\,0\,0\,0\,0\,1\,1\,0\,0\,1\,1\,1\,1\,1\,0\,0\,]^t. \tag{8.41c}$$

For the time-advanced sampling vector we take the one in Eq.(8.40). Note, however, that the sampling and correction times differ from those of Example 8.3. Fig. 8.19 depicts the resulting BDSA spreader and despreader circuits, along with the related timing diagrams.

4.   PARALLEL DSA (PDSA)

Now we introduce the PDSA - - another multi-bit based DSA scheme. If compared to the BDSA scheme, the PDSA scheme is the simpler as the *serial-input parallel-output* (SIPO) and *parallel-input serial-output* (PISO) connectors are not needed and the related timing circuitry is much simplified. This simplification costs complexity increase in implementing the parallel and time-advanced parallel sampling circuits, which however is merely a minor logic-level increase.

4.1   PDSA SYSTEM OPERATION

Fig. 8.20 depicts the functional block diagram of the acquisition-related part of the DS/CDMA system that employs the PDSA scheme. The transmitter part consists of a PDSA-spreader and a *parallel sample*(PS)-spreader, and the receiver part contains their despreading counterparts, that is, a PDSA-despreader and a PS-despreader. The PDSA-spreader/despreader pair take the synchronization function while the PS-spreader/despreader pair take the parallel-sample conveyance function.

The time-advanced parallel sampling block in the PDSA-spreader takes the b state samples $z_{ij}, j = 0, 1, \cdots, b-1$, of the main SRG simultaneously at time $(R + i - 1)N_I$, for a reference value R, which are the b main SRG sequence values to be generated at times $(R+i)N_I + \alpha_j, j = 0, 1, \cdots, b-1$, respectively, for $0 \leq \alpha_0 < \alpha_1 < \cdots < \alpha_{b-1} < N_I$. Then the symbol generation block maps those b samples to the corresponding M-ary ($M = 2^b$) orthogonal symbol $\mathbf{x}_i$ (or to b binary symbols in the case of multi-carrier DS/CDMA systems) and spreads it using the igniter SRG.

The PS-despreader despreads the received state signal and matches it with each M-ary orthogonal symbol (or each binary symbol in the case of multi-carrier systems) to detect the conveyed samples $z_{ij}, j = 0, 1, \cdots, b-1$, passing them to the PDSA-despreader at time $(R+i)N_I$. On the other hand, the PDSA-despreader generates its own b SRG state samples $\hat{z}_{ij}$'s simultaneously at time $(R + i)N_I$, with the jth sample $\hat{z}_{ij}$ corresponding to the main SRG sequence value $\hat{s}_{(R+i)N_I+\alpha_j}$, and then compares them with the conveyed counterpart z_{ij}'s. For each $j=0, 1, \cdots, b-1$, if $\hat{z}_{ij}$ coincides with z_{ij}, no action takes place, but otherwise the jth correction circuit is triggered to correct the main SRG state at time $(R + i)N_I + D_c$, with D_c chosen such that $0 < D_c \leq N_I$.

The state sampling and correction processes are timed by the sampling and correction pulses issued by the igniter SRG. The related timing diagrams are depicted in Fig. 8.21. The acquisition and verification procedure of the PDSA is similar to that of the DSA of Section 1, except that the b state samples are manipulated simultaneously.

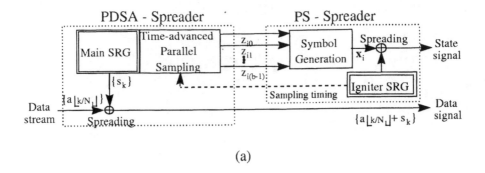

(a)

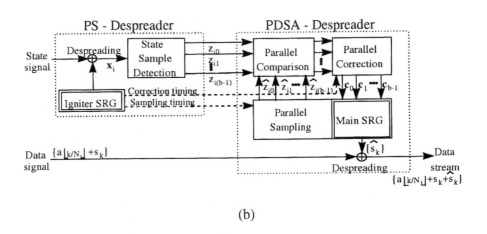

(b)

Figure 8.20.   Functional block diagram of the acquisition-related part of the DS/CDMA system employing the PDSA scheme : (a) Transmitter, (b) receiver. (Interference and thermal noise are not exhibited in the received signals.)

4.2   PDSA SYNCHRONIZATION PARAMETER DESIGN

For a given PDSA system, let $\mathbf{d}_k$ and $\hat{\mathbf{d}}_k$ denote the *state vectors* (L-vectors) of the spreader and despreader SRGs at time k, respectively, $\mathbf{T}$ the *state transition matrix* (an $L \times L$ matrix), $\mathbf{h}$ the *generating vector* (an L-vector) that generates the sequence value s_k out of the state vector $\mathbf{d}_k$ through the relation $s_k = \mathbf{h}^t \cdot \mathbf{d}_k$, $\hat{s}_k = \mathbf{h}^t \cdot \hat{\mathbf{d}}_k$, and $\mathbf{c}_j$, $j = 0, 1, \cdots, b-1$, the *jth correction vector* that corrects the old state vector $\mathbf{d}_{old}$ to a new state vector $\mathbf{d}_{new}$ through the relation $\hat{\mathbf{d}}_{new} = \hat{\mathbf{d}}_{old} + (z_{i0} + \hat{z}_{i0})\mathbf{c}_0 + \cdots + (z_{i(b-1)} + \hat{z}_{i(b-1)})\mathbf{c}_{b-1}$.

In compliance with the PDSA processing, we rearrange the L state samples in $W(=\lceil L/b \rceil)$ bundles of b samples, i.e., $L-1 = b(W-1) + (r-1)$, $1 \le r \le b$, $W \ge 2$. We take samples z_{ij} and $\hat{z}_{ij}$ at *virtually the same* sampling time $(R+i)N_I$, and make corrections at the correction time $(R+i)N_I + D_c$. Note

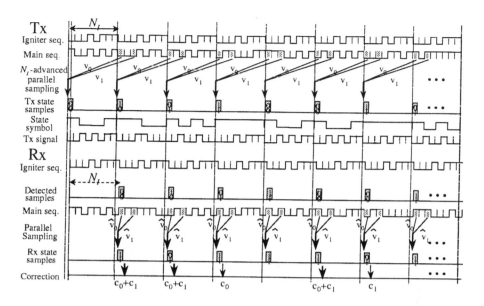

Figure 8.21. System timing diagram (illustrating the case $b{=}2$, $N_I{=}8$). ($\mathbf{v}_0$, $\mathbf{v}_1$, $\hat{\mathbf{v}}_0$, $\hat{\mathbf{v}}_1$ are sampling vectors and $\mathbf{c}_0$, $\mathbf{c}_1$ are correction vectors.)

that the samples z_{ij} and $\hat{z}_{ij}$ are the sequence values that are supposed to be generated at $(R+i)N_I + \alpha_j, 0 \le \alpha_0 < \alpha_1 < \cdots < \alpha_{b-1} < N_I$, respectively, from the transmitter main SRG and receiver main SRG. So, the samples z_{ij} and $\hat{z}_{ij}$ are related to the state vectors by the relations $z_{ij} = s_{(R+i)N_I+\alpha_j} = \mathbf{h}^t \cdot \mathbf{d}_{(R+i)N_I+\alpha_j}, \hat{z}_{ij} = \hat{s}_{(R+i)N_I+\alpha_j} = \mathbf{h}^t \cdot \hat{\mathbf{d}}_{(R+i)N_I+\alpha_j}, i = 0, 1, \cdots, W-1,$ $j = 0, 1, \cdots, b-1,$ and the state vectors at the correction are $\mathbf{d}_{(R+i)N_I+D_c} = \mathbf{T} \cdot \mathbf{d}_{(R+i)N_I+D_c-1}, \hat{\mathbf{d}}_{(R+i)N_I+D_c} = \mathbf{T} \cdot \hat{\mathbf{d}}_{(R+i)N_I+D_c-1} + \sum_{j=0}^{b-1}(z_{ij} + \hat{z}_{ij})\mathbf{c}_j,$ $i = 0, 1, \cdots, W-1.$

Then we can easily derive that the *state distance vector* $\delta_k \stackrel{\triangle}{=} \mathbf{d}_k + \hat{\mathbf{d}}_k$ has the relation $\delta_{(R+W-1)N_I+D_c} = \tilde{\Lambda} \cdot \delta_{RN_I}$ for the $L \times L$ *correction matrix*

$$\tilde{\Lambda} \stackrel{\triangle}{=} (\mathbf{T}^{N_I} + \sum_{j=0}^{b-1} \mathbf{c}_j \cdot \mathbf{h}^t \cdot \mathbf{T}^{\alpha_j+N_I-D_c})^{W-1} \times (\mathbf{T}^{D_c} + \sum_{j=0}^{b-1} \mathbf{c}_j \cdot \mathbf{h}^t \cdot \mathbf{T}^{\alpha_j}). \quad (8.42)$$

In order to achieve the synchronization by applying W parallel corrections, it is necessary to make $\tilde{\Lambda}$ a zero matrix. Therefore, for a given PDSA spreader/despreader pair, the synchronization problem turns to a problem of determining appropriate values of the parameters N_I, α_j, D_c, and $\mathbf{c}_j$ $(j = 0, 1, \cdots, b-1)$ that make the correction matrix null.

For a PDSA system characterized by b, $\mathbf{T}$, $\mathbf{h}$, we define the *discrimination matrix* $\tilde{\Delta}$ to be the $bW \times L$ matrix

$$\tilde{\Delta} \triangleq [(\mathbf{T}^{\alpha_0})^t \cdot \mathbf{h} \cdots (\mathbf{T}^{\alpha_{b-1}})^t \cdot \mathbf{h} \ (\mathbf{T}^{N_I+\alpha_0})^t \cdot \mathbf{h} \cdots (\mathbf{T}^{N_I+\alpha_{b-1}})^t \cdot \mathbf{h} \cdots$$
$$(\mathbf{T}^{(W-1)N_I+\alpha_0})^t \cdot \mathbf{h} \cdots (\mathbf{T}^{(W-1)N_I+\alpha_{b-1}})^t \cdot \mathbf{h}]^t. \tag{8.43}$$

Then we can determine the igniter sequence period N_I and sampling times, α_0, $\alpha_1, \cdots, \alpha_{b-1}$, based on the following theorem.

THEOREM 8.7 (PARALLEL SAMPLING CONDITION) *For a nonsingular state transition matrix* $\mathbf{T}$, *the correction matrix* $\tilde{\Lambda}$ *in Eq.(8.42) can be zero only if the igniter sequence period* N_I *and the sampling times* α_j, $j = 0, 1, \cdots, b-1$, *are chosen such that the rank of the discrimination matrix* $\tilde{\Delta}$ *in Eq.(8.43) is L.*

Proof: We prove that for α_j, $j = 0, 1, \cdots, b-1$, and N_I chosen such that the rank of $\tilde{\Delta}$ in Eq.(8.43) is less than L, the correction matrix $\tilde{\Lambda}$ cannot be zero. To prove this by contradiction, we suppose $\tilde{\Lambda}$ becomes zero for some choice of D_c and $\mathbf{c}_j$, $j = 0, 1, \cdots, b-1$, and choose a nonzero vector $\hat{\delta}$ such that $\tilde{\Delta} \cdot \hat{\delta} = 0$. Then by Eq.(8.43) we have $\mathbf{h}^t \cdot \mathbf{T}^{iN_I+\alpha_j} \cdot \hat{\delta} = 0$ for $i = 0, 1, \cdots, W-1$, $j = 0, 1, \cdots, b-1$. Applying this relation repeatedly to $\tilde{\Lambda} \cdot \hat{\delta}$ for the correction matrix $\tilde{\Lambda}$ in Eq.(8.42), we obtain the equality $\tilde{\Lambda} \cdot \hat{\delta} = \mathbf{T}^{(W-1)N_I+D_c} \cdot \hat{\delta}$ or, equivalently, $\hat{\delta} = \mathbf{T}^{-(W-1)N_I-D_c} \cdot \tilde{\Lambda} \cdot \hat{\delta}$. This implies that $\hat{\delta} = 0$ since $\tilde{\Lambda}$ is a zero matrix, which contradicts the assumption $\hat{\delta} \neq 0$. ∎

According to the theorem, we are supposed to choose the igniter sequence period and the sampling times such that the rank of the relevant discrimination matrix becomes L. This is an expected result because L independent samples are needed to achieve the synchronization and because the sample vector $\mathbf{z} = [z_{00}, \cdots, z_{0(b-1)}, \cdots z_{(W-1)0}, \cdots, z_{(W-1)(b-1)}]^t$ is related to the discrimination matrix $\tilde{\Delta}$ by the equation $\mathbf{z} = \tilde{\Delta} \cdot \mathbf{d}_{RN_I}$. Consequently, the PDSA system becomes synchronizable only when the discrimination matrix includes L independent row vectors.

There are $\binom{bW}{L}$ possible combinations in selecting the L independent samples out of bW received ones and as many possibly different synchronization schemes can exist. Among them we select the igniter sequence period N_I and the sampling times $\alpha_0, \alpha_1, \cdots \alpha_{b-1}$ that make the *PDSA matrix* $\bar{\Delta}$, the $L \times L$ submatrix of $\tilde{\Delta}$ composed of the last L rows, i.e.,

$$\bar{\Delta} \triangleq [(\mathbf{T}^{\alpha_{b-r}})^t \cdot \mathbf{h} \cdots (\mathbf{T}^{\alpha_{b-1}})^t \cdot \mathbf{h} \ (\mathbf{T}^{N_I+\alpha_0})^t \cdot \mathbf{h} \cdots (\mathbf{T}^{N_I+\alpha_{b-1}})^t \cdot \mathbf{h} \cdots$$
$$(\mathbf{T}^{(W-1)N_I+\alpha_0})^t \cdot \mathbf{h} \cdots (\mathbf{T}^{(W-1)N_I+\alpha_{b-1}})^t \cdot \mathbf{h}]^t \tag{8.44}$$

nonsingular. Then we can determine the correction delay D_c, and the correction vectors $\mathbf{c}_j$, $j = 0, 1, \cdots, b-1$, as follows:

THEOREM 8.8 (PDSA CORRECTION DELAY & CORRECTION VECTOR)
For a nonsingular state transition matrix $\mathbf{T}$, *let the igniter sequence period* N_I
and the sampling times α_j, $j = 0, 1, \cdots, b - 1$, *be chosen such that the PDSA*
matrix $\bar{\Delta}$ *is nonsingular. Then the correction matrix* $\tilde{\Lambda}$ *in Eq.(8.42) becomes a*
zero matrix if the correction vectors $\mathbf{c}_j$, $j = 0, 1, \cdots, b - 1$, *take the expressions*

$$
\mathbf{c}_j = \begin{cases} \mathbf{T}^{(W-1)N_I + D_c} \cdot \bar{\Delta}^{-1} \cdot \mathbf{e}_{(W-1)b+j}, \\ \qquad \text{if } r = b, \\ \mathbf{T}^{(W-1)N_I + D_c} \cdot \bar{\Delta}^{-1} \cdot \{ \mathbf{e}_{(W-2)b+r+j} + \mathbf{y}_j \}, \\ \qquad \text{if } 0 < r < b, \end{cases} \tag{8.45a}
$$

$$
\mathbf{y}_j \overset{\triangle}{=} \sum_{k=0}^{\min(r-1, b-r-1)} u_k (\mathbf{h}^t \cdot \mathbf{T}^{\alpha_k} \cdot \bar{\Delta}^{-1} \cdot \mathbf{e}_{(W-2)b+r+j}) \mathbf{e}_k,
$$

$$
j = 0, 1, \cdots, b - 1, \tag{8.45b}
$$

for an arbitrary correction delay D_c *in the interval* $0 < D_c \le N_I$, *where the*
L-vector $\mathbf{e}_i$, $i = 0, 1, \cdots, L - 1$, *denotes the ith standard basis vector whose*
ith element is 1 and the others are 0, and u_k *is a binary number that can be*
arbitrarily set to either 0 or 1 for $k = 0, 1, \cdots, r - 1$.

Proof : We can rewrite the correction matrix of Eq.(8.42) as $\tilde{\Lambda} = \Lambda_*^W \cdot$
$\mathbf{T}^{D_c - N_I}$ for $\Lambda_* \overset{\triangle}{=} \mathbf{T}^{N_I} + \sum_{j=0}^{b-1} \mathbf{c}_j \cdot \mathbf{h}^t \cdot \mathbf{T}^{\alpha_j + N_I - D_c}$. In addition, we expand
the equation $\bar{\Delta} \cdot \bar{\Delta}^{-1} = \mathbf{I}$ to get the relations

$$
\mathbf{h}^t \cdot \mathbf{T}^{\alpha_j + lN_I} \cdot \bar{\Delta}^{-1} = \mathbf{e}_{b(l-1)+r+j}^t,
$$

$$
\begin{cases} l = 0; \quad j = b - r, \cdots, b - 1, \\ l = 1, 2, \cdots, W - 1; \quad j = 0, 1, \cdots, b - 1. \end{cases} \tag{8.46}
$$

We first prove the theorem for the case $r = b$. For the $\mathbf{c}_j$ in the upper line of
Eq.(8.45a), Λ_* takes the expression $\Lambda_* = \mathbf{T}^{(W-1)N_I + D_c} \cdot \mathbf{A}_0 \cdot \mathbf{T}^{-(W-1)N_I - D_c}$,
where $\mathbf{A}_0 \overset{\triangle}{=} \bar{\Delta} \cdot \mathbf{T}^{N_I} \cdot \bar{\Delta}^{-1} + \sum_{j=0}^{b-1} \mathbf{e}_{(W-1)b+j} \cdot \mathbf{h}^t \cdot \mathbf{T}^{\alpha_j + WN_I} \cdot \bar{\Delta}^{-1}$. Applying
the relations in Eq.(8.46) we get $\mathbf{A}_0 = [\, \mathbf{0}_{Wb \times b} \ \mathbf{e}_0 \ \mathbf{e}_1 \ \cdots \ \mathbf{e}_{(W-1)b-1} \,]$, which
is a nilpotent matrix of nilpotency W. Therefore, $\mathbf{A}_0^W = \mathbf{0}$ and $\Lambda_*^W = \mathbf{0}$, and
hence $\tilde{\Lambda} = \mathbf{0}$.

Next, we consider the case $0 < r < b$. Due to the lower line of Eq.(8.45a),
Λ_* takes the expression $\Lambda_* = \mathbf{T}^{(W-1)N_I + D_c} \cdot (\mathbf{A}_1 + \mathbf{B}_1) \cdot \mathbf{T}^{-(W-1)N_I - D_c}$,
for $\mathbf{A}_1 \overset{\triangle}{=} \bar{\Delta} \cdot \mathbf{T}^{N_I} \cdot \bar{\Delta}^{-1} + \sum_{j=0}^{b-1} \mathbf{e}_{(W-2)b+r+j} \cdot \mathbf{h}^t \cdot \mathbf{T}^{\alpha_j + WN_I} \cdot \bar{\Delta}^{-1}$ and $\mathbf{B}_1 \overset{\triangle}{=}$
$\sum_{j=0}^{b-1} \sum_{k=0}^{\min(r-1, b-r-1)} u_k \mathbf{e}_k \cdot (\mathbf{h}^t \cdot \mathbf{T}^{\alpha_k} \cdot \bar{\Delta}^{-1} \cdot \mathbf{e}_{(W-2)b+r+j}) \mathbf{h}^t \cdot \mathbf{T}^{\alpha_j + WN_I} \cdot \bar{\Delta}^{-1}$.
Applying the relations in Eq.(8.46), we get $\mathbf{A}_1 = [\, \mathbf{0}_{\{(W-1)b+r\} \times b} \ \mathbf{e}_0 \ \mathbf{e}_1 \ \cdots$

$\mathbf{e}_{(W-2)b+r-1}$]. On the other hand, by Lemma A.5 in Appendix A, we find that the (k, n) element of $\mathbf{B}_1$ is zero for $k = 0, 1, \cdots, \min(r - 1, b - r - 1)$, $n = 0, 1, \cdots, r-1$, and for $k = \min(r-1, b-r-1)+1, \cdots, (W-1)b+r-1$, $n = 0, 1, \cdots, (W - 1)b + r - 1$.

Then, noting that $\min(r, b - r) \le r \le b$, we can easily get the relations $\mathbf{B}_1^2 = \mathbf{0}$ and $\mathbf{A}_1 \cdot \mathbf{B}_1 = \mathbf{0}$, which together yield the relation

$$(\mathbf{A}_1 + \mathbf{B}_1)^W = \mathbf{A}_1^W + \mathbf{B}_1 \cdot \mathbf{A}_1^{W-1}. \tag{8.47}$$

Furthermore, by raising $\mathbf{A}_1$ to the $(W - 1)$th and the Wth powers, we get the relations $\mathbf{A}_1^{W-1} = \left[\begin{array}{c|c} \mathbf{0}_{r \times (W-1)b} & \mathbf{I}_{r \times r} \\ \hline \mathbf{0}_{(W-1)b \times (W-1)b} & \mathbf{0}_{(W-1)b \times r} \end{array} \right]$ and $\mathbf{A}_1^W = \mathbf{0}$.

Therefore, by Eq.(8.47), we obtain $(\mathbf{A}_1 + \mathbf{B}_1)^W = \mathbf{0}$, which means that $\Lambda_*^W = \mathbf{0}$. Therefore $\tilde{\Lambda} = \mathbf{0}$, and the proof is complete. ∎

We finally consider how to realize the parallel sampling and time-advanced parallel sampling, which can be done by employing new sampling vectors as specified in the following theorem:

THEOREM 8.9 (PARALLEL SAMPLING VECTOR) : *The sample z_{ij} generated at time $(R + i)N_I + \alpha_j$ is identical to the time-advanced sample taken at time $(R + i - 1)N_I$ using the sampling vector $\mathbf{v}_j = (\mathbf{T}^{N_I + \alpha_j})^t \cdot \mathbf{h}$, or to the sample taken at time $(R + i)N_I$ using the sampling vector $\hat{\mathbf{v}}_j = (\mathbf{T}^{\alpha_j})^t \cdot \mathbf{h}$, for $j = 0, 1, \cdots, b - 1$.*

Proof: The theorem can be proved in a similar way to that of Theorem 8.3. ∎

Based on the theorems we have discussed so far, we can design the PDSA synchronization parameters in the following procedure: *Given an SRG structured by the state transition matrix $\mathbf{T}$ and the generating vector $\mathbf{h}$, we first take the igniter sequence period N_I and the sampling times α_j, $j = 0, 1, \cdots, b-1$, such that the PDSA matrix $\tilde{\Delta}$ becomes nonsingular. Then, after taking an arbitrary correction delay D_c that satisfies $0 < D_c \le N_I$, we determine the correction vector $\mathbf{c}_j$, the time-advanced parallel sampling vector $\mathbf{v}_j$, and the parallel sampling vector $\hat{\mathbf{v}}_j$, for $j = 0, 1, \cdots, b - 1$, as specified in Theorem 8.8 and Theorem 8.9, respectively.*

4.3   PDSA SPREADER DESIGN EXAMPLE

EXAMPLE 8.5 For the same SRG as in Example 8.3, we consider the PDSA system for $b=2$ (i.e., the quaternary signaling PDSA system). If we insert $\alpha_0 = 0$, $\alpha_1 = N_I/b = 64$ and $N_I = 128$, along with those $\mathbf{T}$ and $\mathbf{h}$ in Eq.(8.38) into

Eq.(8.44), the resulting matrix $\bar{\Delta}$ becomes nonsingular. In addition, we take $D_c = 1$ for immediate correction [69], and insert $N_I = 128$, $W = 8$, $b = 2$, $r = 1$ and $D_c = 1$ to Eq.(8.45). Then we obtain the following two correction vector sets by assigning $u_0 = 0$ and $u_0 = 1$, respectively:

$$\begin{cases} \mathbf{c}_0 = [\,1\,1\,1\,0\,1\,1\,0\,0\,1\,0\,0\,1\,0\,1\,1\,]^t, \\ \mathbf{c}_1 = [\,1\,1\,1\,1\,0\,0\,0\,1\,0\,0\,0\,1\,0\,1\,1\,]^t, \end{cases} \tag{8.48a}$$

$$\begin{cases} \mathbf{c}_0 = [\,1\,1\,1\,0\,1\,1\,0\,0\,1\,0\,0\,1\,0\,1\,1\,]^t, \\ \mathbf{c}_1 = [\,1\,1\,1\,0\,1\,0\,0\,1\,0\,0\,1\,1\,0\,1\,0\,]^t. \end{cases} \tag{8.48b}$$

On the other hand, by Theorem 8.9, we get the time-advanced parallel sampling vectors and parallel sampling vectors as follows:

$$\begin{cases} \mathbf{v}_0 = [\,1\,0\,1\,1\,0\,1\,1\,1\,0\,0\,0\,0\,0\,1\,0\,]^t, \\ \mathbf{v}_1 = [\,0\,0\,0\,1\,1\,1\,1\,1\,1\,1\,1\,0\,0\,0\,0\,]^t, \end{cases} \tag{8.49}$$

$$\begin{cases} \hat{\mathbf{v}}_0 = [\,1\,0\,0\,0\,0\,0\,0\,0\,0\,0\,0\,0\,0\,0\,0\,]^t, \\ \hat{\mathbf{v}}_1 = [\,1\,0\,0\,0\,1\,1\,0\,1\,1\,1\,1\,0\,0\,0\,1\,]^t. \end{cases} \tag{8.50}$$

Fig. 8.22 (a) depicts the resulting PDSA spreader circuit that employs the sampling vectors in Eq.(8.49); and Fig. 8.22 (b) depicts its counterpart PDSA despreader circuit employing the correction vectors in Eq.(8.48a) and the sampling vectors in Eq.(8.50).

Example 8.6 (Example 8.5 continued) For the same SRG, we consider another signaling system in which *three* bits are conveyed per symbol (i.e., the 8-ary signaling system). Since $b = 3$ divides $L = 15$, we have only one set of correction vectors which can be designed according to the upper line of Eq.(8.45a). From the relation $L = (W - 1)b + r$ $(1 \leq r \leq b)$, we have $W = 5$ and $r = 3$. If we take $\alpha_j = j$ for $j = 0, 1, 2$ and $D_c = 1$, the resulting 15×15 matrix $\bar{\Delta}$ becomes nonsingular. Then, by inserting $N_I = 128$, $W = 5$, $b = 3$, $r = 3$ and $D_c = 1$ to Eq.(8.45a), we can obtain the correction vectors

$$\begin{cases} \mathbf{c}_0 = [\,0\,0\,1\,0\,1\,0\,1\,0\,0\,0\,1\,1\,0\,0\,1\,]^t, \\ \mathbf{c}_1 = [\,1\,0\,1\,0\,1\,0\,0\,0\,0\,0\,0\,1\,1\,0\,0\,]^t, \\ \mathbf{c}_2 = [\,0\,1\,0\,0\,0\,0\,1\,1\,0\,0\,1\,1\,1\,1\,1\,]^t. \end{cases} \tag{8.51}$$

For the time-advanced parallel sampling vectors and the parallel sampling vectors, we get

$$\begin{cases} \mathbf{v}_0 = [\,1\,0\,1\,1\,0\,1\,1\,1\,0\,0\,0\,0\,0\,1\,0\,]^t, \\ \mathbf{v}_1 = [\,0\,1\,0\,1\,1\,0\,1\,1\,1\,0\,0\,0\,0\,0\,1\,]^t, \\ \mathbf{v}_2 = [\,1\,0\,1\,0\,1\,0\,0\,0\,0\,0\,0\,0\,0\,1\,0\,]^t, \end{cases} \tag{8.52}$$

$$\begin{cases} \hat{\mathbf{v}}_0 = [\,1\,0\,0\,0\,0\,0\,0\,0\,0\,0\,0\,0\,0\,0\,0\,]^t, \\ \hat{\mathbf{v}}_1 = [\,0\,1\,0\,0\,0\,0\,0\,0\,0\,0\,0\,0\,0\,0\,0\,]^t, \\ \hat{\mathbf{v}}_2 = [\,0\,0\,1\,0\,0\,0\,0\,0\,0\,0\,0\,0\,0\,0\,0\,]^t, \end{cases} \tag{8.53}$$

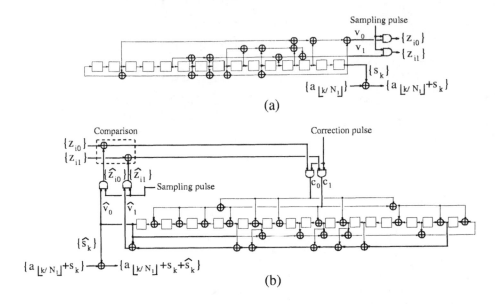

Figure 8.22.   PDSA spreader(a) and despreader(b) circuits designed in Example 8.5.

respectively. Fig. 8.23 depicts the resulting PDSA spreader and despreader circuit set.

5.   DIFFERENTIAL DSA (D²SA)

The original DSA scheme was designed to convey only one state sample in an igniter sequence period, but this restriction was lifted afterwards in the BDSA and the PDSA schemes, where multiple samples are conveyed in the form of *state symbol*. Such symbol based multiple sample conveyances enabled a flexible design of signaling schemes suitable for the pilot power budget in the noisy environment, and also enabled the reuse of the modem equipment readily deployed for data transmission for sample transmission as well (in ad hoc networks), but it accompanied inefficiency in the despreading process. For a mobile station to acquire the orthogonally modulated igniter sequence in the first stage of the synchronization process, 2^b noncoherent correlator branches are to be equipped in the synchronization block, as b state samples are simultaneously conveyed in the form of an orthogonal symbol. In this section we introduce the DPSK signaling scheme as the state sample conveyance means to resolve the above multiple correlator problems. The DPSK based state sample conveyance, however, accompanies a critical problem of phase ambiguity [71] in the channel

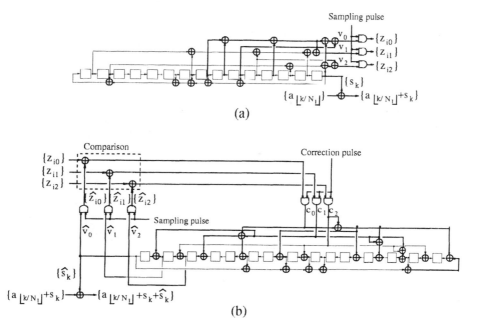

Figure 8.23.   PDSA spreader(a) and despreader(b) circuits designed in Example 8.6.

estimation respect, for which we are going to introduce the concept of the *pre-rotation* of data constellation in this section.

5.1   PRINCIPLES OF THE D²SA

In the DSA family introduced in Sections 1, 3, and 4, $b(\geq 1)$ state samples are simultaneously conveyed in the form of 2^b-ary orthogonal symbols spread by the igniter sequence whose period is equal to the symbol period. The orthogonal modulation of the igniter sequence, however, has the problem of exponentially increasing noncoherent correlators: In order to acquire the igniter sequence modulated by 2^b-ary orthogonal symbols, the DSA receiver (mobile station) must be equipped with 2^b noncoherent correlators and a decision logic that compares the 2^b correlator outputs with a threshold value, as shown in Fig. 8.24. This multiple correlator-based acquisition scheme imposes a big burden on the mobile station in terms of hardware complexity, and also degrades the acquisition time performance, increasing the false acquisition probability. In order to resolve these problems, we introduce the D²SA scheme in the following:

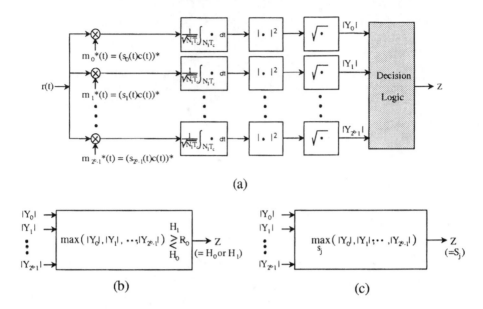

(a)

(b)   (c)

Figure 8.24. Structure of parallel sample despreader of the orthogonal signaling PDSA or BDSA system: (a) Operational structure; (b) decision logic in the igniter sequence acquisition mode; (c) decision logic in the sample detection mode.

5.1.1   D²SA SYSTEM ORGANIZATION AND OPERATION

Fig. 8.25 depicts the functional block diagram of the acquisition-related part of the DS/CDMA system that employs the D^2SA scheme. The basic structure of the D^2SA scheme which consists of PDSA-spreader/despreader and PS-spreader/despreader is similar to those of the existing orthogonal-signaling DSA schemes, except that the PS-spreader/despreader employs DPSK for state sample conveyance and the DPSK state symbols of the pilot channel are provided to all traffic channels (see the dashed lines), which is to pre-rotate the constellation of the data symbols. [29]

Now the main SRG in the BS (or, BS main SRG) generates the *main sequence* $\{s_m\}$ of period $N_M \equiv 2^L - 1$, which is used for data scrambling and whose fast acquisition is our ultimate goal. On the other hand, the igniter SRG in the BS (or, BS igniter SRG) generates the *igniter sequence* $\{c_m\}$ of short period N_I. We assume that $\{s_m\}$ and $\{c_m\}$ are the normalized complex sequences taking the value $(\pm 1 \pm j)/\sqrt{2}$ and that one-to-one correspondence exists between the in-phase (I-) and quadrature (Q-) SRG states [10, 106]. Then, once the in-phase

---

[29]The pre-rotation technique of the data constellation will be investigated in detail in relation to the channel estimation function of the state signal in the next section.

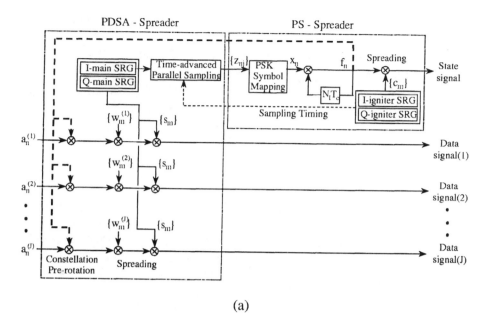

(a)

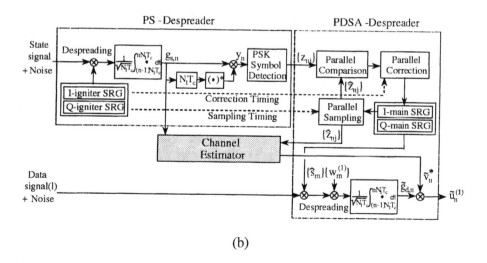

(b)

Figure 8.25.   Functional block diagram of the the D^2SA-based DS/CDMA system: (a) Base station, (b) mobile station.

SRGs are synchronized, the quadrature SRGs can be immediately synchronized according to this one-to-one relation. [30] The time-advanced parallel sampling block takes the b state samples z_{ij}, $j = 0, 1, \cdots, b - 1$, of the I-main SRG at time $(R + i - 1)N_I$, for a reference value R, and provides them to the 2^b-ary DPSK modulator. [31] The DPSK modulator maps the b state samples to the corresponding PSK *state symbol* x_i and produces the DPSK *pilot symbol* f_i by adding the phase of x_i to the phase accumulated up to the previous time slot. The resulting pilot symbol is spread by a period of the igniter sequence and transmitted through the pilot channel in the interval $[(R+i-1)N_I, (R+i)N_I)$. On the other hand, each user's M-ary PSK data $a_i^{(l)}$ is multiplied by the 2^b-ary pilot symbol f_i, spread by one of the orthogonal Walsh sequences $\{w_m^{(l)}\}$ and scrambled by the common main sequence $\{s_m\}$, and then transmitted in the interval $[(R + i - 1)N_I, (R + i)N_I)$. [32] Note that once the main sequence is acquired the Walsh sequence boundary (i.e., the data symbol boundary) can be determined immediately [10]. The state signal (or, pilot signal) and the data signals go through the same multipath fading channel to arrive at the mobile station.

The MS first acquires the DPSK modulated igniter sequence employing the simple noncoherent threshold detector shown in Fig. 8.26 (a) [7, 45]. We assume that the simple serial search method is applied to the igniter sequence acquisition. Then, the noncoherent delay-locked loop shown in Fig. 8.26 (b), whose use is optional, tracks the accurate chip timing [4, 45]. After the timing synchronization of the igniter sequence, the MS despreads the received state signal and differentially detect the conveyed samples z_{ij}, $j = 0, 1, \cdots, b - 1$, which are passed to the PDSA-despreader at time $(R + i)N_I$. On the other hand, the PDSA-despreader generates its own b SRG state samples $\hat{z}_{ij}$'s simultaneously at time $(R+i)N_I$ from the I-main SRG, and compares them with the conveyed counterparts z_{ij}'s. For each $j = 0, 1, \cdots, b - 1$, if $\hat{z}_{ij}$ coincides with z_{ij}, no action takes place, but otherwise the jth correction circuit is triggered to correct the I-main SRG state at time $(R+i)N_I + D_c$, with D_c chosen such that $0 < D_c \leq N_I$. If we employ the sampling and correction circuits designed according to the theorems in Section 4, the MS I-main SRG gets synchronized to

---

[30] We assume complex sequences for their popularity in commercial systems [8, 9, 10]. However, all the discussions in the paper can equally be applied to real sequence-based systems with a trivial modification of the system structure and some relevant approximations related to system performance analysis.

[31] In this section, we describe the system on a discrete time basis, setting the unit time to be the chip duration T_c.

[32] The data symbol period may take any divisor of the state symbol period N_I. We assume that the data symbol period is equal to the state symbol period for notational convenience. The data symbol set size M need not be equal to 2^b.

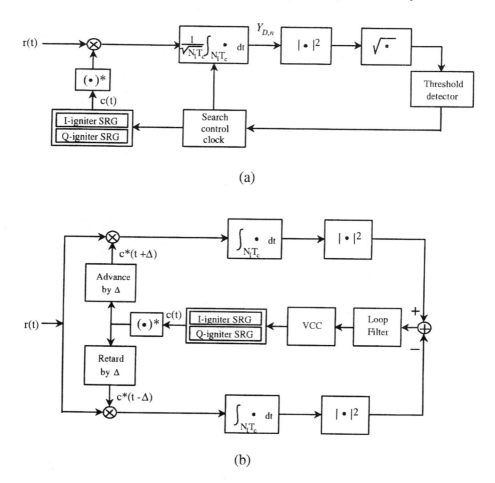

Figure 8.26. Structure of the igniter sequence synchronization block: (a) Acquisition detector, (b) noncoherent DLL for sequence tracking (optional).

the BS I-main SRG after $W \stackrel{\Delta}{=} \lceil L/b \rceil$ comparison-correction operations when no detection error occurs, where L denotes the length of the main SRG. [33]

For safety purpose against detection error, we install a verification process that checks whether or not the conveyed and receiver-generated symbols coincide V more times after the W comparison-correction operations. While verifying the synchronization of the I-main SRGs for the interval $[(R+W)N_I, (R+W+V-1)N_I]$, we also investigate the mapping table to get the corresponding Q-main SRG state at the end of the interval. If all the V symbol sets (or, the

---

[33] In this section, we take the PDSA approach in relation to the main SRG synchronization function. But the BDSA approach is equally applicable.

bV sample sets) coincide, then the MS loads the obtained Q-main SRG states and declares completion of synchronization of the main sequence, beginning to track and estimate the channel gain and carrier phase. If there happens any discrepancy in V symbol sets, the mobile station resumes the initial igniter sequence search mode and repeats the acquisition procedure. (Refer to Fig. 8.3.) Despite the above verification process, there still exists a possibility of false acquisition, no matter how small it may be. If a false acquisition is really detected in this stage, the MS recalls the declaration of the synch-completion and restarts the initial igniter sequence searching process, which may be modeled by a delay of KN_I time units for a very large number K.

After the synchronization process, the MS despreads the data signal by multiplying the conjugate of the synchronized main sequence and the corresponding Walsh sequence, then coherently demodulates the despread data by using the channel estimation technique to be described below.

5.1.2   D²SA CHANNEL ESTIMATION

Most DS/CDMA receivers maximize the signal-to-noise ratio by taking the RAKE structure that combines each multipath data signal coherently. The pilot channel signal which usually takes the form of unmodulated PN sequence helps the MS estimate the channel characteristics of each path, based on which the RAKE receiver accomplishes the *maximal-ratio combining* (MRC) of the incoming data signals [7]. In the case of the DSA family including the D²SA, the state signal, that is, the modulated igniter sequence, should render the channel estimation reference as well as the timing acquisition/tracking reference as it is the only pilot channel signal of the system. Note that, in the existing DSA schemes, once the main sequence as well as the igniter sequence is synchronized, the orthogonally modulated igniter sequence can be autonomously regenerated in the MS. This implies that once the synchronization process is completed the incoming orthogonally modulated igniter sequence is eligible for being the channel estimation reference. However, the DPSK-modulated igniter sequence in the D²SA scheme cannot take this role due to the phase ambiguity problem inherent in the DPSK signaling [71]. We investigate this problem more specifically referring to the D²SA channel estimator shown in Figs. 8.25 and 8.27.

In the nth symbol interval, the BS pilot symbol f_n and its input symbol x_n take the complex values

$$f_n = \exp(j\theta_n) = \exp\{j(\theta_0 + \sum_{l=1}^{n} \Delta\theta_l)\}, \tag{8.54}$$

$$x_n = \exp(j\Delta\theta_n), \tag{8.55}$$

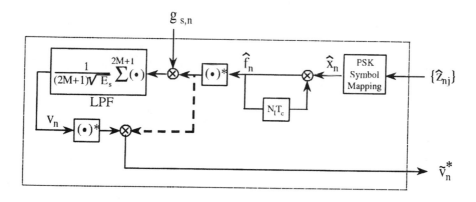

Figure 8.27.   Structure of the D^{2}SA channel estimator.

where θ_0 denotes the initial phase of the BS DPSK modulator and $\Delta\theta_n$ is determined by the b main SRG state samples z_{nj}, $j = 0, 1, \cdots, b-1$. The state signal constructed by spreading and power-adjusting the pilot symbol goes through the unknown channel and arrives at the receiver. The nth received state signal takes the expression

$$r_s^{(n)}(t) = H_n e^{j\phi_n} \sqrt{P_s} e^{j(\theta_0 + \sum_{l=1}^{n} \Delta\theta_l)}$$
$$\times \sum_{m=0}^{N_I-1} c_m p_{T_c}(t - [m + (n-1)N_I]T_c) + N_s(t), \qquad (8.56)$$

where $H_n e^{j\phi_n}$ denotes the channel gain during the transmission of the nth state signal, P_s the pilot channel power, and $p_{T_c}(t)$ the chip pulse shaping filter whose amplitude is 1 in the interval $[0, T_c]$ and 0, elsewhere. The pilot channel noise $N_s(t)$ contains the traffic channel inteference, multipath interference, other cell interference and the thermal noise, and thus is assumed to be a complex white gaussian random process whose inphase and quadrature components have the power spectral density of $N_0/2$, respectively. [34] The MS despreads the incoming state signal with the synchronized igniter sequence and integrate it for the symbol period $[(n-1)N_I T_c, nN_I T_c]$ to produce the nth sufficient statistic

$$g_{s,n} = \frac{1}{\sqrt{N_I T_c}} \int_{(n-1)N_I T_c}^{n N_I T_c} r_s^{(n)}(t) \sum_{m=0}^{N_I-1} c_m^* p_{T_c}(t - [m + (n-1)N_I]T_c) dt$$

---

[34]Since the spread spectrum rake receivers process each multipath channel independently in relation to the synchronization and the channel estimation, we regard the other multipath signals except the one we are focusing on simply as a component of the pilot channel noise, and deal with the received signal as if it had gone through a single path channel [7], which is reflected in Eq.(8.56).

$$= \sqrt{E_s} H_n e^{j\phi_n} e^{j(\theta_0 + \sum_{l=1}^{n} \Delta\theta_l)} + N_{s,n}, \tag{8.57}$$

where $E_s \overset{\triangle}{=} P_s N_I T_c$ and $N_{s,n}$ is the zero mean circularly-symmetric complex gaussian noise [100, 133] with $E\{[Re\{N_{s,n}\}]^2\} = E\{[Im\{N_{s,n}\}]^2\} = N_0/2$. By multiplying $g_{s,n}$ by $g_{s,n-1}^*$, we get the nth conveyed state symbol (corrupted by the channel noise)

$$y_n = E_s H_n H_{n-1} e^{j(\Delta\phi_n + \Delta\theta_n)} + \tilde{N}_{s,n}, \tag{8.58}$$

where $\Delta\phi_n \overset{\triangle}{=} \phi_n - \phi_{n-1}$ and the differential demodulator output noise takes the expression $\tilde{N}_{s,n} = N_{s,n} N_{s,n-1}^* + \sqrt{E_s} H_n e^{j(\phi_n + \theta_n)} N_{s,n-1}^* + \sqrt{E_s} H_{n-1} \cdot e^{-j(\phi_{n-1} + \theta_{n-1})} N_{s,n}$. In the slowly fading channels, the channel characteristic remains nearly constant for several symbol periods, and thus we have $\Delta\phi_n \approx 0$ in Eq.(8.58). Finally, after the magnitude normalization, the 2^b-ary PSK slicer restores b state samples $\hat{z}_{nj}$, $j = 0, 1, \cdots, b - 1$, from y_n, which are used to synchronize the MS I-main SRG to the BS I-main SRG in the DSA-despreader. Once the synchronization of the igniter SRG and the I-main SRG is completed, the MS starts to regenerate the DPSK-modulated pilot symbol by sampling the MS I-main sequence, which is represented by

$$\hat{f}_{n_0+n} = \exp(j\hat{\theta}_{n_0+n}) = \exp\{j(\hat{\theta}_{n_0} + \sum_{l=n_0+1}^{n_0+n} \Delta\hat{\theta}_l)\}, \tag{8.59}$$

where n_0 denotes the time when the MS starts to regenerate the pilot symbols after completing the I-main SRG synchronization, and $\hat{\theta}_{n_0}$ denotes the initial phase of the MS DPSK-modulator. Note that the BS and the MS generate the same differential-phase symbols after the synchronization of the I-main SRGs, but that they cannot have the same initial phase. That is,

$$\Delta\hat{\theta}_l = \Delta\theta_l, l = n_0 + 1, n_0 + 2, \cdots, \tag{8.60a}$$

$$\hat{\theta}_{n_0} \neq \theta_{n_0}, \tag{8.60b}$$

in general. Due to the discrepancy of these initial phases, we cannot apply the general low pass filtering approach to get the channel estimate, which, for example, takes the expression $(n > M)$ [7]

$$v_{n_0+n} = \frac{1}{(2M+1)\sqrt{E_s}} \sum_{l=n-M}^{n+M} g_{s,n_0+l} \, \hat{f}_{n_0+l}^* \tag{8.61a}$$

$$= \frac{1}{(2M+1)} \sum_{l=n-M}^{n+M} H_{n_0+l} e^{j\phi_{n_0+l}} e^{j\{\theta_{n_0} - \hat{\theta}_{n_0} + \sum_{m=n_0+1}^{n_0+l}(\Delta\theta_m - \Delta\hat{\theta}_m)\}}$$

$$+ \hat{N}_{s,n_0+n} \tag{8.61b}$$

$$\approx H_{n_0+n} e^{j\{\phi_{n_0+n} + (\theta_{n_0} - \hat{\theta}_{n_0})\}} + \hat{N}_{s,n_0+n}, \tag{8.61c}$$

where $2M + 1$ denotes the number of averaging symbol periods during which the channel characteristic is assumed to remain nearly constant. The averaged complex gaussian noise $\hat{N}_{s,n_0+n} \triangleq \frac{1}{2M+1} \sum_{l=n-M}^{n+M} N_{s,n_0+l} \exp(-j\hat{\theta}_{n_0+l})$ is also zero-mean, circularly symmetric, and $E\{[\,Re\{\hat{N}_{s,n_0+n}\}]^2\} = E\{[Im\{\hat{N}_{s,n_0+n}\}]^2\} = \frac{N_0/2}{(2M+1)}$. [35]

On the other hand, the data signal arrives at the receiver and is despread by the main sequence and the corresponding Walsh sequence. It is then integrated for the period $[(n_0 + n - 1)N_I, (n_0 + n)N_I]$, producing the $(n_0 + n)$th sufficient statistic

$$g_{d,n_0+n} = \sqrt{E_d}H_{n_0+n}e^{j\phi_{n_0+n}}a_{n_0+n} + N_{d,n_0+n}, \qquad (8.62)$$

where $E_d \triangleq P_d N_I T_c$, P_d the corresponding data channel power, and N_{d,n_0+n} the data channel noise. From Eq.(8.61c) and Eq.(8.62), we obtain the MRC input component for the current path

$$
\begin{aligned}
u_{n_0+n} &= g_{d,n_0+n}\, v^*_{n_0+n} &\qquad (8.63a) \\
&\approx \sqrt{E_d}H^2_{n_0+n}e^{j(\hat{\theta}_{n_0}-\theta_{n_0})}a_{n_0+n} + \tilde{N}_{d,n_0+n}, &\qquad (8.63b)
\end{aligned}
$$

which will generally cause a wrong detection of the MPSK data symbol a_{n_0+n} due to the initial phase discrepancy $\hat{\theta}_{n_0} - \theta_{n_0}$, even when the noise component $\tilde{N}_{d,n_0+n} \equiv N_{d,n_0+n}\hat{N}^*_{s,n_0+n} + N_{d,n_0+n}\,H_{n_0+n}e^{-j\{\phi_{n_0+n}+(\theta_{n_0}-\hat{\theta}_{n_0})\}} + \hat{N}^*_{s,n_0+n}\sqrt{E_d}\,H_{n_0+n}\,e^{j\phi_{n_0+n}}a_{n_0+n}$ becomes very small.

5.1.3   PRE-ROTATION OF DATA CONSTELLATION

The invalidity of the simple low pass filtering-based channel estimation in the D^2SA originates from the fundamental difference of the signaling (or, modulation / demodulation) methods between the pilot channel and the traffic channel. Since the pilot channel synchronization must be completed before the channel estimation, we cannot but take a noncoherent signaling to convey the synchronization information (or, the state samples) over the pilot channel, while we prefer a coherent signaling for the traffic channel data in favor of better error performance. As an embodiment of this noncoherent pilot signaling and coherent traffic signaling, the D^2SA employs the noncoherent DPSK scheme for the pilot channel and the coherent MPSK scheme for the traffic channel. The noncoherent DPSK was chosen in consideration of its advantage over the noncoherent orthogonal signaling in terms of system complexity and acquisition

---

[35]This is for the case of a simple moving average of the channel characteristic. More refined low pass filtering is also applicable for the estimation [134].

time performance, which unfortunately brought about the critical problem of initial phase discrepancy between the BS and the MS DPSK modulators.

In order to resolve the phase discrepancy problem, we employ the technique of pre-rotation of data constellation. [36] From Eq.(8.59), Eq.(8.60a), and Eq.(8.61c), we see that the MS can easily generate the (normalized) pilot signal distorted by the channel, which is represented by

$$\tilde{v}_{n_0+n}$$

$$= v_{n_0+n}\hat{f}_{n_0+n} \tag{8.64a}$$

$$\approx H_{n_0+n}e^{j(\phi_{n_0+n}+\theta_{n_0}+\sum_{l=n_0+1}^{n_0+n}\Delta\theta_l)} + \hat{N}_{s,n_0+n}\cdot e^{j(\hat{\theta}_{n_0}+\sum_{l=n_0+1}^{n_0+n}\Delta\hat{\theta}_l)} \tag{8.64b}$$

$$= H_{n_0+n}e^{j(\phi_{n_0+n}+\theta_{n_0+n})} + \hat{N}_{s,n_0+n}\cdot e^{j\hat{\theta}_{n_0+n}}. \tag{8.64c}$$

Now, comparing Eq.(8.62) and Eq.(8.64c), we can deduce the fact that the desired channel compensation is realized if the transmitted data symbol a_{n_0+n} is pre-rotated by θ_{n_0+n} (or, equivalently, pre-multiplied by $e^{j\theta_{n_0+n}}$) in the BS. More specifically, if the MS despreads and integrates the $(n_0 + n)$th *pre-rotated* data signal, producing the sufficient statistic

$$\tilde{g}_{d,n_0+n} = \sqrt{E_d}H_{n_0+n}e^{j(\phi_{n_0+n}+\theta_{n_0+n})}a_{n_0+n} + N_{d,n_0+n}, \tag{8.65}$$

then the corresponding MRC input component becomes

$$\tilde{u}_{n_0+n} = \tilde{g}_{d,n_0+n}\,\tilde{v}^*_{n_0+n} \tag{8.66a}$$

$$\approx \sqrt{E_d}H^2_{n_0+n}a_{n_0+n} + \check{N}_{d,n_0+n}, \tag{8.66b}$$

where the noise component is given by $\check{N}_{d,n_0+n} = e^{-j\hat{\theta}_{n_0+n}}\{\,N_{d,n_0+n}\,\hat{N}^*_{s,n_0+n}$
$+N_{d,n_0+n}\,H_{n_0+n}\,e^{-j\{\phi_{n_0+n}+(\theta_{n_0}-\hat{\theta}_{n_0})\}} + \hat{N}^*_{s,n_0+n}\sqrt{E_d}\,H_{n_0+n}\,e^{j(\phi_{n_0+n}+\theta_{n_0+n})}\,\cdot$
$\cdot a_{n_0+n}\}$. Note that $\check{N}_{d,n_0+n}$ has the same statistical characteristic as $\tilde{N}_{d,n_0+n}$ in Eq.(8.63b). This means that the D^2SA system incorporated with the pre-rotation and modified channel estimation achieves the same data error performance as the conventional coherent MPSK receiver having no intial phase discrepancy problem. The pre-rotation operation in the BS and the generation of the modified estimate $\tilde{v}_{n_0+n}$ in the MS are incorporated in Figs. 8.25 (a) and 8.27, respectively. (See the dashed lines in the figures.)

5.2   MEAN ACQUISITION TIME ANALYSIS

Now we analyze the mean acquisition time of the D^2SA scheme referring to the analysis of that of the DSA in Section 1. In relation to the mean acquisition

---

[36]The approach of modifying the transmitted symbol constellation can also be found in the Tomlison-Harashima precoding scheme for the channel pre-equalization [135, 136] or in the derivation of the carrierless AM/PM (CAP) scheme [137].

time analysis, we assume that the channel characteristic remains stationary throughout the acquisition process [7, 14, 15], that is, $H_n e^{j\phi_n} = H e^{j\phi}$, for any n. Though the analysis under this AWGN channel assumption may be somewhat different from what we get under correlated Rayleigh or Rician fading channels, it should be sufficient to clarify the advantage of the D^2SA over the orthogonal signaling DSA or the conventional serial search acquisition methods.

We first derive the mean acquisition time formula of the 2^b-ary DSA familiy when the *detection probability per run* $P_{d,r}$, the *false acquisition probability per cell* $P_{fa,c}$, and the *probability of correct symbol decision* P_c are given. (Refer to Section 1 for the definition of these probability parameters in the binary signaling case.) The derivation procedure is just the same as that of Eq.(8.19), except that the number of comparison-correction operations is $W (= \lceil L/b \rceil)$, not L, and $P_f = 1/2^b$, not $1/2$. The resulting formula is [37]

$$
\begin{aligned}
E\{T_{acq}\} = {} & \frac{1}{P_{d,r} P_c^{W+V}} \Big[1 + P_{d,r} \Big\{ W + 2 + \frac{1 - P_c^W}{2^b - 1} + \frac{P_c - P_c^V}{1 - P_c} P_c^W \\
& + \frac{(K - \frac{2^b}{2^b - 1})(1 - P_c^W)}{2^b V} \Big\} + (N_I - 1)\Big\{ 1 + P_{fa,c}(W + \frac{2^b}{2^b - 1} \\
& + \frac{K - \frac{2^b}{2^b - 1}}{2^b V})\Big\}(1 - \frac{P_{d,r} P_c^{W+V}}{2}) \Big] \cdot \tau_D,
\end{aligned}
\tag{8.67}
$$

where τ_D denotes the symbol integration period $N_I T_c$ and the uniform prior probability distribution is assumed for the initial code time-shift difference between the MS and the BS SRGs [57]. Note that Eq.(8.67) reduces to Eq.(8.19b) when $b = 1$.

To complete the analysis, we focus on determining various probability parameters involved in the above formula for both the orthogonal signaling parallel DSA scheme and the D^2SA.

5.2.1 PROBABILITY PARAMETERS: ORTHOGONAL SIGNALING

Let $m_j(t)$, $j = 0, 1, \cdots, 2^b - 1$ denote the orthogonal state symbol waveform $s_j(t)$ spread by the igniter sequence $c(t) \triangleq \sum_{m=0}^{N_I - 1} c_m p_{T_c}(t - mT_c)$, that is,

$$
m_j(t) = s_j(t)c(t),
\tag{8.68}
$$

$$
\frac{1}{N_I T_c} \int_0^{N_I T_c} m_j(t) m_k^*(t) dt = \delta_{jk}, \quad j, k = 0, 1, \cdots, 2^b - 1,
\tag{8.69}
$$

---

[37] We assume for simplicity that the chip-timing fine-alignment (tracking) operation does not exist between the igniter sequence acquisition and the state symbol detection. The overall synchronization performance will improve if the tracking procedure is placed before the state symbol detection when implementing the D^2SA system.

where $\delta_{jk}=1$ if $j = k$ and 0 otherwise. We assume, without loss of generality, that $m_0(t)$ is transmitted in the nth interval. Then the received signal takes the baseband expression

$$r_s^O(t) = m(t + lT_c + \eta T_c) + N_s(t), \tag{8.70}$$

where, $m(t) \triangleq \sum_{k=-\infty}^{\infty} \sqrt{P_s} H e^{j\phi} m_{j(k)}(t-[k-1]N_I T_c)$, $j(k) \in \{0,1,\cdots,2^b -1\}$, $j(n) = 0$, and P_s denotes the pilot channel power and $N_s(t)$ the complex white gaussian noise whose in-phase and quadrature components both have the power spectral density $N_0/2$. Parameters l and η respectively denote the integer and fractional part of the phase difference between the incoming and receiver-generated igniter sequences normalized by T_c.

Referring to Fig. 8.24, we can represent nth matched components $Y_{j,n}$ ($j=0,1,\cdots,2^b$-1) by

$$\begin{align}
Y_{j,n} &= \frac{1}{\sqrt{N_I T_c}} \int_{(n-1)N_I T_c}^{nN_I T_c} r_s^O(t) m_j^*(t - [n-1]N_I T_c) dt \tag{8.71a} \\
&= X_{j,n} + N_{j,n} \tag{8.71b} \\
&= \bar{X}_{j,n} + (X_{j,n}^r + N_{j,n}), \tag{8.71c}
\end{align}$$

where

$$X_{j,n} \triangleq \frac{1}{\sqrt{N_I T_c}} \int_{(n-1)N_I T_c}^{nN_I T_c} m(t + lT_c + \eta T_c) m_j^*(t - [n-1]N_I T_c) dt, \tag{8.72a}$$

$$\bar{X}_{j,n} \triangleq \begin{cases} \sqrt{E_s} H e^{j\phi}(1 - |\eta|)\delta_{0j}, & \text{if } \Omega_1(\text{in-phase}), \\ 0, & \text{if } \Omega_0(\text{out-of-phase}), \end{cases} \tag{8.72b}$$

$$X_{j,n}^r \triangleq X_{j,n} - \bar{X}_{j,n}, \tag{8.72c}$$

for $E_s \triangleq P_s N_I T_c$, and $N_{j,n}$ is the zero mean circularly-symmetric complex gaussian noise [100, 133] with $E\{[Re\{N_{j,n}\}]^2\} = E\{[Im\{N_{j,n}\}]^2\} = N_0/2$. Note that $\bar{X}_{j,n}$ and $X_{j,n}^r$ respectively represent the deterministic and random parts [15] of the correlation value $X_{j,n}$ multiplied by the channel characteristic $He^{j\phi}$. Since $X_{j,n}^r$ and $N_{j,n}$ are both random variables, we integrate these terms into a composite noise term $N_{j,n}^c \triangleq X_{j,n}^r + N_{j,n}$. Then, approximating $X_{j,n}^r$, $j = 0, 1, \cdots, 2^b - 1$, to the zero mean circularly-symmetric uncorrelated complex gaussian random variables, we may regard $N_{j,n}^c$, $j = 0, 1, \cdots, 2^b - 1$, as the zero mean circularly-symmetric complex gaussian random noises having the second-order statistic

$$E\{N_{j,n}^c N_{k,n}^{c\,*}\} = \begin{cases} N_0[1 + \gamma_c|\eta|^2]\delta_{jk}, & \text{if } \Omega_1(\text{in-phase}), \\ N_0[1 + \gamma_c(1 - 2|\eta| + 2|\eta|^2)]\delta_{jk}, & \text{if } \Omega_0(\text{out-of-phase}), \end{cases} \tag{8.73}$$

for $j, k = 0, 1, \cdots, 2^b - 1$, where $\gamma_c \triangleq P_s H^2 T_c / N_0$ denotes the pilot channel chip-SNR at the MS.

Now noting that the magnitude of the zero-mean circularly-symmetric complex gaussian random variable has a Rayleigh distribution and becomes Rician distributed when a non-zero value is added, we can determine the probabilities that $|Y_{j,n}|$ exceeds the given threshold R_0 under the hypothesis Ω_i, $P_{ji} \triangleq Pr\{|Y_{j,n}| > R_0 | \Omega_i\}$, $j = 0, 1, \cdots, 2^b - 1$, $i = 0, 1$, as follows:

$$P_{ji}(|\eta|) = \begin{cases} Q\left(\sqrt{\frac{2N_I \gamma_c (1-|\eta|)^2}{1+\gamma_c|\eta|^2}}, \sqrt{\frac{c}{1+\gamma_c|\eta|^2}}\right) \triangleq P_1, & j = 0, \ i = 1, \\ \exp\{-\frac{c}{2[1+\gamma_c|\eta|^2]}\} \triangleq P_2, & j = 1, 2, \cdots, 2^b - 1, \ i = 1, \\ \exp\{-\frac{c}{2[1+\gamma_c(1-2|\eta|+2|\eta|^2)]}\} \triangleq P_3, & j = 0, 1, \cdots, 2^b - 1, \ i = 0, \end{cases}$$
(8.74)

where c and $Q(\alpha, \beta)$ respectively denote the *normalized threshold* and the *Marcum's Q-function* [23] defined by $c \equiv \frac{2R_0^2}{N_0}$ and $Q(\alpha, \beta) \equiv \int_\beta^\infty x \exp\{-\frac{1}{2}(x^2 + \alpha^2)\} I_0(\alpha x) dx$ for the 0th-order modified Bessel function $I_0(x)$.

Since the igniter sequence false acquisition happens when any of $|Y_{j,n}|$'s exceeds R_0 under H_0, we can determine the *false acquisition probability per cell*, $P_{fa,c}(|\eta|)$, as follows:

$$P_{fa,c}(|\eta|) = 1 - \prod_{j=0}^{2^b-1} (1 - P_{j0}(|\eta|)) \tag{8.75a}$$

$$= 1 - [1 - \exp\{-\frac{c}{2[1 + \gamma_c(1 - 2|\eta| + 2|\eta|^2)]}\}]^{2^b} \tag{8.75b}$$

Thus, given the desired false acquisition probability, the threshold c should be set to

$$c = -2[1 + \gamma_c(1 - 2|\eta| + 2|\eta|^2)] \log(1 - [1 - P_{fa,c}(|\eta|)]^{1/2^b}). \tag{8.76}$$

On the other hand, the *detection probability for a cell* with the phase difference $|\eta|$ under Ω_1 is given by,

$$Q_d(|\eta|) = 1 - \prod_{j=0}^{2^b-1} (1 - P_{j1}(|\eta|)) \tag{8.77a}$$

$$= 1 - \left[1 - Q\left(\sqrt{\frac{2N_I \gamma_c(1 - |\eta|)^2}{1 + \gamma_c|\eta|^2}}, \sqrt{\frac{c}{1 + \gamma_c|\eta|^2}}\right)\right]$$

$$\times \left[1 - e^{-\frac{c}{2[1+\gamma_c|\eta|^2]}}\right]^{2^b-1}. \tag{8.77b}$$

Since we assume that the local phase advances by the step size T_c, there exist two cells in the in-phase region whose phase differences are $|\eta|$ and $1 - |\eta|$, respectively. Therefore, the *detection probability per run*, $P_{d,r}(|\eta|)$, is determined by

$$P_{d,r}(|\eta|) = Q_d(|\eta|) + (1 - Q_d(|\eta|))Q_d(1 - |\eta|). \tag{8.78}$$

Finally, assuming that the symbol decision is made with the smaller phase difference out of $|\eta|$ and $1 - |\eta|$ if both cells in the in-phase region are detected, we can determine the *probability of correct symbol decision (under Ω_1)*

$$
\begin{aligned}
P_c(|\eta|) \approx \; & \frac{Q_d(|\eta|)(1 - Q_d(1 - |\eta|))\hat{P}_c(|\eta|)}{Q_d(|\eta|) + Q_d(1 - |\eta|) - Q_d(|\eta|)Q_d(1 - |\eta|)} \\
& + \frac{(1 - Q_d(|\eta|))Q_d(1 - |\eta|)\hat{P}_c(1 - |\eta|)}{Q_d(|\eta|) + Q_d(1 - |\eta|) - Q_d(|\eta|)Q_d(1 - |\eta|)} \\
& + \frac{Q_d(|\eta|)Q_d(1 - |\eta|) \cdot \max\{\hat{P}_c(|\eta|), \hat{P}_c(1 - |\eta|)\})}{Q_d(|\eta|) + Q_d(1 - |\eta|) - Q_d(|\eta|)Q_d(1 - |\eta|)},
\end{aligned} \tag{8.79}
$$

where $\hat{P}_c(|\eta|)$ denotes the *correct symbol detection probability (under $|\eta|$)* that the symbol s_j is detected when the state symbol s_j is actually transmitted and the fractional chip phase difference is $|\eta|$, which is determined in Appendix B.1 to be

$$\hat{P}_c(|\eta|) = \sum_{n=0}^{2^b - 1} \frac{(-1)^n}{n+1} \binom{2^b - 1}{n} \exp\left\{-\frac{n}{n+1} \frac{N_I \gamma_c (1 - |\eta|)^2}{(1 + \gamma_c |\eta|^2)}\right\}. \tag{8.80}$$

5.2.2   PROBABILITY PARAMETERS: DPSK SIGNALING

When we employ 2^b-ary DPSK signaling instead of the orthogonal signaling, the received signal can be represented by

$$r_s^D(t) = \sum_{k=-\infty}^{\infty} \sqrt{P_s} H e^{j(\phi + \theta_k)} c(t + [l + \eta]T_c - [k-1]N_I T_c) + N_s(t), \tag{8.81}$$

where $\theta_k \in \{\frac{2\pi m}{2^b} : m = 0, 1, \cdots, 2^b - 1\}$.

Referring to Fig. 8.26 (a), we can represent the nth matched components $Y_{D,n}$ by

$$
\begin{aligned}
Y_{D,n} &= \frac{1}{\sqrt{N_I T_c}} \int_{(n-1)N_I T_c}^{n N_I T_c} r_s^D(t) c^*(t - [n-1]N_I T_c)dt && \text{(8.82a)} \\
&= X_{D,n} + N_{D,n} && \text{(8.82b)} \\
&= \bar{X}_{D,n} + \underbrace{(X_{D,n}^r + N_{D,n})} && \text{(8.82c)} \\
&= \bar{X}_{D,n} + N_{D,n}^c, && \text{(8.82d)}
\end{aligned}
$$

where $X_{D,n}$ and $N_{D,n}$ respectively denote the matched signal and AWGN components, and the signal component is composed of the deterministic part

$$\bar{X}_{D,n} \triangleq \begin{cases} \sqrt{E_s}He^{j(\phi+\theta_n)}(1-|\eta|), & \text{if } \Omega_1(\text{in-phase}), \\ 0, & \text{if } \Omega_0(\text{out-of-phase}), \end{cases} \quad (8.83)$$

and the random part $X_{D,n}^r \triangleq X_{D,n} - \bar{X}_{D,n}$ with the variance

$$E\{|X_{D,n}^r|^2\} = P_s H^2 T_c |\eta|^2. \quad (8.84)$$

The composite noise $N_{D,n}^c$ is defined in the same way as in Section 5.2.1, that is, the sum of $X_{D,n}^r$ and $N_{D,n}$. Then, we may approximate $N_{D,n}^c$ to the zero mean circularly-symmetric complex gaussian random noise having the second-order statistic

$$E\{|N_{D,n}^c|^2\} = \begin{cases} N_0[1+\gamma_c|\eta|^2], & \text{if } \Omega_1(\text{in-phase}), \\ N_0[1+\gamma_c(1-2|\eta|+2|\eta|^2)], & \text{if } \Omega_0(\text{out-of-phase}). \end{cases} \quad (8.85)$$

Now taking the same procedure as was done in Section 5.2.1, and noting that the igniter sequence false acquisition happens when $|Y_{D,n}|$ exceeds the threshold R_0 under H_0, we can determine

$$P_{fa,c}(|\eta|) = P_3 = \exp\{-\frac{c}{2[1+\gamma_c(1-2|\eta|+2|\eta|^2)]}\}. \quad (8.86)$$

Thus, for the desired false acquisition probability, the threshold c should be set to

$$c = -2[1+\gamma_c(1-2|\eta|+2|\eta|^2)]\log(P_{fa,c}(|\eta|)). \quad (8.87)$$

On the other hand, the detection probability for a cell with the phase difference $|\eta|$ under Ω_1 is given by

$$Q_d(|\eta|) = P_1 = Q(\sqrt{\frac{2N_I\gamma_c(1-|\eta|)^2}{1+\gamma_c|\eta|^2}}, \sqrt{\frac{c}{1+\gamma_c|\eta|^2}}), \quad (8.88)$$

and the *detection probability per run*, $P_{d,r}(|\eta|)$, can be determined by inserting Eq.(8.88) to Eq.(8.78).

The final step is to determine the correct detection probability that the symbol $e^{j\Delta\theta_n}$ is detected when the state symbol $e^{j\Delta\theta_n}$ is actually transmitted. To solve this problem, we depicted the relations between the $(n-1)$th matched filter output $Y_{D,n-1}$ and the nth output $Y_{D,n}$ in Fig. 8.28. Note that the matched signal components $X_{D,n-1}$ and $X_{D,n}$ have the same magnitude $\sqrt{E'}$ and the angle difference between them is just the transmitted information angle $\Delta\theta_n$, which is true because the igniter sequence repeats in every state symbol period.

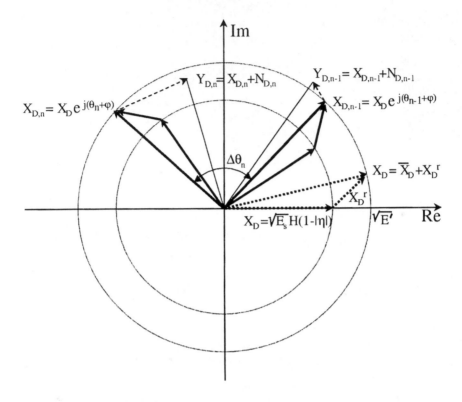

Figure 8.28.   Relations between the $(n-1)$th and the nth igniter-matched filter output symbols.

Then, by Appendix B.2, we can obtain the correct symbol detection probability (under $|\eta|$)

$$\hat{P}_c(|\eta|) = 1 - \frac{\sin\frac{\pi}{2^b}}{2\pi} \int_{-\pi/2}^{\pi/2} \frac{\exp\{-\frac{N_I\gamma_c(1-|\eta|)^2(1-\cos(\pi/2^b)\cos\theta)}{1+\gamma_c|\eta|^2(1-\cos(\pi/2^b)\cos\theta)}\}}{(1-\cos\frac{\pi}{2^b}\cos\theta)[1+\gamma_c|\eta|^2(1-\cos\frac{\pi}{2^b}\cos\theta)]}\, d\theta,$$

(8.89)

and the eventual *probability of correct symbol decision (under Ω_1) P_c* can be determined by inserting Eq.(8.89) to Eq.(8.79).

5.3   PERFORMANCE EVALUATION

Now based on the analysis results in the last section, we evaluate the acquisition time performance of the D^2SA. We consider an m-sequence of period $2^{15} - 1$ ($L=15$) as the main sequence and an extended sequence (e.g., extended m-sequence, or extended Gold-sequence) of period 256 ($N_I=256$) as the igniter sequence. The false alarm penalty factor is 1,000 and the fractional chip misalignment η is assumed to be 0.25 (except for Fig. 8.31). The gaussian approximation is applied to the multiple access interference and the channel noise.

Fig. 8.29 shows the detection probability per run $P_{d,r}$ and the correct symbol decision probability (under Ω_1) P_c under the threshold settings that produce the false acquisition probability per cell of 0.1. Figs. 8.29 (a)–(c) exhibit the probabilities for the chip SNR of -20dB through 0dB when b is 1, 2, and 3 (or, 2-ary, 4-ary, and 8-ary signaling cases). Solid and dashed lines respectively represent the D^2SA and the DSA performances. We observe that when $b=1$ the D^2SA has an SNR gain of about 2.5dB for P_c and larger than 1dB for $P_{d,r}$ over the DSA in low SNR ranges. When $b=2$, the gain for P_c reduces to smaller than 1dB while the gain for $P_{d,r}$ increases. However, the performance of the D^2SA is much inferior to that of the DSA for P_c when $b=3$. This phenomenon can be interpreted as follows: As the number of bits per symbol increases, the minimum Euclidean distance among the signal symbols decreases in the case of DPSK while it remains the same in the orthogonal signaling case. The minimum Euclidean distance of the 4-ary DPSK symbols is the same as that of the orthogonal signaling symbols having the same energy. On the contrary, $P_{d,r}$ of the orthogonal signaling DSA degrades as b increases, because the igniter sequence is modulated by the increased number of candidate symbols, which makes it difficult to acquire the sequence itself.

Fig. 8.30 exhibits the mean acquisition time performance of the conventional serial search acquisition (SSA) [14], DSA, and D^2SA schemes for varying threshold value c. The chip-SNR is set to -15dB and the number of verification steps for each scheme is set to a value that produces the smallest acquisition time. The mean acquisition time of the SSA scheme is evaluated on the basis of the analysis result of Eq.(8.20b). Fig. 8.30 (a) and (b) respectively show the results for $b=1$ and $b=2$. In the binary signaling case, the D^2SA completes the acquisition about 5 times faster than the DSA and about 70 times faster than the SSA. In the quaternary signaling case, however, such superiority of acquisition time performance of the D^2SA over the DSA diminishes significantly, while it still outperforms the SSA by about 15 times faster. We observed that in the case of the 8-ary or higher-order D^2SA schemes the acquisition time performance is poorer than the DSA counterpart. [38] Nevertheless this is no problem because in practical cellular systems the BS will not employ the 8-ary or higher order signaling schemes to broadcast the control information. From the perspective of system complexity, on the other hand, the D^2SA scheme always requires only one correlator for the igniter sequence acquisition, while the DSA requires 2^b correlators.

---

[38] In fact, we may improve the D^2SA acquisition time performances much better than those exhibited in Figs. 8.30 and 8.31 by modifying the overall acquisition procedure more elaborately. Refer to [66].

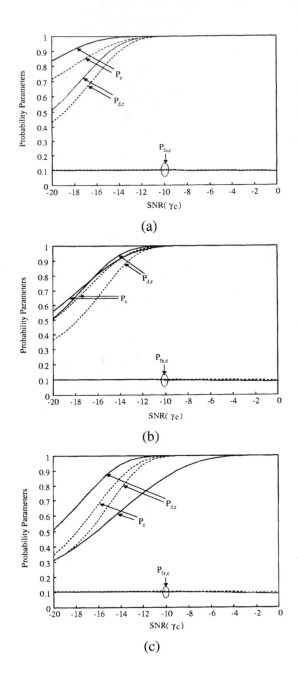

Figure 8.29. Probability parameters vs. chip-SNR (γ_c) ($P_{d,r}$: detection probability per run, P_c: correct symbol decision probability (under Ω_1), $P_{fa,c}$: false acquisition probability per cell (=0.1), N_I=256, η=0.25.): (a) When b=1, (b) when b=2, (c) when b=3. Solid lines: D^2SA, dashed lines: (orthogonal signaling) DSA.

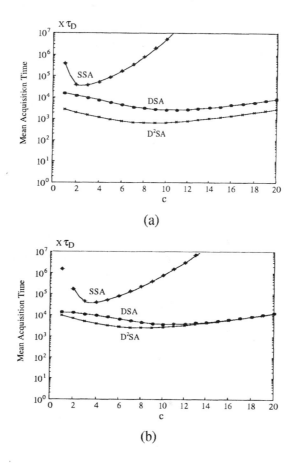

Figure 8.30.   Mean acquisition time vs. normalized threshold c. (γ_c=-15dB, $L = 15$, $N_I =$ 256, $K = 1000$, η=0.25, uniform prior probability): (a) When b=1 (verification step size = 9), (b) when b=2 (verification step size = 4).

Fig. 8.31 plots the acquisition times for varying chip-SNR: Figs. 8.31 (a)– (b) and Figs. 8.31 (c)–(d) plot the mean acquisition times for b=1 and b=2, respectively, with the threshold set such that the false acquisition probability becomes 0.25, 0.1 and 0.01. The D^{2}SA and the DSA both exhibit their best performances when the threshold is set such that $P_{fa,c} = 0.01$ while the SSA has different optimal thresholds depending on the SNR. Figs. 8.31 (a) and (c) plot them when the chip misalignment η is 0.25, and Figs. 8.31 (b) and (d) when η is 0. The verification step size for each scheme is set to a value that produces the smallest acquisition time. We see in Figs. 8.31 (a) and (b) that the acquisition times of the D^{2}SA and the DSA both reduce to about a hundredth of the SSA acquisition time if the chip-SNR is larger than -11dB when $\eta = 0.25$ and larger than -14dB when $\eta = 0$. The D^{2}SA can keep this reduction ratio

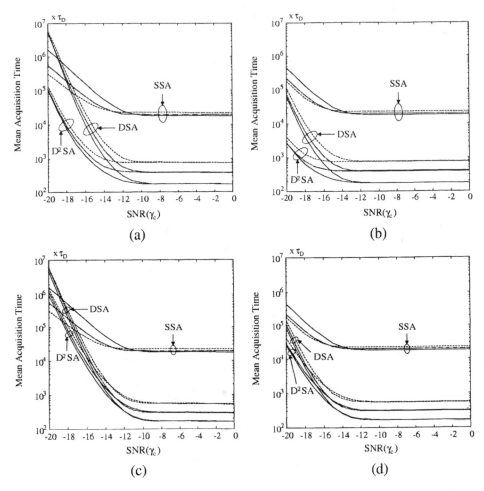

Figure 8.31.   Mean acquisition time vs. chip-SNR (γ_c) ($L = 15$, $N_I = 256$, $K = 1000$, uniform prior probability): (a) When $b=1$, $\eta=0.25$; (b) when $b=1$, $\eta=0$; (c) when $b=2$, $\eta=0.25$; (d) when $b=2$, $\eta=0$. Solid lines: $P_{fa,c}=0.01$, dotted lines: $P_{fa,c}=0.1$, dashed lines: $P_{fa,c}=0.25$.

up to about -14dB when $\eta = 0.25$ and -17dB when $\eta = 0$, while the DSA exhibits rapid performance degradation if the chip-SNR is smaller than -11dB when $\eta = 0.25$ and -14dB when $\eta = 0$. In the low SNR ranges, the binary D^2SA has the SNR gain of about 2.5dB over the binary DSA.

In the quaternary signaling cases (Fig. 8.31 (c) and (d)), the D^2SA has only a small gain over the DSA, and the performances of both systems degrade rapidly when the chip-SNR becomes smaller than -11dB when $\eta = 0.25$ and smaller than -14dB when $\eta = 0$. For performance improvement, the D^2SA can easily reduce the chip misalignment η to a small value by incorporating a simple delay-locked tracking loop (see Fig. 8.26 (b)) before the state symbol decision.

However, the DSA has to design somewhat complex tracking system [7] in order to overcome the orthogonal modulation effect of the igniter sequence and expedite the tracking process.

Chapter 9

CORRELATION-AIDED DSA (CDSA) TECHNIQUES

The DSA techniques acquire code synchronization relying on the state samples that are conveyed over the igniter sequence, or the pilot channel. The conveyed state samples are compared with the receiver-generated state samples and the resulting discrepancy is used in correcting the state of the main SRG in the receiver. When the channel condition is good, the synchronization can be done extremely fast, as the correct conveyance of L samples suffices to acquire the synchronization. In very poor channel condition, however, it is highly probable that the state samples get errored during the conveyance, consequently delaying the code synchronization substantially. Therefore, it is very much demanding to devise some means of improving the acquisition performance in such poor channel condition.

In this chapter, we introduce the *correlation-aided DSA* (CDSA) technique that has been presented as a solution to the above problem. We first discuss the principles of the CDSA and analyze its acquisition time performance. Then we consider the applications of the CDSA to the inter-cell synchronous and asynchronous DS/CDMA systems respectively. Similar to the DSA case, the CDSA technique is new and not yet widely referenced, so we provide in-depth discussions on the theoretical background and applications.

1. PRINCIPLES OF THE CDSA

The acquisition performance improvement of the DSA scheme over the conventional *serial search acquisition* (SSA) or the *parallel search acquisition* (PSA) scheme is significant in high or moderately low SNR environment. However, there often exists a BS-to-MS radio link that is in extremely poor channel state, which occurs when the MS is located in a shadowed or deep fading area, or when a considerable frequency offset exists between the BS and the MS

261

oscillators [13]. Unfortunately, the DSA-based acquisition scheme cannot operate successfully in such poor channel environments. When the SNR becomes very low due to poor channel state, the probability to get correct detection in $L + V$ successive samples drops very low, thus making DSA-based receiver spend tremendously long time in acquiring the PN sequence. [1]

In order to keep the short acquisition time even in this worst-case environment, we discuss in this section how to incorporate a state symbol correlation-based acquisition process in the existing DSA scheme. Since the received state symbol streams are a kind of pseudo-random sequences having good correlation property [69], the correlation of the despread state symbol sequence and the symbol sequence previously stored in the MS memory can aid the synchronization process. As the state symbol-SNR is relatively high, the state symbol correlation process enables reliable synchronization even in very low chip-SNR environment. As it will become clear in later sections, the resulting CDSA scheme maintains good performance even in the fading channels with large frequency offset.

1.1   CDSA SYSTEM ORGANIZATION AND OPERATION

The CDSA is a variation of the DSA that modifies the acquisition procedure in the receiver only. Fig. 9.1 depicts the functional block diagrams of the acquisition-related part of a receiver (MS). The DSA system in Fig. 9.1 (a) is a simplified version of that shown in Fig. 8.1. We observe that the CDSA system in Fig. 9.1 (b) is an outgrowth of the DSA system in Fig. 9.1 (a). It contains two SRGs - - the igniter SRG of length S and the main SRG of length L - - as the basic building blocks and also contains the comparison-correction based acquisition blocks. The added blocks, shown in shade, are only those for storing and correlating the "soft" state symbols. From this block diagram, we can easily imagine that the CDSA is basically the same as the DSA but is additionally equipped with the function of storing and correlating the "soft" state symbols in compensation for the performance degradation.

The functional block diagram is a generic one that can take different forms depending on the signalling scheme. For example, in case the DPSK-signalling is employed, the SRG may be composed of two parts - - in-phase SRG and quadrature SRG - - for both the igniter and the main SRGs. This change accompanies additional functional blocks for pre-rotating processing of data constellation. (Refer to Section 5 of Chapter 8.)

The Operation of the CDSA begins with the igniter sequence acquisition. This first stage processing is basically identical to that of the DSA, except that

---

[1]Note that L and V respectively denote the SRG length and the verification step size.

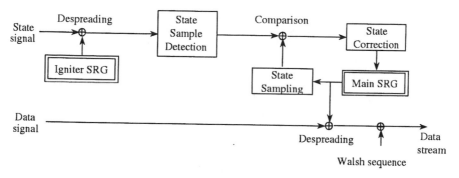

(a)

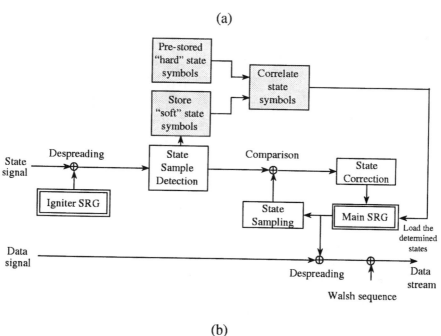

(b)

Figure 9.1. Functional block diagram of the acquisition-related part of the receiver (MS): (a) DSA system, (b) CDSA system.

the igniter SRG may be divided into in-phase and quadrature SRGs if DPSK-signalling is used. In terms of the igniter sequence period N_I and the chip time T_c, we may assume that is takes $S_I N_I T_c$ for the igniter acquisition and $V_I N_I T_c$ for its verification.

Once the igniter sequence acquisition is completed, the main SRG acquisition begins. This second stage processing now consists of two steps - - the comparison-correction based first step and the correlation based second step. The first step is essentially identical to the DSA case: The receiver-generated state samples are compared with the igniter-conveyed state samples, and the comparison result is fedback to the main SRG for state correction. Assuming 2^b-ary signalling, it takes $W \stackrel{\triangle}{=} \lceil L/b \rceil$ symbol conveyances for the main sequence acquisition and V symbol comparisons for its verification. Therefore, in terms of the $N_I T_c$ unit, the overall comparison-correction based acquisition time becomes $(W + V)N_I T_c$.

The second step processing of the main SRG acquisition is a supplementary processing that is triggered only when the first step processing does not terminate within the pre-specified limit. In preparation for this, the MS stores the "soft" state symbols in the received signal during comparison-correction processing. By the "soft" state symbol we mean the received raw data out of which a "hard" state symbol is to be determined. In addition, the MS keeps a pre-determined set of "hard" state symbols that encompasses all the 2^b-ary PSK symbols uniquely determined by the b binary numbers. If the first step processing does not terminate within F units of $(W + V)N_I T_c$, the MS begins to cross-correlate the "soft" state symbols with the "hard" state symbols, determining the "hard" state symbol that produces the maximum value. Once the "hard" state symbol is loaded to the main SRG, the second stage processing, as well as the second step processing, terminates. We may assume that this second step process takes D_{cor} units of $N_I T_c$.

The auxiliary processing of confirming the possibility of false synchronization or false alarm is maintained in the same way as in the DSA. We set the false alarm penalty time of $KN_I T_c$ as before, for a large number K.

Fig. 9.2 depicts the flow diagram that outlines the CDSA process we have discussed so far. Fig. 9.3 is the synchronization timing diagram that illustrates the time-requirements for each of the two stages.

1.2   MEAN ACQUISITION TIME ANALYSIS

We analyze the mean acquisition time of the CDSA-based cellular DS/CDMA system referring to Section 2 of Chapter 8. In the analysis we assume the AWGN channel for simplicity [7, 15]. The mean acquisition time performance in the Rayleigh fading channel will be exhibited through computer simulation, in the subsequent sections.

1.2.1   SIMPLIFIED FLOW DIAGRAM

The acquisition process may be arranged into two stages: The 1st stage includes all the processing from the start of the search to the completion of the igniter code and timing acquisition, and the 2nd stage includes all the processing from the start of the main SRG correction to the completion of the main sequence timing acquisition. In order to analyze the mean acquisition time in line with this arrangement, we rearrange the overall flow diagram in Fig. 9.2 accordingly.

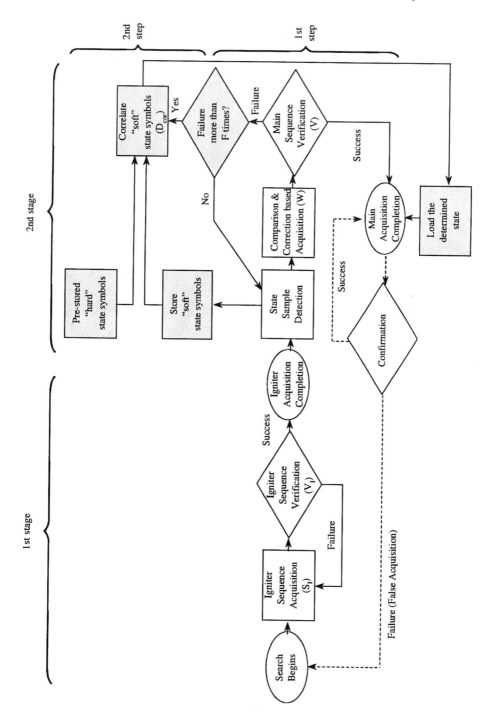

Figure 9.2. Flow diagram outlining the CDSA process.

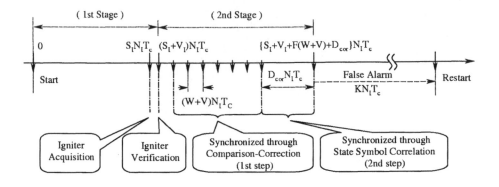

Figure 9.3.   Synchronization timing diagram of the CDSA scheme.

In this rearrangement we simplify the overall flow diagram by assuming that the igniter search procedure restarts with a randomly distributed igniter code and the timing difference between the BS and MS if the main sequence false acquisition occurs under the igniter true acquisition. [2] Under this assumption, we may deal with the igniter sequence acquisition process separately from the remaining main sequence acquisition process, deriving the *equivalent igniter false acquisition penalty time*. This penalty time is obtained by calculating the mean time that the MS wastes after its false declaration of the igniter acquisition until it returns to the next igniter phase search step, which is determined to be $K_I N_I T_c$. [3] Based on this equivalent igniter false acquisition time, we can redraw the flow diagram in Fig. 9.2 into the simplified one in Fig. 9.4 (a). In the new flow diagram, the 2nd stage begins only after the igniter sequence is truly acquired, and the whole process restarts when the main sequence acquisition

---

[2]The assumption for simplification results in an overestimation of the overall CDSA acquisition time because the 1st stage acquisition process would resume in the igniter in-phase state if the igniter true acquisition were kept, which means that the 1st stage process could be completed immediately after returning from the false acquisition. However, as the tracking loop may not keep the original chip timing during the main sequence false acquisition state, we may simply assume a random shift of the igniter code phase.

[3]In the case of the inter-cell synchronous DS/CDMA systems (such as IS-95 and cdma2000) having the minimum PN sequence offset 2^D, we can determine

$$
\begin{aligned}
K_I \;\approx\; & \sum_{n=0}^{F-1} (n+1)(W+V)P_{MVS,F}(1-P_{MVS,F})^n \\
& + \{(W+V)F + D_{cor}\}(1-P_{MVS,F})^F + K \\
=\; & \frac{(W+V)\{1-(1-1/2^{bV+D})^F\}}{1/2^{bV+D}} + D_{cor}(1-1/2^{bV+D})^F + K,
\end{aligned}
$$

where $P_{MVS,F} \triangleq (1/2^b)^V \cdot 1/2^D$ is the *false main verification success probability* that the state validity check and the coincidence test turn out the success result when the main sequence is not truly acquired.

declaration turns out false. This simplified flow diagram enables us to determine the overall acquisition time after analyzing the 1st and the 2nd stage acquisition times, separately.

Now in order to obtain the overall mean acquisition time by taking the moment generating function approach of [14], we denote by $H_I(z)$ the transfer function from the *1st stage beginning state* to the *2nd stage beginning state*, which is related to the mean igniter sequence acquisition time $E[T_I]$ through the relation $E[T_I] = H_I'(1)$ (see Section 2 of Chapter 8.). [4] Then, we divide the *acquisition completion declaration state* to two sub-states - - one through comparison-correction and the other through symbol correlation, and denote by $H_{CC}(z)$ and $H_{SC}(z)$ the transfer functions from the *2nd stage begin state* to these two sub-states, respectively. This division is necessary for the analysis, as the probabilities to declare acquisition completion for the two cases, namely, P_{TCC} and P_{TSC}, are different in general.

The function $H_{CC}(z)$ is determined to be

$$
\begin{aligned}
H_{CC}(z) &= \sum_{n=0}^{F-1}(1 - P_{MVS})^n P_{MVS} \cdot z^{(n+1)(W+V)} \\
&= P_{MVS} \cdot z^{W+V} \cdot \frac{1 - (1 - P_{MVS})^F \cdot z^{(W+V)F}}{1 - (1 - P_{MVS}) \cdot z^{W+V}}, \quad (9.1)
\end{aligned}
$$

where P_{MVS} is the *main verification success probability* that the state validity check and the coincidence test turn out successful result. [5] On the other hand, the function $H_{SC}(z)$ can be easily determined to be

$$
H_{SC}(z) = (1 - P_{MVS})^F \cdot z^{(W+V)F+D_{cor}}. \quad (9.2)
$$

The probability P_{TCC} and P_{TSC} vary depending on the fractional chip alignment offset η and the minimum PN sequence offset 2^D. In the case of the P_{TCC}, we can easily get

$$
P_{TCC}(|\eta|) = \frac{P_{NB}P_c(|\eta|)^{W+V}}{(1/2^{bV+D})\{1 - P_{NB}P_c(|\eta|)^W\} + P_{NB}P_c(|\eta|)^{W+V}}, \quad (9.3)
$$

---

[4]The transfer function $H_{ij}(z)$ from *state-i* to *state-j* is defined as $H_{ij}(z) \triangleq \sum_{n=0}^{\infty} p_{ij}(n)z^n$ for the probability $p_{ij}(n)$, which is the probability that the system leaves state-i and reach state-j in n time units (see Section 2 of Chapter 8.). The time unit is $N_I T_c$ throughout this section, and $H_{ij}'(1) \equiv \frac{d}{dz}H_{ij}(z)|_{z=1}$.

[5]In the case of the inter-cell synchronous DS/CDMA systems having the minimum PN sequence offset of 2^D, P_{MVS} is given by

$$
P_{MVS}(|\eta|) = (1/2^{bV+D})\{1 - P_{NB}P_c(|\eta|)^W\} + P_{NB}P_c(|\eta|)^{W+V},
$$

for the fractional chip alignment offset η. In the above equation $P_{NB} \triangleq 1 - W/2^{L-S}$ denotes the probability that the frame boundary does not hit during the main SRG correction stage [66], and P_c denotes the *correct state symbol detection probability*. (See Eq.(8.98) in Section 5 of Chapter 8.)

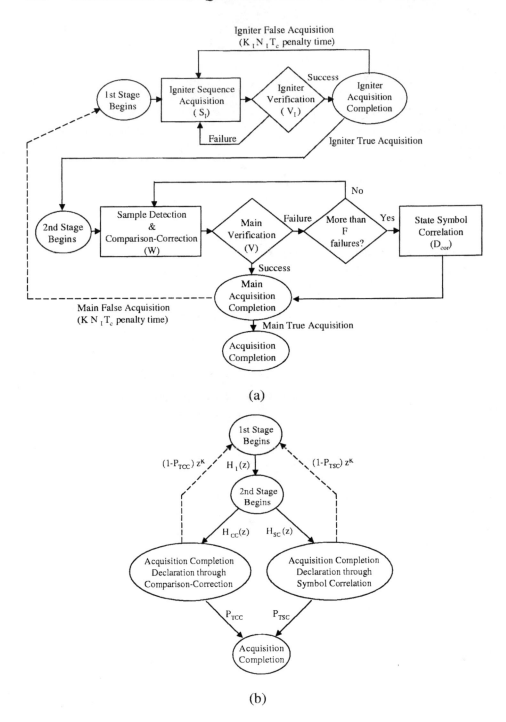

Figure 9.4.   Simplified flow diagrams for the CDSA process shown in Fig. 9.1: (a) Simplified flow diagram, (b) state transition diagram.

where $P_{NB} \triangleq 1 - W/2^{L-S}$. Derivation is rather complex in the case of P_{TSC}. According to the derivation given in Appendix B.3, we get

$$P_{TSC}(|\eta|) =$$

$$
\begin{cases}
\sum_{n=0}^{2^{L-D}-1}(-1)^n \binom{2^{L-D}-1}{n} \dfrac{(1+N_I\gamma_c)^2 \exp\{-\frac{nBN_I^2\gamma_c^2}{(1+N_I\gamma_c)^2+n(4-2/B)N_I\gamma_c+n}\}}{\sqrt{(1+N_I\gamma_c)^2+2nN_I\gamma_c/B+n}} \\
\quad\times \dfrac{1}{\sqrt{(1+N_I\gamma_c)^2+n(4-2/B)N_I\gamma_c+n}}, & |\eta| = 0, \\[2em]
\sum_{n=0}^{2^{L-D}-1}(-1)^n \binom{2^{L-D}-1}{n} \displaystyle\int_0^\infty \dfrac{(1+\gamma_c y)^2 \exp\{-\frac{nB\gamma_c^2 y^2}{(1+\gamma_c y)^2+n(4-2/B)\gamma_c y+n}\}}{\sqrt{\{(1+\gamma_c y)^2+2n\gamma_c y/B+n\}}} \\
\quad\times \dfrac{1}{|\eta|^2\sqrt{(1+\gamma_c y)^2+n(4-2/B)\gamma_c y+n}} \exp\{-\dfrac{N_I(1-|\eta|)^2+y}{|\eta|^2}\} \\
\quad\times I_0(2\dfrac{(1-|\eta|)}{|\eta|^2}\sqrt{N_I \cdot y})\,dy, & |\eta| \neq 0,
\end{cases}
$$

$$\text{(9.4a)}$$

where $I_0(x)$ is the 0th order modified Bessel function and γ_c denotes the pilot chip-SNR. This applies to all values of b, providing an approximate value for the case $b = 1$. When $b = 1$, the exact P_{TSC} is lower-bounded by

$$P_{TSC}(|\eta|) \geq$$

$$
\begin{cases}
\sum_{n=0}^{2^{L-D}-1} \dfrac{(-1)^n \binom{2^{L-D}-1}{n}\{(N_I\gamma_c)^2+(1+N_I\gamma_c)^2\}}{\sqrt{(N_I\gamma_c)^2+(1+N_I\gamma_c)^2+2nN_I\gamma_c/B+n}} \\
\quad\times \dfrac{\exp\{-\frac{nBN_I^2\gamma_c^2}{(N_I\gamma_c)^2+(1+N_I\gamma_c)^2+n(4-2/B)N_I\gamma_c+n}\}}{\sqrt{(N_I\gamma_c)^2+(1+N_I\gamma_c)^2+n(4-2/B)N_I\gamma_c+n}}, & |\eta| = 0, \\[2em]
\sum_{n=0}^{2^{L-D}-1} \displaystyle\int_0^\infty \dfrac{(-1)^n \binom{2^{L-D}-1}{n}\{(\gamma_c y)^2+(1+\gamma_c y)^2\}}{\sqrt{(\gamma_c y)^2+(1+\gamma_c y)^2+2n\gamma_c y/B+n}} \\
\quad\times \dfrac{\exp\{-\frac{nB\gamma_c^2 y^2}{(\gamma_c y)^2+(1+\gamma_c y)^2+n(4-2/B)\gamma_c y+n}\}}{|\eta|^2\sqrt{(\gamma_c y)^2+(1+\gamma_c y)^2+n(4-2/B)\gamma_c y+n}} \exp\{-\dfrac{N_I(1-|\eta|)^2+y}{|\eta|^2}\} \\
\quad\times I_0(2\dfrac{(1-|\eta|)}{|\eta|^2}\sqrt{N_I \cdot y})\,dy, & |\eta| \neq 0,
\end{cases}
$$

$$\text{(9.4b)}$$

which will provide an upper bound of the mean acquisition time of the binary signaling CDSA system. Based on these transfer function parameters, we can redraw Fig. 9.4 (a) into the state transition diagram shown Fig. 9.4 (b). The mean acquisition time can be obtained by analyzing the overall transfer function from the *1st state beginning state* to *acquisition completion state* in Fig. 9.4 (b).

1.2.2 MEAN ACQUISITION TIME

The overall transfer function $H(z)$ in Fig. 9.4 (b) is easily determined to be

$$H(z) = \frac{P_{TCC}H_{CC}(z) + P_{TSC}H_{SC}(z)}{1/H_I(z) - z^K\{(1 - P_{TCC})H_{CC}(z) + (1 - P_{TSC})H_{SC}(z)\}},$$

$$\text{(9.5)}$$

and the overall mean acquisition time $E[T_{CDSA}]$ can be obtained through the relation $E[T_{CDSA}] = H'(1)$. Therefore, by using Eq.(9.1), Eq.(9.2), Eq.(9.5), and the facts $H_I(1) = 1$ and $H_I'(1) = E[T_I]$, we get the mean acquisition time

$$E[T_{CDSA}] = (N_{CDSA}/D_{CDSA}) \times N_I T_c, \qquad (9.6a)$$

for

$$
\begin{aligned}
N_{CDSA} &= [\, P_{TCC}(W+V)\{1 - (1 + FP_{MVS})(1 - P_{MVS})^F\}/P_{MVS} \\
&+ P_{TSC}\{(W+V)F + D_{cor}\}(1 - P_{MVS})^F \,] \\
&+ (P_{TSC} - P_{TCC}) \times [\, (1 - P_{MVS})^F(W+V) \\
&\times \{1 - (1 + FP_{MVS})(1 - P_{MVS})^F\}/P_{MVS} \\
&- \{1 - (1 - P_{MVS})^F\}\{(W+V)F + D_{cor}\}(1 - P_{MVS})^F \,] \\
&+ [\, P_{TCC}\{1 - (1 - P_{MVS})^F\} + P_{TSC}(1 - P_{MVS})^F \,] \\
&\times [\, E[T_I] + K \times [\, (1 - P_{TCC})\{1 - (1 - P_{MVS})^F\} \\
&+ (1 - P_{TSC})(1 - P_{MVS})^F \,]\,],
\end{aligned}
$$

$$(9.6b)$$

$$
\begin{aligned}
D_{CDSA} &= [\, 1 - (1 - P_{TCC})\{1 - (1 - P_{MVS})^F\} \\
&- (1 - P_{TSC})(1 - P_{MVS})^F \,]^2.
\end{aligned}
$$

$$(9.6c)$$

The remaining work to complete Eq.(9.6) is the determination of the mean igniter sequence acquisition time $E[T_I]$. In analyzing the mean acquisition time, we consider the two extremal cases only - - when the fractional chip alignment offset is $0.25T_c$ (i.e., the worst-case scenario) and 0 (i.e., the best-case scenario) [41], assuming that the chip advancement step in the acquisition stage is $0.5T_c$. For simple analysis, we assume that all cells in the *in-phase* (H_1) *state* [14] have a chip alignment offset smaller than $0.5T_c$, while all cells in the *out-of-phase* (H_0) *state* have an offset larger than T_c [41]. Under this assumption, the cells of actual offset $0.75T_c$ and $0.5T_c$ are regarded as having an offset larger than T_c, and there are two H_1 state cells of offset $0.25T_c$ for the worst-case scenario and a single H_1 state cell of offset 0 for the best-case scenario, respectively. Finally, as the fractional chip alignment offset of the H_0 state cells affect little the acquisition performance in the practical low-SNR environment, we simply regard that all H_0 state cells have the same fractional offset of $|\eta|T_c$. [6]

Under the above assumption, we can easily determine the mean igniter sequence acquisition time by employing the state transition diagram shown in

---

[6] As the chip advancement step is $0.5T_c$, a half of the H_0 state cells actually have the fractional chip alignment offset of $|\eta|T_c$ while the other half have the offset of $(\frac{1}{2} - |\eta|)T_c$.

[14, Fig.3] and the resulting mean acquisition time formula [14, Eq.(9)]

$$E[T_I] = \frac{1}{H_D(1)} \{H'_D(1) + H'_M(1) + (\nu - 1)H'_0(1)(1 - \frac{H_D(1)}{2})\}, \quad (9.7)$$

where ν denotes the number of cells in the uncertainty region ($\approx 2N_I$), and $H_0(z)$, $H_D(z)$, and $H_M(z)$ are the branch gains defined in [14, Fig.3]. According to the derivation given in Appendix B.4, the three branch gain functions take the following forms in the DPSK signalling environment:

$$
\begin{aligned}
H_0(z) &= (1 - P_{fa})z + P_{fa}(1 - P_{VSI,H_0})z^{1+V_I} \\
&\quad + P_{fa}P_{VSI,H_0}z^{1+V_I+K_I},
\end{aligned}
\tag{9.8a}
$$

$$
H_D(z) =
\begin{cases}
P_d P_G z^{1+V_I}, & |\eta|=0, \\
P_d P_G z^{1+V_I}[1 + (1 - P_d)z + P_d(1 - P_G)z^{1+V_I} \\
\quad + P_d(P_{VSI,H_1} - P_G)z^{1+V_I+K_I}], & |\eta|=0.25,
\end{cases}
\tag{9.8b}
$$

$$
H_M(z) =
\begin{cases}
(1 - P_d)z + P_d(1 - P_{VSI,H_1})z^{1+V_I} \\
\quad + P_d(P_{VSI,H_1} - P_G)z^{1+V_I+K_I}, & |\eta|=0, \\
[(1 - P_d)z + P_d(1 - P_{VSI,H_1})z^{1+V_I} \\
\quad + P_d(P_{VSI,H_1} - P_G)z^{1+V_I+K_I}]^2, & |\eta|=0.25,
\end{cases}
\tag{9.8c}
$$

where P_d denotes the detection probability per H_1 state cell, P_{fa} the false acquisition probability per H_0 state cell, P_{VSI,H_1} the igniter verification success probability under H_1 state, P_{VSI,H_0} the igniter verification success probability under H_0 state, and P_G the probability that the correlator branch corresponding to the igniter code of the current cell actually produces the maximum accumulated energy in the igniter verification stage and the square-root of the value exceeds the threshold under H_1 state. Inserting these relations back to Eq.(9.7), we get the mean igniter acquisition time (normalized by $N_I T_c$)

$$
\begin{aligned}
E[T_I] &= \frac{1}{P_d P_G} [1 + V_I P_d + K_I P_d (P_{VSI,H_1} - P_G) \\
&\quad + (2N_I - 1)\{1 + P_{fa}(V_I + K_I P_{VSI,H_0})\}(\frac{1}{F_d} - \frac{P_d P_G}{2})],
\end{aligned}
\tag{9.9}
$$

where $F_d=1$ if $|\eta|=0$, and $F_d=2 - P_d P_G$ if $|\eta|=0.25$. Finally, inserting Eq.(9.9) into Eq.(9.6), we obtain the mean acquisition time of the CDSA scheme. [7]

---

[7]The mean acquisition time of the inter-cell synchronous DS/CDMA systems employing 2^b-ary DPSK signalling and active correlator (but not the CDSA) is derived in Appendix B.5.

2.   APPLICATION TO INTER-CELL SYNCHRONOUS SYSTEMS

The CDSA technique introduced in the previous section is a generic one that can be applied to the inter-cell synchronous as well as the inter-cell asynchronous DS/CDMA systems. In this section, we consider how the CDSA techniques can be used in the code synchronization in the inter-cell synchronous environment, such as the IS-95 and cdma2000 systems. As it will become clear through performance evaluations, the CDSA based code acquisition substantially outperforms the DSA based code acquisition in the low-SNR and fading channel environment.

2.1   CDSA-BASED SYSTEM ORGANIZATION

Fig. 9.5 depicts the functional block diagram of the DPSK-modulated CDSA scheme containing the state symbol sequence correlation block. The main shift register generators, I-main SRG and Q-main SRG, both of length L, generate the complex *m-sequence* of period $2^L - 1$ and extend it, by zero insertion, to the complex *extended m-sequence* $\{s_m^{[0]}\} \equiv \{s_{I,m}^{[0]} + js_{Q,m}^{[0]}\}$ of period $N_M = 2^L$ [9, 10]. For the minimum PN phase offset 2^D, the BS of the kth cell uses one of 2^{L-D} shifted versions of the extended sequence

$$\{s_m^{[k]} : s_m^{[k]} \stackrel{\triangle}{=} s_{m+k2^D} \equiv s_{I,m+k2^D} + js_{Q,m+k2^D},$$

$$m = \cdots, 0, 1, \cdots, 2^L - 1, \cdots\}, \quad k = 0, 1, \cdots, 2^{L-D} - 1, (9.10)$$

as its data scrambling sequence [9, 10]. The fast acquisition of the scrambling sequence (or, main sequence) of the corresponding cell is our ultimate goal. The inserted 0+j0 bit is defined as the *start bit of the main sequence* regardless of the amount of shifts. Note that one-to-one correspondence exists between the states of the I-main SRG and the Q-main SRG [9, 10], which enables us to acquire the entire complex sequence once we acquire the I-main sequence.

On the other hand, the igniter shift register generators, Igniter SRG-1 and Igniter SRG-2, both of length S, can possibly generate the complex *igniter sequences* $\{c_m^{[l]}\} = \{c_{I,m}^{[l]} + jc_{Q,m}^{[l]}\}, l = 0, 1, \cdots, 2^{S-1} - 1$, of period $N_I = 2^S$. We assume that N_I is larger than or equal to 2^D. The $\{c_{I,m}^{[l]}\}$ and $\{c_{Q,m}^{[l]}\}$ are a pair of orthogonal Gold sequences formed by the binary addition of two m-sequences generated from the igniter SRG-1 and the igniter SRG-2 along with the 0 insertion to extend the resulting sequence period to 2^S. The number of possible orthogonal Gold sequences generated from the two igniter SRGs of length S is 2^S, but the cellular system uses just $2\hat{R}$ sequences (or, $\hat{R}$ complex sequences) for $\hat{R} \equiv \lceil R/2^{S-D} \rceil$, where R denotes the igniter sequence reuse factor [6] of the cellular system. Using R shifted sequences out of $\hat{R}2^{S-D}$ available sequences, which are composed of $\hat{R}$ complex orthogonal Gold sequences and

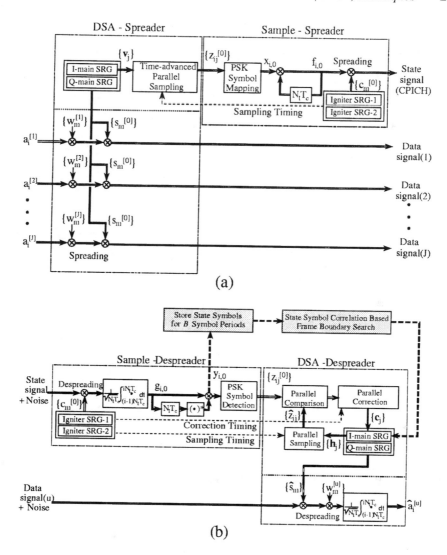

Figure 9.5. Functional block diagram of the acquisition-related part of the inter-cell synchronous DS/CDMA system employing the CDSA scheme: (a) 0th base-station (transmitter), (b) mobile-station (receiver).

their 2^{S-D} shifted versions $\{c^{[l]}_{m+q2^D}\}, l = 0, 1, \cdots, \hat{R}, q = 0, 1 \cdots, 2^{S-D} - 1$, we can implement the inter-cell synchronous cellular system with the igniter sequence reuse factor R. The *start bit of the igniter sequence* is defined to be the inserted $0+j0$ bit regardless of the amount of shifts as in the definition of the main sequence start bit. Each cell of the cellular system belongs to one of the R groups and each group uses one of the R shifted igniter sequences. The cells using the same shifted igniter sequence are spatially separated to minimize co-igniter code interference.

We assume that the 0th cell of the system uses the *zero-offset main sequence* $\{s_m^{[0]}\}$ as its main sequence and the 0th *zero-offset igniter sequence* $\{c_m^{[0]}\}$ as its igniter sequence. The start bits of the zero-offset main sequence are generated at times $0, 2^L, 2 \cdot 2^L, 3 \cdot 2^L, \cdots$, while those of the lth ($l = 0, 1, \cdots, \hat{R}$) zero-offset igniter sequence are generated at times $0, 2^S, 2 \cdot 2^S, 3 \cdot 2^S, \cdots$. [8] The main and the igniter sequences of each cell are chip-shifted versions of the zero-offset main sequence and the (corresponding) zero-offset igniter sequence, with the shifts being integer multiples of 2^D. (Note that 2^S is divided by 2^D.) Therefore, the I-main sequence samples of the kth cell, taken when every αth ($0 \leq \alpha < N_I$) chip of its igniter sequence is generated, form the n_kth *state sample sequence* of period 2^{L-S}

$$\{\tilde{s}_{m,\alpha}^{[n_k]} : \tilde{s}_{m,\alpha}^{[n_k]} \triangleq s_{I,m2^S+n_k2^D+\alpha}, \; m = \cdots, 0, 1, \cdots, 2^{L-S} - 1, \cdots\}, \quad (9.11)$$
$$n_k = 0, 1, \cdots, 2^{S-D} - 1,$$

where n_k is determined by the difference (modulo N_I) between the start bit generation time of the main sequence and that of the igniter sequence in the kth cell. The exemplary timing relations of the main and the igniter sequences as well as the igniter reuse pattern are depicted in Fig. 9.6 for L=15, S=8, D=6, and R=7. The 7 igniter sequences used in the system are $\{c_m^{[0]}\}$, $\{c_{m+64}^{[0]}\}$, $\{c_{m+128}^{[0]}\}$, $\{c_{m+192}^{[0]}\}$, $\{c_m^{[1]}\}$, $\{c_{m+64}^{[1]}\}$, and $\{c_{m+128}^{[1]}\}$, where $\{c_m^{[0]}\}$ and $\{c_m^{[1]}\}$ can be simultaneously generated from the igniter SRGs.

2.2   CDSA-BASED SYSTEM OPERATION

Based on the CDSA-based inter-cell synchronous system organization we have discussed so far, we now consider the operation of the system. As the operation differs depending on the communication site, we consider operation at the BS first, and then at the MS. At the BS, state symbol generation and broadcasting are the major functions. In contrast, acquisition of the igniter and main sequences are the major functions at the MS.

2.2.1   OPERATION OF THE BS

We focus on the BS operation in the 0th cell, as its generalization to the kth cell is straightforward. The BS time-advanced sampling block takes the b state samples

$$z_{ij}^{[0]} \triangleq s_{I, (r+i)2^S+\alpha_j}^{[0]}, \quad j = 0, 1, \cdots, b - 1, \quad (9.12)$$

---

[8] In this section, we describe the system discrete-time based, taking the unit time of chip duration T_c.

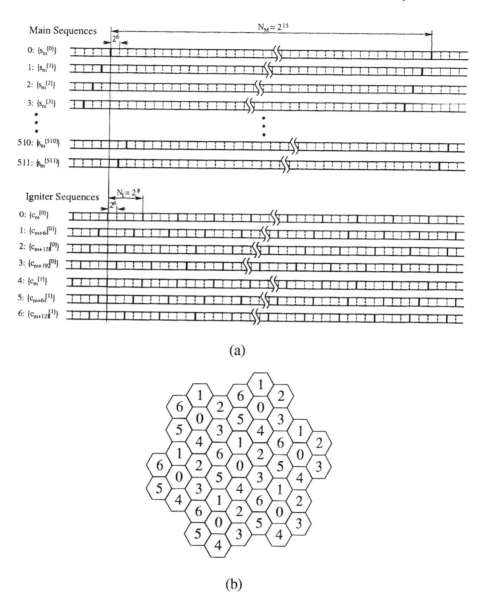

Figure 9.6. Timing diagram (a) and igniter sequence reuse pattern (b) of the 2^{L-D} main sequences and R igniter sequences in the inter-cell synchronous DS/ CDMA system. Each cell employs a main sequence and an igniter sequence out of them. ($L = 15, S = 8, D = 6, R = 7$.)

of the I-main SRG at time $(r + i - 1)N_I$, for a reference value r and the jth sample delay α_j ($0 \le \alpha_0 < \alpha_1 < \cdots < \alpha_{b-1} < N_I$), and provides them to the 2^b-ary DPSK modulator. [9] Note that, for a given α_j, the sequence $\{z_{ij}^{[k]}\}$ is just a shifted version of the sequence $\{\bar{s}_{i,\alpha_j}^{[n_k]}\}$ in Eq.(9.11), and thus its period is 2^{L-S}. The DPSK modulator maps the b state samples to the corresponding 2^b-ary PSK state symbol $x_{i,0} \equiv e^{j\Delta\theta_{r+i,0}}$ and produces the DPSK pilot symbol $f_{i,0} \equiv e^{j\theta_{r+i,0}}$ by adding the phase of $x_{i,0}$ to the phase accumulated up to the previous symbol time. The resulting pilot symbol is spread by a period of the igniter sequence and transmitted through the pilot channel for the time interval $[(r + i - 1)N_I, (r + i)N_I)$. Note that *the state sampling times and the pilot symbol boundaries of an arbitrary cell always correspond to the times when the start bit of its igniter sequence (i.e., $c_0^{[l]}$) is generated.* On the other hand, each user's M-ary PSK data $a_i^{[u]}$ is spread by one of the orthogonal Walsh sequences $\{w_m^{[u]}\}$ and scrambled by the cell-specific main sequence $\{s_m^{[0]}\}$, and then transmitted in the interval $[(r + i - 1)N_I, (r + i)N_I)$. [10] Once the main sequence is acquired, the Walsh sequence boundary(i.e., the data symbol boundary) can be determined immediately [9, 10, 106]. The state signal and the data signals go through the same channel to arrive at the mobile station.

2.2.2   OPERATION OF THE MS

A. Igniter Sequence Acquisition: The MS first acquires the DPSK modulated igniter sequence employing the serial or parallel noncoherent acquisition detector shown in Fig. 9.7. We assume that parallel correlator scheme is employed for the igniter sequence acquisition. The $\hat{R}$ correlators corresponding to the complex orthogonal Gold sequences employed in the system operate in parallel and the maximum value of $\hat{R}$ correlator outputs is compared with the predetermined threshold. Whenever the threshold overtaking occurs, a verification test that compares the energy of V_I noncoherently accumulated symbols with another threshold follows to confirm the true acquisition. In this stage, the MS first accomplishes igniter code identification by determining the igniter code that produces the maximum accumulated energy, and then compares the maximum energy with the threshold.

B. Comparison-Correction based Main Sequence Acquisition: Once the igniter code and timing of the current cell is acquired (i.e., the maximum accumulated energy overtakes the verification threshold), the comparison-correction based main sequence acquisition process begins. The MS despreads the re-

---

[9]Refer to Chapter 8 for the state sample selection, sampling/correction time determination, and the related circuit design issues of the 2^b-ary signaling DSA system.

[10]In fact, $a_i^{[u]}$ is multiplied by the pilot symbol $f_{i,0}$ before spreading, which is needed when we coherently demodulate the data symbol using the state signal as a channel estimation reference (see Section 5 of Chapter 8.).

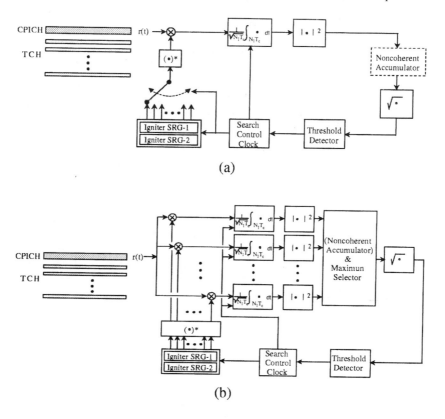

Figure 9.7.   Igniter sequence acquisition and verification detectors: (a) Pure serial correlator implementation, (b) parallel correlator implementation. Noncoherent accumulator is employed only for verification.

ceived state signal using the acquired igniter sequence and differentially detect the conveyed samples $z_{ij}^{[0]}$, $j = 0, 1, \cdots, b - 1$, which are passed to the DSA-despreader at time $(r + i)N_I$. On the other hand, the DSA-despreader generates its own state samples $\hat{z}_{ij}$'s simultaneously at time $(r + i)N_I$ from the MS I-main SRG, and compares them with the conveyed counterparts $z_{ij}^{[0]}$'s. When we denote by $\{\hat{s}_{I,m}\}$ the extended m-sequence currently generated from the MS I-main SRG, the sample $\hat{z}_{ij}$ is defined to be

$$\hat{z}_{ij} \triangleq \hat{s}_{I,(r+i)N_I + \alpha_j}, \quad j = 0, 1, \cdots, b - 1. \tag{9.13}$$

If $\hat{z}_{ij}$ does not coincide with $z_{ij}^{[0]}$, the jth correction circuit is triggered to correct the state of the I-main SRG at time $(r + i)N_I + D_c$, with D_c chosen such that $0 < D_c \leq N_I$. Otherwise, no action takes place. If we employ the sampling and correction circuits designed according to the theorems in Section 3 of Chapter 8, the MS I-main SRG gets synchronized to the BS I-main SRG after $W \stackrel{\triangle}{=} \lceil L/b \rceil$ state symbol conveyance if no detection error is involved.

For fear that there should be a detection error, we install a main sequence verification process that checks if the conveyed and receiver-generated symbols coincide for additional V symbol comparisons after the synchronization. While verifying the synchronization of the I-main SRG through the coincidence test, we also investigate the mapping table to check the validity of the corrected I-main SRG state and get the corresponding Q-main SRG state that will appear at the end of the verification process. Note that there are $2^{S-D} \times 2^{L-S}$ valid I-main SRG states each of which is mapped to each shifted version of the state symbol sequences (see Eq.(9.14)). The valid I-main SRG states (i.e., L-bit shift register internal values) together with the corresponding Q-main SRG states must be stored in advance in the memory. The required MS memory is nevertheless minimal in practical system implementations. [11] On the other hand, during the main SRG correction and verification processes, i.e., for $(W + V)$ symbol transmission periods, the MS stores the *soft* state symbols $\mathbf{y}_{i,0}$'s in the memory (refer to Fig. 9.5 (b)). If the conveyed and receiver-generated symbols indeed coincide for all V symbol pairs and the I-main SRG corrected state coincides with a valid state listed in the memory, then the MS loads the corresponding Q-main SRG state and declares completion of synchronization of the main sequence, beginning to track and estimate the channel gain and carrier phase. If less than V symbol-pairs coincide or the corrected I-main SRG state does not coincide with any of the valid states, the MS goes back to the main SRG correction stage and detects another set of $(W + V)$ state symbols for SRG correction, verification, and soft symbol storage. Unless the verification test succeeds (i.e., V symbol-pairs coincide in the verification stage and the corrected state is valid), this procedure continues until the number of stored symbols exceeds a predetermined number B, which is assumed to be 2^{L-S} (i.e., the period of the state symbol sequence) in this chapter. [12]

C. Correlation based Main Sequence Acquisition: If 2^{L-S} state symbols have been received after initiation of the comparison-correction process and the main SRG synchronization is not yet declared, the comparison-correction based acquisition process stops and the *state symbol correlation process* begins. In the mean time $F \stackrel{\triangle}{=} \lceil 2^{L-S}/(W+V) \rceil$ trials of comparison-correction process are completed only to fail. For the state symbol correlation process, the MS stores in the memory, in advance, 2^{S-D} sets of possible PSK state symbol sequences of length 2^{L-S}, $\{ \{\tilde{x}_{i,n_k} : i = 0, 1, \cdots, 2^{L-S} - 1\} : n_k = 0, 1, \cdots, 2^{S-D} - 1$

---

[11] When L=15, S=8, and D=6, the required memory size is about 16kbits.

[12] We set B to be 2^{L-S} for realizing the full-period correlation of the state symbol sequence. However, in general, we can achieve the same goal by employing partial correlation when B is not equal to 2^{L-S}.

}, for the symbol $\tilde{x}_{i,n_k}$ defined by

$$\tilde{x}_{i,n_k} \overset{\triangle}{=} S_{2^b PSK}(\tilde{s}_{i,\alpha_0}^{[n_k]}, \tilde{s}_{i,\alpha_1}^{[n_k]}, \cdots, \tilde{s}_{i,\alpha_{b-1}}^{[n_k]}) \equiv e^{j\Delta\theta_{i,n_k}}, \qquad (9.14)$$

where $S_{2^b PSK}(\nu_0, \nu_1, \cdots, \nu_{b-1})$ denotes the 2^b-ary PSK symbol on the unit circle that is uniquely determined by b binary numbers ν_0, ν_1, $\cdots$, ν_{b-1}. In the state symbol correlation process, the 2^{S-D} *hard* state symbol sequences $\{\tilde{x}_{i,n_k}\}$, $n_k = 0, 1, \cdots, 2^{S-D} - 1$, and their cyclically shifted versions are correlated one by one with the received *soft* state symbol sequence $\{y_{i,0}\}$ of length 2^{L-S}, and then the shifted sequence which produces the maximum correlation energy is determined to be the truly transmitted sequence $\{x_{i,0}\}$ (refer to Fig. 9.5). Note that as the sequences obtained by decimating an m-sequence by the decimation factor 2^S is just a shifted version of the m-sequence [69], the state symbol sequences which are generated on the basis of the decimated sequences generally keep good pseudo-random property. Furthermore, the state symbol-SNR is relatively high due to the high processing gain (or, 2^S) even in very low chip-SNR range. Therefore, the state symbol correlation process will provide a very high synchronization success probability even in poor channel environments. As the processing time required for the state symbol correlation process is constant (we denote it by $D_{cor}N_I$ time units), we can map all candidate sequences $\{\tilde{x}_{i,n_k}\}$ and their shifted versions with the corresponding I- and Q- main SRG states that will appear after the processing time has elapsed. Fig. 9.8 depicts the resulting overall acquisition process (see the solid line part). The required time for each sub-process (i.e., igniter sequence acquisition and verification, comparison-correction based main sequence acquisition, and the state symbol correlation processes) is as illustrated in Fig. 9.3. If the overall acquisition process turns out to be a false alarm in the tracking stage, the MS recalls the declaration of the synch-completion and goes back to the igniter search stage, which may be modeled by a delay of KN_I time units for a very large number K (see the dashed line part and refer to Section 1 of Chapter 8.).

After the synchronization process, the MS despreads the data signal by multiplying the conjugate of the synchronized main sequence and the corresponding Walsh sequence, and then coherently demodulates the despread data using the channel estimation results. [13]

2.3 ACQUISITION TIME EVALUATIONS

We compare the acquisition time performance of the CDSA, the DPSK-signaling DSA, and the conventional PSA schemes when they are applied to acquiring the IS-95 PCS forward link scrambling sequence (1.2288Mcps,

---

[13]Detailed discussions on the channel estimation of the DPSK-modulated DSA scheme can be found in Section 5 of Chapter 8.

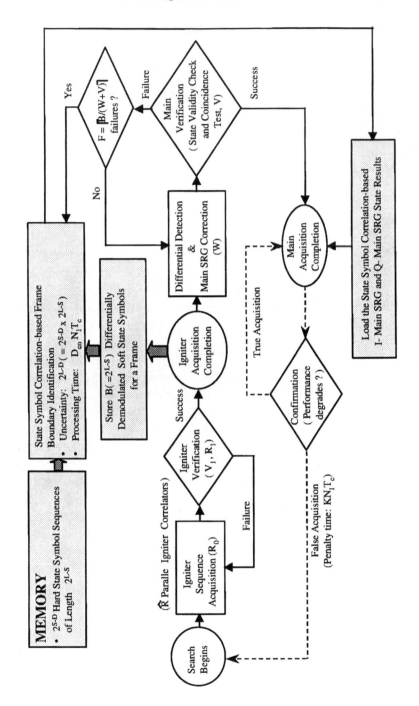

Figure 9.8.   Flow diagram outlining the CDSA process for inter-cell synchronous DS/CDMA systems.

$L = 15$, $D = 6$). The igniter sequence period N_I and reuse factor R are respectively assumed to be 256 (S=8) and 16, which means 4 different complex igniter sequences are needed for the entire cellular system($\hat{R} = 4$). For the numerical evaluations, we took the false alarm penalty factor K of $1,280$, which corresponds to 10 frame periods, and the dwell time τ_D of $256T_c$. We applied the Gaussian approximation to the multiple access interference and the channel noise, thus integrating all the noise processes to a single complex white Gaussian random process whose in-phase and quadrature components have the power spectral density of $N_o/2$. We take the igniter verification step size V_I of 10, and the coincidence test step size V of 4 for $b = 1$ and 2 for $b = 2$ respectively. We selected the normalized threshold ζ_0 such that P_{fa} becomes 0.05, and the threshold ζ_1 such that P_{VSI,H_0} becomes 0.001. Finally, the correlation speed in the state symbol correlation stage is assumed to be the same to that in the igniter sequence acquisition stage, which means that we spend $128T_c$ in obtaining $\hat{R}$(=4) correlation results between soft and hard state symbol sequences of length 2^{L-S}(=128). As 512 correlation results must be obtained per state symbol correlation process, D_{cor} becomes 64. [14]

Fig. 9.9 plots the overall mean acquisition times of the three schemes in the AWGN channel for $|\eta| = 0.25$. Fig. 9.9 (a) exhibits the performances when DBPSK signaling is employed (b=1) for state sample transmission, while Fig. 9.9 (b) does the same when DQPSK signaling is employed (b=2). The mean acquisition time of the DBPSK signaling CDSA exhibited in Fig. 9.9 (a) is the upper bound [Eq.(9.4b)], [15] while in Fig. 9.9 (b) the exact mean acquisition time of the DQPSK signaling CDSA [Eq.(9.4a)] is exhibited. In Fig. 9.9 (a) and (b) we observe that the overall acquisition times of the DSA and the CDSA schemes are shorter than that of the PSA scheme in the chip-SNR range above -25dB, but the DSA performance begins to degrade rapidly as the chip-SNR becomes lower than -21dB for $b = 1$ and -16dB for $b = 2$. Unfortunately, in the case b=1, we observe that the performance improvement of the CDSA over the DSA is not significant in the given SNR range. To explicate this, we also plotted the igniter sequence acquisition time components (or, $E[T_I] \times N_I T_c$ and $E[T_1] \times N_I T_c$) together with the rest components given by subtracting them from the overall acquisition times (or, $E[T_{CDSA}]$-$E[T_I] \times N_I T_c$ and $E[T_{DSA}]$-$E[T_1] \times N_I T_c$) in Fig. 9.9 (c) and (d). The rest components reflect the overall acquisition time increase due to the 2nd stage processing. As the chip-SNR decreases, the igniter sequence acquisition time components increase rapidly, but it also accompanies the increase of the rest components. When b=1 (Fig. 9.9 (c)),

---

[14]In practice, the MS is likely to employ a high-speed processor to manipulate the stored data, significantly reducing the state symbol correlation time. In this paper, for fair comparisons, we kept the CDSA computation complexity always lower than that of the $\hat{R}$ parallel correlators operating at the chip rate.

[15]The approximation based on Eq.(9.4a) provides nearly the same plot as the upper bound does in the given SNR range.

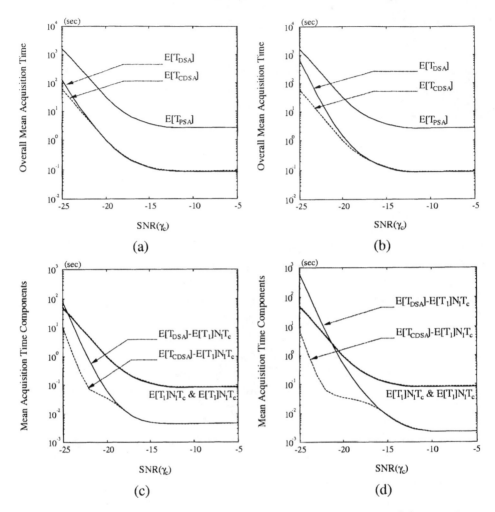

Figure 9.9.   The mean acquisition times vs. chip-SNR(γ_c) in AWGN channel ($|\eta|$=0.25): Overall mean acquisition times for (a) b=1 and (b) b=2; igniter sequence acquisition time components and the rest components for (c) b=1 and (d) b=2.

the dominant component of the overall acquisition time is the igniter sequence acquisition time component in the given SNR range, which leaves little room for improvement from the adoption of the state symbol correlation process working in the 2nd stage (see Fig. 9.9 (a)). However, when $b = 2$ (Fig. 9.9 (d)), the growth rate of the rest component of the DSA scheme is so high that it becomes the dominant factor of the overall acquisition time at low SNR parts in the given range, which significantly reduces the acquisition time performance gap between the DSA and the PSA. In the case of the CDSA scheme the state symbol correlation process effectively suppresses the growth of the rest component such that it cannot become the dominant factor, thus contributes to keeping the large acquisition time reduction ratio in any environment. If we plot the acquisition time performance in the SNR range below -25dB, we can observe the performance gap between the CDSA and the DSA in the case $b=1$ as well. The CDSA scheme enables about 30times faster acquisition than the PSA scheme over the entire SNR range shown in Fig. 9.9 (a) and (b).

Then, we carried out computer simulations to examine the system performances when the arriving signals suffer from fading and frequency offset, which are inevitable in the practical code acquisition environment. For the simulations, we assumed the DBPSK signaling $(b=1)$ for state sample transmission, and we took the single-path Rayleigh fading channel with low average chip-SNR, as the usual multipath signals except the focused path can be regarded as the interference components during the path acquisition process. We used the JTC classic (U-shape) spectrum [138] of maximum Doppler frequency 100Hz (except for Fig. 9.10 (b)) to generate the fading process, and set the fractional chip alignment offset $|\eta|$ to 0 and the chip advancement step to $0.5T_c$.

Fig. 9.10 (a) plots the mean acquisition times of the three schemes in the Rayleigh fading channel with respect to the average chip-SNR ($\gamma_c \equiv E[H^2]P_sT_c$ /No) varying from -25dB to -5dB. We observe that at the chip-SNR of -21dB, the acquisition times are about 1(sec) for CDSA, 5(sec) for DSA, and 40(sec) for PSA. The performance of the DSA scheme degrades rapidly in the lower SNR range. In the high-SIR range, the mean acquisition time reduces to about 0.1(sec) in the CDSA and the DSA cases, while it maintains the long acquisition time of 3(sec) in the PSA case. When compared with the AWGN performances, the performance degradation of the DSA scheme begins at a much higher chip-SNR value in the fading channel (though $|\eta|=0$ in Fig. 9.10), which reflects the significance of incorporating the state symbol correlation process for the practical DSA scheme. Fig. 9.10(b) compares the mean acquisition times of the CDSA and the DSA at the chip-SNR of -15dB as the Doppler fading rate varies from 100Hz to 500Hz. We observe that the CDSA scheme maintains its low acquisition time performance in the high Doppler environment, while the DSA scheme degrades rapidly as the fading rate increases, which is a critical weakness in the mobile wireless communications.

Fig. 9.11 plots the acquisition time degradations caused by frequency offset between the BS and the MS. In normal operation, code acquisition is done before the frequency discrepancy between BS and MS reduces (through phase-locked loop (PLL) operation) to a minimal level, so the robustness to frequency offset is critically important in acquisition. We took the same fading channel as in Fig. 9.10. Fig. 9.11(a) shows the acquisition time performances of the CDSA and the DSA when the frequency offset is 1kHz(or, 0.5ppm for 2GHz carrier frequency). We find that the performance of the DSA scheme degrades seriously (compare with Fig. 9.10 (a)) when the frequency offset exists, which results from

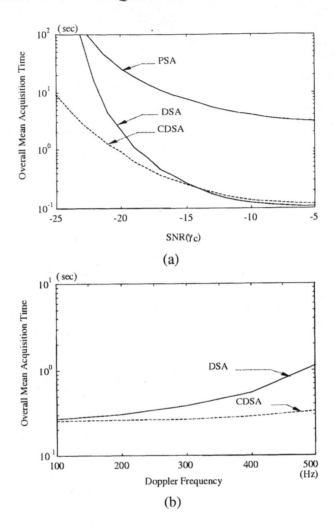

Figure 9.10. The overall mean acquisition times in Rayleigh fading channels ($b = 1$, $|\eta| = 0$): (a) vs. chip-SNR(γ_c), 100Hz Doppler, (b) vs. Doppler frequency, -15dB chip-SNR. The classic Doppler spectrum of the JTC fading model is employed.

the frequent DPSK detection errors in the frequency offset environment. The DSA acquisition time becomes longer than 10(sec) at the chip-SNR of about -13dB. However, the CDSA scheme degrades little thanks to the robustness of the state symbol correlation process to the frequency offset. Fig. 9.11 (b) shows the acquisition time performances with respect to the frequency offset that varies from 0 to 4kHz(2ppm) for a fixed chip-SNR of -10dB. We observe that the DSA acquisition time increases rapidly as the frequency offset becomes larger than 1kHz, which means that the DSA synchronization scheme may malfunction

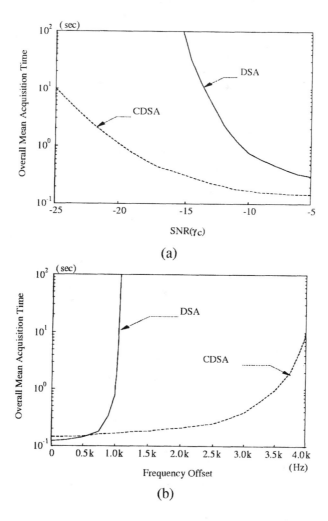

Figure 9.11. Overall mean acquisition times in Rayleigh fading channels with frequency offset ($b = 1$, $|\eta| = 0$): (a) vs. chip-SNR(γ_c), 1kHz frequency offset, 100Hz Doppler, (b) vs. frequency offset, -10dB chip-SNR, 100Hz Doppler.

even in the moderate SNR range. In contrast, the CDSA scheme maintains short acquisition time in stable and gradual manner.

3.   APPLICATION TO INTER-CELL ASYNCHRONOUS SYSTEMS

Now we consider how to apply the CDSA technique to cell search in the inter-cell asynchronous IMT-2000 W-CDMA systems. It is a promising challenge to use the igniter sequence and state symbol-based CPICH for efficient cell search in this new inter-cell asynchronous cellular systems. This CDSA technique should prove itself to be a fast and robust two-stage cell search scheme for the IMT-2000 W-CDMA systems. As will soon become clear through extensive performance comparisons, the CDSA substantially outperforms the 3GPP three-step approach in terms of code acquisition time and system power consumption.

3.1   CDSA-BASED CELL SEARCH IN IMT-2000 W-CDMA SYSTEMS

In the IMT-2000 W-CDMA systems, a cell (or sector) may use multiple scrambling codes to enhance the downlink code capacity [8]. Included among them is the primary scrambling code which is uniquely determined by the cell-number and used to scramble the primary CPICH and the primary CCPCH. In the initial cell search process an MS identifies and synchronizes the primary scrambling code of the cell currently located. The primary scrambling code allocated to each BS is a segmented *complex* Gold sequence of length 38,400. In fact, the I-phase primary scrambling sequence is this segmented Gold sequence which is obtained by binary-adding two sets of the first 38,400 m-sequence segments respectively generated by two different shift register generators (SRG) having the characteristic polynomials $\Psi_1(x) = x^{18} + x^7 + 1$ and $\Psi_2(x) = x^{18} + x^{10} + x^7 + x^5 + 1$ respectively.

We assume that the cellular system is composed of 512 cells numbered from 1 to 512 and that the 1st m-sequence generator (or, main SRG-1) of each BS is initially loaded with the cell-number and the 2nd m-sequence generator (or, main SRG-2) with all-1 string. We decimate the two segmented m-sequences by the decimation period of 256 chips to get two *state sample sequences* of period 150 each, with each sequence corresponding to a different segment of the original m-sequences. Then we get 512 different sequences of period 150, $\{z_{i,k}^{(1)} : i = 0, 1, \cdots, 149, \cdots\}$, $k = 1, 2, \cdots, 512$, for the 1st m-sequences of 512 BS's, and one sequence $\{z_i^{(2)} : i = 0, 1, \cdots, 149, \cdots\}$ for the 2nd m-sequences. These 513 available sample sequences of length 150 each are stored in advance in the memory of each MS.

On the other hand, the Q-phase primary scrambling sequence is obtained by taking the first 38,400 sequence values of the time-lagged version (lagged by 131,072 chips [8]) of the Gold sequence used for I-phase primary scrambling code generation. According to the shift-and-add property of the constituent m-sequences, the Q-phase sequence can be easily generated from the main SRG-1 and main SRG-2 by employing the corresponding *sequence generating vectors* [4]. Since the system chip rate is 3.84Mcps [139], the period of the primary scrambling sequence is commensurate with 10ms, which corresponds to the frame time of the transmitted signals. The frame is composed of 15 slots of length 2,560 chips each.

In the following we detail the CDSA scheme for the DQPSK modulation system.

Fig. 9.12 depicts the functional block diagram of the BS and MS in the kth cell ($k = 1, 2, \cdots, 512$) of the cellular system that employs the DQPSK-based CDSA scheme. The time-advanced sampling block of the BS takes the *state samples* $z_{r+i,k}^{(1)}$ and $z_{r+i}^{(2)}$ out of the main SRG-1 and the main SRG-2 at time $(r + i - 1) \times 256T_c$ employing the sampling vectors $\mathbf{v}_1$ and $\mathbf{v}_2$, for a reference value r, and supplies them to the DQPSK modulator. The main SRG-1 and the main SRG-2 are the m-sequence generators that respectively generate the 1st and the 2nd segmented m-sequences of length 38,400. The DQPSK modulator first maps the two state samples to the I-phase and the Q-phase of the QPSK constellation to generate the *state symbol* $x_{r+i,k}$, and then produces the differentially encoded symbol $f_{r+i,k}$ by adding the phase of $x_{r+i,k}$ to the phase accumulated up to the previous symbol time. As the period of the primary scrambling code (38,400 chips) is a multiple of the sampling period (256 chips), the resulting state symbol sequence $\{x_{r+i,k}\}$ also becomes periodic, with the period being 150 symbols. The differentially encoded symbol $f_{r+i,k}$ is broadcast through the primary CPICH channel in the time interval $[(r + i - 1) \times 256T_c, (r + i) \times 256T_c)$.

The spreading sequence of the primary CPICH channel $\{c_{m,k}\}$, which we call the *igniter sequence* of the cell, is one of the 16 short sequences of length 256. The short sequences used here are the complex orthogonal Gold sequences [53] but may be changed into another set of sequences for improved performances (For example, the set of 16 Hadamard-modulated Golay codes used for secondary SCH of the 3GPP three-step approach [8] may be used as the igniter sequence set.). The multiple short sequences are necessary for the spatial separation of co-code cells, each of which has a one-to-one correspondence with each cell group composed of 32 cells.

In parallel with the primary CPICH transmission, the PSC symbols spread by the GHG code of length 256 are broadcast through the SCH at the beginning of each slot for fast synchronization of the slot boundary. During the period when the the PSC symbols are not transmitted, the broadcast channel (BCH) symbols are transmitted through the primary CCPCH. However, the SSC is not transmitted in the proposed CDSA scheme, thereby making all the SCH power be used for slot boundary acquisition only, without being divided into two for slot boundary acquisition and frame/group identification as is done in the 3GPP 3-step scheme case. On the other hand, each user's M-ary PSK data $\{a_{r+i,k}^{[l]}\}$ is spread by an orthogonal Walsh sequence $\{w_m^{[l]}\}$ and scrambled by the primary

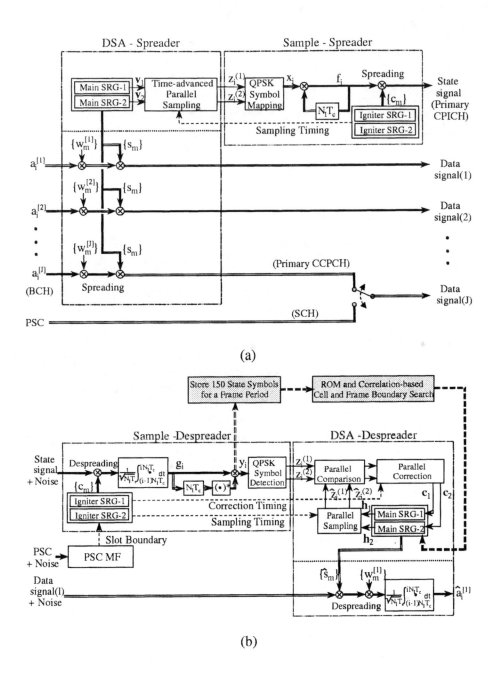

Figure 9.12. Functional block diagram of the acquisition-related part of the IMT-2000 W-CDMA system employing the DQPSK-based CDSA scheme: (a) Base-station (transmitter), (b) mobile-station (receiver).

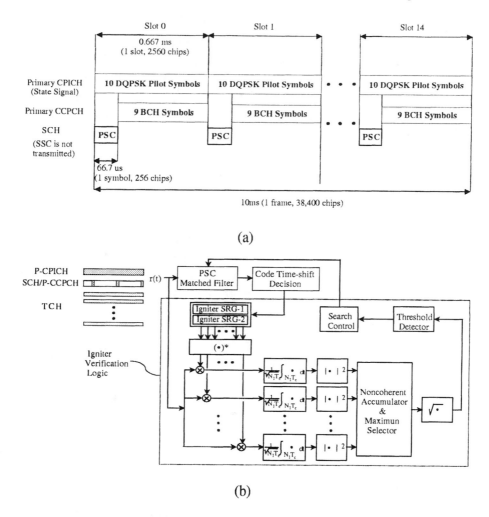

Figure 9.13. Channel structure and receiver structure of passive matched filter based CDSA system: (a) Synchronization-related BS channel structures (BCH:Broadcast channel), (b) igniter sequence acquisition and verification related receiver structure.

or one of the secondary scrambling sequences of the cell, and then transmitted as the data signal.

Fig. 9.13 depicts the channel structure and the receiver structure of the passive matched filter-based CDSA system. The primary CPICH, the SCH, the primary CCPCH, and the other data and control signals go through the same channel to the MS. We assume the single path Rayleigh fading channel, which is the worst-case channel model as far as the initial synchronization process is concerned [54].

3.1.1   (1ST STAGE) SLOT BOUNDARY ACQUISITION AND IGNITER SEQUENCE IDENTIFICATION

The MS first acquires the slot boundary by using the PSC, for which we take the 1st step of the 3GPP three-step approach, i.e., the matched filtering with the Golay correlator. More specifically, the MS first takes a discrete sample stream by sampling the incoming analog signal at every half chip interval. Then, it stores 5,120 correlation values per slot obtained by correlating the sample stream with the PSC by employing the Golay correlator of length 256 [52]. Finally, it acquires the slot boundary by taking the sample position that produces the maximum correlation energy out of the 5,120 candidate positions. For an improved acquisition performance in the fading channel, we usually take the slot-wise energy accumulation throughout S slots before taking the sample position having the maximum correlation energy [54].

After acquiring the slot boundary, the MS identifies the igniter sequence broadcast in the current cell by correlating the incoming primary CPICH signal with 16 candidate igniter sequences in parallel and then determining the sequence having the maximum correlation energy. We take the symbol-wise energy accumulation throughout V_I primary CPICH symbols for an improved identification performance. By verifying whether or not the square-rooted maximum correlation energy overtakes a pre-specified threshold value R_I, we can also confirm whether or not the slot boundary is truly synchronized. If the square-rooted maximum energy turns out to overtake R_I, the 1st step terminates, but otherwise, the slot boundary acquisition step resumes. Once one of the 16 igniter sequences is determined, the cell uncertainty reduces from 512 to 32, according to the mapping relation between the igniter sequences and the co-code cells.

3.1.2   (2ND STAGE) MAIN SRG SYNCHRONIZATION AND STATE SYMBOL CORRELATION

A. Comparison-correction Based Main SRG Synchronization: Once the slot boundary is acquired, the MS then acquires the primary scrambling code through the *comparison-correction based main SRG synchronization* process. For this, the MS first despreads the received primary CPICH signal $r(t)$ using the acquired igniter sequence to get the *soft CPICH symbol* g_{r+i} and then multiplies it to the conjugate of the previously obtained symbol g_{r+i-1}^* to get the *soft state symbol* y_{r+i}. More specifically, if we denote by P_s the primary CPICH signal power, by $f_{n,k} \equiv e^{j\theta_{n,k}}$ the nth DQPSK symbol, by $c_k(t) \stackrel{\triangle}{=} \sum_{m=0}^{255} c_{m,k} p_{T_c}(t - mT_c)$ the igniter sequence of the kth BS, [16] by $N(t)$ the

---

[16]We define $p_w(t)$ to be the pulse shaping function which is 1 in $[0, w]$ and 0 elsewhere.

channel noise, by H the channel gain, by ω_o the frequency alignment offset, by ϕ the carrier phase, and by ηT_c the fractional chip alignment offset, then we get the expressions

$$r(t) = \sum_{n=-\infty}^{\infty} \sqrt{P_s} H e^{j(\omega_o t + \phi + \theta_{n,k})} \times c_k(t + \eta T_c - [n-1] \times 256 T_c)$$

$$+ N(t), \tag{9.15}$$

$$g_{r+i} = \frac{1}{\sqrt{256 T_c}} \times \int_{256(r+i-1)T_c}^{256(r+i)T_c} r(t) c_k^*(t - [r+i-1] \times 256 T_c) dt, \tag{9.16}$$

and

$$y_{r+i} = g_{r+i} g_{r+i-1}^*. \tag{9.17}$$

The channel noise is assumed complex white Gaussian distributed, with its I-phase and Q-phase components both having the power spectral density of $N_0/2$.

The DPSK symbol detection block then determines the conveyed state samples $z_{r+i,k}^{(1)}$ and $z_{r+i}^{(2)}$, which are then passed to the DSA-despreader at time $(r+i) \times 256 T_c$. On the other hand, the DSA-despreader generates its own state samples $\hat{z}_{r+i}^{(1)}$ and $\hat{z}_{r+i}^{(2)}$ at time $(r+i) \times 256 T_c$ from the MS main SRG-1 and main SRG-2, and compares them with the conveyed counterparts $z_{r+i,k}^{(1)}$ and $z_{r+i}^{(2)}$. If $\hat{z}_{r+i}^{(1)}$ and/or $\hat{z}_{r+i}^{(2)}$ do not coincide with $z_{r+i,k}^{(1)}$ and/or $z_{r+i}^{(2)}$ respectively, the corresponding jth $(j = 1, 2)$ correction circuit c_j is triggered to correct the state of the main SRG-j at time $(r+i) \times 256 T_c + D_c T_c$, with D_c chosen such that $0 < D_c \leq 256$. Otherwise, no action takes place. If we employ a pair of sampling and correction circuits designed for main SRG-1 and main SRG-2 according to Section 3.2, the MS main SRGs will get synchronized to the BS main SRGs whenever 18 successive state symbols are conveyed without any detection errors.

For fear that there may occur one or more detection errors while detecting the 18 symbols, we observe $V=7$ additional symbols to verify the synchronization. Since the main SRGs in the BS and the MS generate the same sequence samples after the true synchronization, we can verify the synchronization by comparing the observed samples with the MS generating samples. If the conveyed and receiver-generated symbols indeed coincide for all V symbol pairs, then the MS declares completion of synchronization of the primary scrambling sequence, beginning to track and estimate the channel gain and carrier phase. If less than V symbol-pairs coincide, the MS restarts the comparison-correction based main SRG synchronization using another set of 18 successive state symbols. Therefore, it nominally takes 25 state symbol periods (i.e., 1.66ms) to complete a run of this main SRG synchronization process of the CDSA, but it could take multiple runs in poor channel environments as detection errors would occur more frequently.

B. State Symbol Correlation: To prevent unbounded runs going in the main SRG synchronization process, we initiate a synch-termination process after detecting 150 soft state symbols, by correlating them with the pre-stored state sample sequences. This state symbol correlation process is composed of two phases - - frame boundary detection and cell number detection. For frame boundary detection, we first correlate the soft state symbol sequence $\{y_{10\tilde{r}+i} : i = 0, 1, \cdots, 149\}$, for an arbitrary integer $\tilde{r}$, with the 15 shifted versions of the pre-stored 2nd state sample sequence $\{w^{(2)}_{10n+i} : i = 0, 1, \cdots, 149\}$, $n = 0, 1, \cdots, 14$, where $w^{(2)}_i \triangleq (-1)^{z^{(2)}_i}$. Note that the timing reference r is set to $10\tilde{r}$, which indicates that the soft state symbol collection begins at the start of a slot, taking advantage that the slot boundaries are readily acquired in the slot identification step. Then we compare the resulting 15 correlation energy values

$$E_{n,frame} = |\frac{1}{\sqrt{150}} \sum_{i=0}^{149} y_{10\tilde{r}+i} w^{(2)}_{10n+i}|^2, \quad n = 0, 1, \cdots, 14, \qquad (9.18)$$

and determine the time shift number $n = n_0$ at which the correlation energy becomes maximum as the frame boundary. For cell number detection, we correlate the soft state symbol sequence $\{y_{10\tilde{r}+i} : i = 0, 1, \cdots, 149\}$ with the 32 candidate (1st state sample) sequences obtained at the end of the igniter sequence identification process. More specifically, if the acquired igniter sequence $\{c_k(t)\}$ is mapped to the 32 co-code cells using the 1st state sample sequences $\{z^{(1)}_{i,q}\}$, $q = 1, 2, \cdots, 32$, we compare the 32 correlation energy values

$$E_{q,cell} = |\frac{1}{\sqrt{150}} \sum_{i=0}^{149} y_{10\tilde{r}+i} w^{(1)}_{10n_0+i,q}|^2, \quad q = 1, 2, \cdots, 32, \qquad (9.19)$$

where $w^{(1)}_{i,q} \triangleq (-1)^{z^{(1)}_{i,q}}$, and determine the cell number $q=q_0$ at which the correlation energy becomes maximum. Now that the frame boundary information n_0 and the cell number information q_0 are both available, we can complete the cell search by loading the corresponding initial state values (or, the cell number information) to the 1st main-SRG and loading the all-1 string to the 2nd main-SRG at the moment the next frame boundary begins. This completes the 2nd stage synchronization process.

In the state symbol correlation process, we deliberately use the *differentially demodulated* state symbols to enhance the robustness of the scheme in the frequency offset environment. If there were no frequency offset at all, the CPICH symbols $\{g_i\}$ would yield better correlation performance than the state symbols $\{y_i\}$ does, as the state symbols accompany higher noise variances due to the differential demodulation process. In reality, however, frequency offset is unavoidable in any initial synchronization process, and CPICH symbol based correlation degenerates even at a small amount of frequency offset because the symbols are collected over relatively long period (i.e., a frame time, 10ms). By using differentially demodulated symbols for the correlation, we can maintain normal correlation performances unless the frequency offset becomes exceedingly high to cause a considerable phase angle rotation during the symbol interval.

3.1.3   SUPPLEMENTAL PROCESSINGS

Once the cell search is completed by way of the comparison-correction based SRG synchronization process, we declare synchronization, but at the same time we carry out a validity test by checking whether or not the state of the 1st m-sequence generator takes one of the 32 valid cell numbers when the state of the 2nd m-sequence generator takes all-1 string. This test can be done within a frame time. If this validity test fails, we recall the synchronization declaration and load the main SRGs with the initial state values determined through the state symbol correlation process. [17] Fig. 9.14 depicts the resulting overall cell search process (see the solid line part). Despite these verification and validity tests, there still exists a small probability of false acquisition, no matter how small it may be. Thus we keep monitoring the data decoding performance even after the acquisition completion declaration. If a false acquisition is really detected in this stage, the MS recalls the declaration of the synch-completion and restarts the slot boundary search stage, which may be modeled by a delay of K frame times, i.e., $K \times 10$ms (see the dashed line part).

Once the synchronization process is done, with the MS igniter and main SRGs synchronized to those in the BS, the MS first despreads the data signal by multiplying the conjugate of the corresponding primary (or secondary) scrambling sequence and the corresponding Walsh sequence. It then regenerates the broadcasted primary CPICH signal (only with an initial phase difference involved in the DQPSK modulator), and estimates the channel characteristics in a way similar to that of the conventional DS/CDMA cellular systems employing the unmodulated pilot sequences [7]. Finally it coherently demodulates the despread data using the channel estimation results. The phase ambiguity problem

---

[17]If the all-1 state does not appear in the main SRG-2 by the frame boundary time determined by the state symbol correlation process, the synchronization declaration is regarded as a false alarm, and the initial states obtained in the state symbol correlation process are loaded. The state validity check can be done even before the all-1 state appears in the main SRG-2 at the cost of increased memory use by pre-storing more state mapping data between the two main SRGs of the 512 cells.

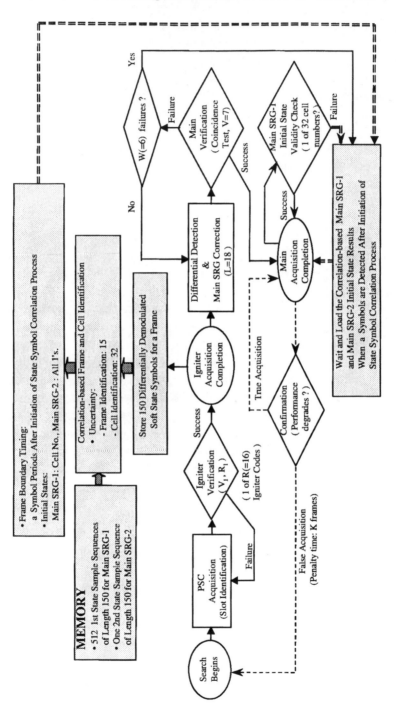

Figure 9.14. Flow diagram outlining the CDSA-based cell-search process for IMT-2000 W-CDMA systems. (The blocks in shade are those related to state symbol correlation process.)

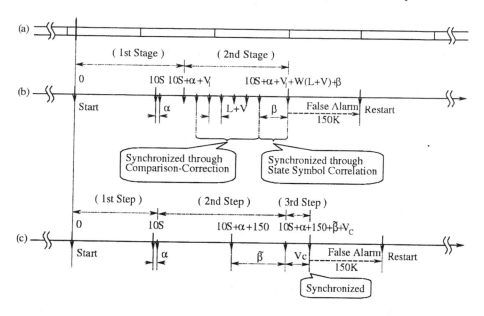

Figure 9.15. Synchronization timing diagram of CDSA and 3GPP 3-step schemes: (a) Primary CPICH frames, (b) CDSA scheme, (c) 3GPP 3-step scheme (see Section 6.2.3 for the definition of parameters).

of the CPICH introduced by employing the differential encoding technique can be resolved by observing the phase of the *dedicated* pilot symbols transmitted in the traffic/control channels or by employing the data constellation pre-rotation technique introduced in Section 5 of Chapter 8.

Fig. 9.15 depicts the synchronization timing diagram of the DQPSK-based CDSA scheme in comparison with that of the 3GPP 3-step scheme: Fig. 9.15 (a) shows the frame boundaries of the primary CPICH signal arriving at the MS, with each block indicating 38,400 chip time (10ms). Fig. 9.15 (b) and (c) respectively show the overall synchronization timing of the CDSA scheme and the 3GPP 3-step scheme, with unit time indicating 256 chips (66.7us). It is assumed in calculating both synchronization times that the data processing time is negligibly small (by dint of high-speed processors) when compared with the data collection time. The parameters α and β [or $\tilde{\beta}$] in Fig. 9.15 (b) and (c) are the random variables that denote the time interval between the slot boundary determination and the slot boundary appearance and the time interval between the frame boundary determination and the frame boundary appearance for the CDSA scheme [or the 3GPP 3-step scheme] respectively. Both random parameters are uniformly distributed, with α being in $[0, 10)$ and β or $\tilde{\beta}$ being in $[0, 150)$. In addition, we denote by $S(=15)$ the number of slots accumulated for the slot boundary acquisition, $V_I(=50)$ the number of accumulated symbols for the igniter sequence identification, $L + V(=25)$ the number of symbols used for a run of the comparison-correction, $W(=6)$ the number of comparison-correction based synchronization processes tried before the initiation of the state symbol correlation process, $K(=1)$ the number of frames lost by the false acquisition declaration, and $V_c(=50)$ the number of accumulated symbols for the cell-specific scrambling sequence identification. We observe that the expectation of the nominal cell search time is 430 symbol time (28.66ms) for both schemes if the CDSA synchronization is completed after the state symbol correlation process. However, in the CDSA case, it reduces to 205 symbol time (13.66ms) in high-SNR environments where the comparison-correction process satiates the synchronization. Note that in the comparison-correction based synchronization, six possible chances are given to acquire the synchronization in a frame time of 10ms long, [18] and the state symbol correlation process is initiated only when the six chances are all missed.

3.2   SYNCHRONIZATION PARAMETER DESIGN

In this section we focus on the design of the main sequence sampling and correction circuitry as well as the multiple igniter sequence generation circuitry for W-CDMA system applications. We apply the DSA design method developed for m-sequences independently to get the jth time-advanced sampling vector $\mathbf{v}_j$ and the jth correction vector $\mathbf{c}_j$.

According to the synchronization parameter design procedure introduced in Section 1 of Chapter 8, we first define the jth *discrimination matrix* $\Delta_{\mathbf{T}_j, \mathbf{h}_j}$ to be the $L \times L$ matrix

$$\Delta_{\mathbf{T}_j, \mathbf{h}_j} \stackrel{\triangle}{=} \left[\mathbf{h}_j \quad (\mathbf{T}_j^{N_I})^t \cdot \mathbf{h}_j \quad (\mathbf{T}_j^{2N_I})^t \cdot \mathbf{h}_j \quad \cdots \quad (\mathbf{T}_j^{(L-1)N_I})^t \cdot \mathbf{h}_j \right]^t,$$
$$j = 1, 2, \quad (9.20)$$

where $\mathbf{T}_j$ and $\mathbf{h}_j$ respectively denote the *state transition matrix* and the *sequence generating vector* of the main SRG-j. Then by Corollary 10.1 both $\Delta_{\mathbf{T}_1, \mathbf{h}_1}$ and $\Delta_{\mathbf{T}_2, \mathbf{h}_2}$ become nonsingular if *we choose N_I such that it is relatively prime to* $2^L - 1$. Once the period of the igniter sequence is chosen to meet this condition, we can determine, by Theorem 8.2 and Theorem 8.3, the correction vector $\mathbf{c}_j$ and the time-advanced sampling vector $\mathbf{v}_j$

$$\mathbf{c}_j = \mathbf{T}_j^{(L-1)N_I + D_c} \cdot \Delta_{\mathbf{T}_j, \mathbf{h}_j}^{-1} \cdot \mathbf{e}_{L-1}, \quad (9.21)$$

$$\mathbf{v}_j = (\mathbf{T}_j^{N_I})^t \cdot \mathbf{h}_j, \quad (9.22)$$

where the L-vector $\mathbf{e}_i$, $i = 0, 1, \cdots, L - 1$, denotes the ith standard basis vector whose ith element is 1 and the others are 0.

---

[18] Only five out of six chances are normally available unless the comparison-correction process begins at the frame boundary. Refer to [66] for details.

The simultaneous synchronization of the main SRG-1 and the main SRG-2, which is acquired by conveying the samples $z_i^{(1)}$ and $z_i^{(2)}$ simultaneously in the form of a state symbol and then applying the DSA synchronization process, enables the MS to identify the current cell and acquire the scrambling code timing in very efficient manner.

EXAMPLE 9.1 We assume that the cellular system is composed of 512 cells and each cell belongs to one of the 7 cell groups ($R=7$) according to the igniter sequence reuse pattern shown in Fig. 9.6 (b). The 512 I-phase main sequences of the system are the 38,400 chip segments of a set of Gold sequences generated by binary-adding the two m-sequences whose characteristic polynomials are $\Psi_1(x)=x^{18}+x^7+1$ and $\Psi_2(x)=x^{18}+x^{10}+x^7+x^5+1$, respectively. The transition matrices T_1, T_2 and the sequence generating vectors h_1, h_2 respectively of the main SRG-1 and the main SRG-2 are given by

$$T_j = \begin{bmatrix} 0 & I_{17\times17} \\ 1 & t_j \end{bmatrix}, \ j = 1, 2, \tag{9.23a}$$

$$t_1 = [\,0\,0\,0\,0\,0\,0\,1\,0\,0\,0\,0\,0\,0\,0\,0\,0\,0\,],$$

$$t_2 = [\,0\,0\,0\,0\,1\,0\,1\,0\,0\,1\,0\,0\,0\,0\,0\,0\,0\,], \tag{9.23b}$$

$$h_1 = h_2 = [\,1\,0\,0\,0\,0\,0\,0\,0\,0\,0\,0\,0\,0\,0\,0\,0\,0\,0\,]^t. \tag{9.23c}$$

The I-phase main sequence of the kth cell ($k=0,1,\cdots$, 511) base station is generated in such a way that a 18-bit string of the cell number (e.g., binary expression of $k+1$) is loaded in the main SRG-1 and a 18-bit string of 1's is loaded in the main SRG-2 as the initial state, and the SRG states advance according to the BS system clock. The Q-phase main sequence of each cell is just the time-lagged version of the I-phase main sequence by the amount of 131,072 chips [8], which can be easily generated from the main SRG-1 and main SRG-2 by employing the *time-advanced generating vectors* $\tilde{h}_j = (T_j^{131,072})^t \cdot h_j$, $j=1,2$. The resulting Q-phase generating vectors are:

$$\tilde{h}_1 = [\,000010100000000100\,]^t, \ \tilde{h}_2 = [\,000001101111111100\,]^t. \ (9.24)$$

The states of the main SRG-1 and main SRG-2 are reset to their initial values (i.e., the cell number and all-1 string, respectively) after advancing by 38,400 chips.

On the other hand, the 7 pairs of I-phase and Q-phase igniter sequences are taken out of a set of 256 orthogonal Gold sequences of period 256, each of which is generated by binary-adding two m-sequences of period 255 and then inserting an extra 0 bit at the sequence head. The characteristic polynomials of the component m-sequences are $I_1(x)=x^8+x^4+x^3+x^2+1$ and $I_2(x)=x^8+x^6+x^5+x^3+1$, and the initial states of the igniter SRG-1 and igniter SRG-2 are set to "00000001" and "11111111" respectively. The 0 bit insertion is done whenever the igniter

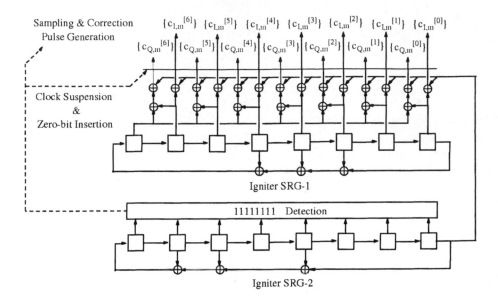

Figure 9.16.   Igniter sequence generator: Orthogonal Gold sequences ($R=7$).

SRG-2 reaches the state "11111111". To get 7 pairs of different complex orthogonal Gold sequences, 14 arbitrary generating vectors (or, 7 pairs of vectors) take as many time-shifted m-sequences from the igniter SRG-1 and add (in binary operation) them to an m-sequence taken from the igniter SRG-2. The resulting example structure is shown in Fig. 9.16.

For the main SRG of length $L=18$ and the igniter sequence of period $N_I=256$, the sampling interval 256 is relatively prime to $2^{18} - 1$. So the discrimination matrices $\Delta_{\mathbf{T}_j,\mathbf{h}_j}$, $j=1,2$, determined by Eq.(9.20) become nonsingular. Therefore, if we take the correction delay D_c of 1, then, by Eqs.(9.21) and (9.22), we get the following correction and time-advanced sampling vectors: [19]

$$\mathbf{c}_1 = [\,0\,0\,1\,0\,1\,0\,1\,0\,0\,0\,0\,1\,1\,1\,0\,1\,0\,0\,]^t,$$
$$\mathbf{c}_2 = [\,0\,0\,1\,0\,1\,1\,1\,1\,0\,1\,0\,1\,0\,1\,0\,1\,0\,0\,]^t, \qquad (9.25a)$$
$$\mathbf{v}_1 = [\,0\,0\,0\,1\,1\,0\,0\,1\,1\,0\,0\,0\,0\,0\,0\,0\,0\,0\,]^t,$$
$$\mathbf{v}_2 = [\,1\,1\,1\,0\,1\,0\,1\,1\,1\,0\,0\,0\,0\,0\,1\,0\,1\,1\,]^t. \qquad (9.25b)$$

Fig. 9.17(a) and (b) depict the resulting DSA spreader and despreader circuits that incorporate the designed sampling and correction functions. The two cir-

---

[19]Note that the last (i.e., 150th) time-advanced samples to be transmitted in a frame is not the 38,400th but the 0th sequence values of the two component m-sequences.

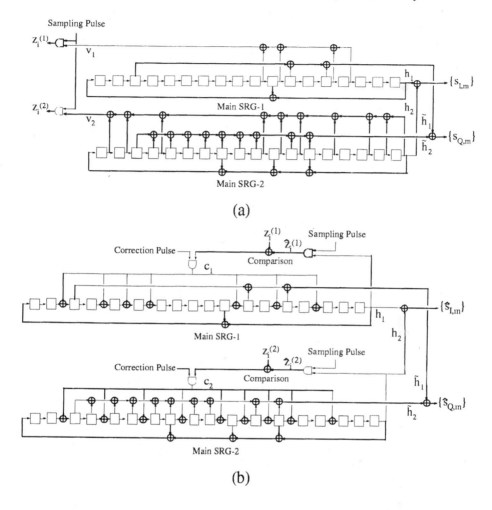

Figure 9.17. Main sequence generator with the sampling and correction vectors incorporated: (a) Base station, (b) mobile station.

cuits exactly fit the DSA spreader and despreader blocks (i.e., the portion above the dotted line) in Fig. 9.12 (a) and (b).

3.3 PERFORMANCE EVALUATIONS

We examine the performances of the CDSA scheme within the IMT-2000 W-CDMA system in terms of mean acquisition time and sytem complexity, and compare them with the 3GPP 3-step synchronization case.

3.3.1   MEAN ACQUISITION TIME

For the simulations, we take the single-path Rayleigh fading channel, which is the worst-case channel model in regard to code acquisition process [54]. We take the JTC flat and classic (U-shape) spectra of maximum Doppler frequencies 5Hz and 200Hz to generate the Rayleigh fading channels for the low-speed and high-speed mobile users respectively. We take the coefficients of the IIR spectral filters of order 32 that are normalized to make the average power of the output samples unity [138]. We set the false alarm penalty factor K to 1 (i.e., 10ms penalty time) as we can easily confirm the synchronization by observing one frame of the primary CCPCH signal. In calculating the total acquisition time, we do not take into account the MS processing capability dependent factors such as the processing time for determining the maximally correlated code or timing in each step. We set the chip rate to 3.84Mcps [139] and process two samples per chip for baseband simulations.

Fig. 9.18 shows how the power allocation and interference generation are arranged for the performance simulations. Note that both the PSC and the SSC are broadcast through the SCH in the 3GPP 3-step approach while only the PSC is broadcast in the CDSA approach. In the 3GPP 3-step case, the SCH (PSC and SSC)/primary CCPCH and the primary CPICH respectively use 10%, of the total BS power, with the remaining 80% being devoted to the other control and traffic channels (i.e., $\alpha=0.1$). [20] In the CDSA case, we consider two different power allocations - - one when the SCH(PSC)/primary CCPCH and the primary CPICH respectively use 10% (i.e., $\alpha=0.1$) of the BS total power and the other when they use only 5% (i.e., $\alpha=0.05$): The former is for a fair comparison of cell search performance with the 3GPP 3-step synchronization case and the latter for investigation of possible reduction of the BS power portion for cell search. Other cell interference signals including the background noise are added to the intra-cell signal after the Rayleigh channel multiplication. The intra-cell control/traffic signals which are not related to the acquisition and other cell interferences are approximated to complex white gaussian random processes in the simulation.

For data processing in the MS, we take the following arrangements: In the 3GPP 3-step case we noncoherently accumulate 15 PSC slots (or, 1 frame) for the 1st step synchronization, 15 SSC slots for the 2nd step synchronization, and 50 primary CPICH symbols (or, 5 slots) for the 3rd step synchronization. [21] When grouping the 512 cells we consider the case of 32 cell-groups having 16

---

[20] We assume that the BS power allocated to the SCH is equally divided for PSC and SSC transmissions in the 3GPP scheme.

[21] After the introduction of the CPICH within the IMT-2000 W-CDMA system, the primary CPICH has been considered as a possible means for the 3rd step synchronization. In real implementations, however, we may use either the primary CPICH or the primary CCPCH for the 3rd step.

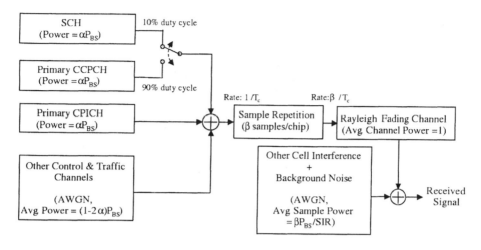

Figure 9.18. Power allocation and interference generation employed in performance simulations. (α=0.1 or 0.05, β=2.)

cells each and the case of 64 cell-groups having 8 cells each [8, 53]. In the CDSA case, we accumulate 15 PSC slots for slot boundary identification (i.e., S=15) and 50 CPICH symbols (or, 5 slots) for igniter sequence identification and verification (i.e., V_I=50) in the 1st stage; and maximally 150 CPICH symbols (or, 1 frame) for the comparison-correction based main SRG synchronization and state symbol correlation processes in the 2nd stage. In support of the igniter sequence identification process, we carry out a slot boundary verification test with respect to the threshold R_I set such that the false alarm probability in the verification stage becomes 0.01.

The simulation results thus carried out are shown in Fig. 9.19 through Fig. 9.21. Fig. 9.19 plots the mean acquisition times of the CDSA and the 3GPP 3-step schemes with respect to the signal-to-interference ratio (SIR, defined by the ratio of the average intra-cell total signal energy per chip to the other cell interference spectral density) varying from -15dB to +10dB. Fig. 9.19 (a) is the case when 10% of the BS power is allocated to both SCH and primary CPICH channels in both the CDSA and 3GPP 3-step schemes, and Fig. 9.19 (b) is the case when the power allocation drops to 5% in the CDSA scheme and is kept at 10% in the 3GPP 3-step scheme. We observe that in the low-SIR range the CDSA scheme has an SIR gain of about 7dB over the 3GPP 3-step scheme in the Fig. 9.19 (a) case of equal power allocation. The CDSA scheme yields much shorter acquisition time than the 3GPP 3-step scheme even in the Fig. 9.19 (b) case. In the high-SIR range, the mean acquisition time reduces to less than 20ms in the CDSA case, while it maintains the nominal acquisition time of

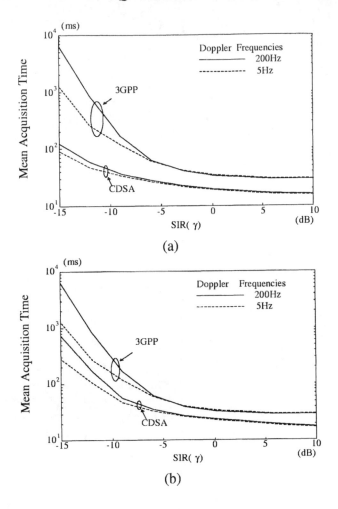

Figure 9.19. Overall mean acquisition time performances with respect to SIR (γ) change in Rayleigh fading channel (32 groups for 3GPP): (a) $\alpha = 0.1$ for both schemes, (b) $\alpha = 0.05$ for CDSA scheme and 0.1 for 3GPP 3-step scheme.

28.6ms in the 3GPP 3-step case. In terms of Doppler frequency, we observe that the acquisition time performance exhibits similar trends in both schemes with the slowly varying indoor channel (5Hz, flat spectrum) outperforming the fast varying outdoor channel (200Hz, classic spectrum) in the low SIR range, but comparable in the high SNR range.

Fig. 9.20 plots the mean acquisition time degradations caused by frequency offset between the BS and the MS. In normal operation, code acquisition is done before the frequency discrepancy between BS and MS reduces (through phase-locked loop (PLL) operation) to a satisfactory level, so the robustness to frequency offset is critically important in acquisition. Shown in the figure is the case of the outdoor channel of Doppler frequency of 200Hz, but a similar performance degradation was observed for the indoor channel of 5Hz Doppler frequency. The case of 10% power allocation is considered for both CDSA and 3GPP 3-step schemes. Fig. 9.20 (a) shows the performance degradation due to the frequency offset of 5kHz (or, 2.5ppm for 2GHz carrier frequency) with reference to the no-offset case. We find that at the SIR of -6dB, mean acquisition time is about 30ms for the CDSA scheme (i.e., little degradation) but increases above 100ms for the 3GPP 3-step scheme, which is twice as long as that for the no-offset case. We also observe in the CDSA case that the mean acquisition time is under-bounded by 28.6ms even in the high SIR range, which implies that acquisition is mostly completed through the state symbol correlation process when the frequency offset is large. Fig. 9.20 (b) shows the performance degradation with respect to the frequency offset that varies from 0 to 10kHz (5ppm) for a fixed value of SIR of -3dB. We observe that acquisition time increases exceedingly rapidly as the frequency offset increases, which warns that the 3GPP 3-step synchronization scheme may malfunction even in the moderate SIR range. In contrast, the CDSA scheme maintains short acquisition time in stable and gradual manner. Notice that the acquisition time performance degradation due to the frequency offset becomes a serious problem in critical regions such as shadowed areas or cell boundaries, where SIR becomes very low and thus acquisition time increases rapidly.

Fig. 9.21 compares the mean acquisition time performance of the 3GPP scheme for different cell groups. We consider the numbers 32 [53] and 64 [8, 110] for this as they have been proposed as the candidate group numbers of the 3GPP 3-step scheme. Fig. 9.21 (a) shows the mean acquisition time performances with respect to the SIR that varies from -15dB to +10dB at no frequency offset, and Fig. 9.21 (b) with respect to the frequency offset that varies from 0 to 10kHz for the SIR of -3dB. We observe that both cell groups exhibit almost the same acquisition time performances. This signifies that the performance comparisons given in Figs. 9.19 and 9.20 also apply to the 3GPP 3-step scheme having 64 cell groups.

A potential disadvantage of the CDSA scheme using a short-period pilot sequence and a long-period traffic scrambling sequence is the non-orthogonality problem between the pilot and the traffic signals. An efficient solution to this problem can be found in the pilot sequence cancellation technique [140], which gets rid of the known pilot components from the received signal before detecting the traffic data. We took a very simple pilot cancellation approach, where each finger independently subtracts the igniter sequence multiplied by the estimated path gain from the received chip value. This technique can be realized without adding any additional complex processing as the channel estimation information of each signal path is readily available for the data demodulation purpose. Fig. 9.22 plots the frame error rate (FER) performances of the 3GPP W-CDMA

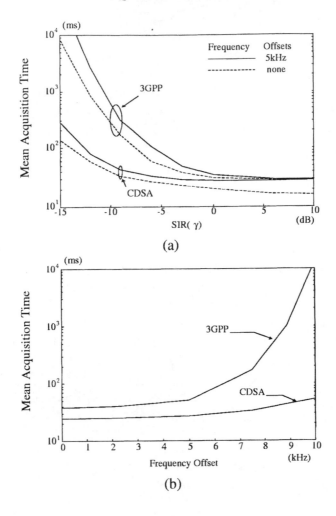

Figure 9.20. Mean acquisition time degradation due to frequency offset in Rayleigh fading channel (32 groups for 3GPP): (a) Overall mean acquisition times vs. SIR (γ) for 5kHz frequency offset and no-offset, (b) overall mean acquisition times vs. frequency offset for -3dB SIR.

system, the DSA system with no cancellation, and the DSA system with pilot cancellation, for the 9.6kbps traffic channels coded by 1/3 rate convolutional coder with the constraint length 9. The ITU vehicular-B model is employed for the multipath channel [141], and the spreading factor of the traffic channels is set to 128. The parameter E_c and I_{or} respectively denote the transmitted chip energy of the reference traffic channel and the total BS signal power spectral density measured in the BS. The figure shows that the FER performance of the

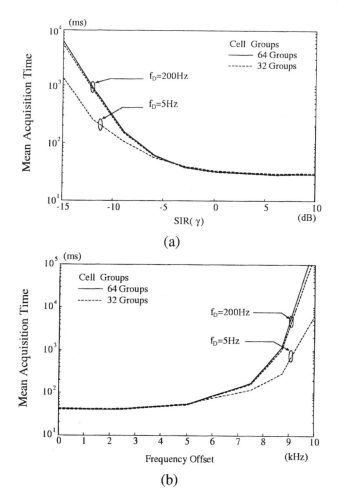

Figure 9.21. Mean acquisition times for the 3GPP 3-step schemes in Rayleigh fading channel: (a) Mean acquisition times vs. SIR (γ) for 0 frequency offset, (b) mean acquisition times vs. frequency offset for -3dB SIR.

DSA scheme is slightly marginal, but the pilot cancellation technique helps to improve it to nearly the same level. [22]

Another issue to examine in realizing the DSA-based cellular system is the determination of the igniter code reuse factor R. A small value of R reduces the system complexity, but may cause the performance degradation during the

---

[22]Each finger can implement the pilot cancellation technique if it has just an additional subtractor, operating at the chip rate, to subtract the igniter sequence modulated by the path gain from the received noisy chip value. If the channel estimate is updated at the pilot symbol rate, the subtractor can operate at the symbol rate.

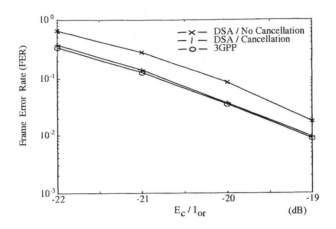

Figure 9.22.   Frame error rate performances of the traffic channels. (1-tap IIR filter with forgetting factor 1/256 operating at the chip rate used for channel estimation. 3 RAKE fingers and ITU vehicular-B/JTC-fading channel model with 200Hz Doppler frequency used. $\hat{I}_{or}/(I_{oc} + N_o)$ set to 9dB.)

Table 9.1.   Mean acquisition time performance degradations due to the igniter code phase collision. (15 SCH/P-CCPCH slots accumulated, 16 igniter sequence correlators at maximum, hexagonal cells, 5Hz single-path/classic Doppler spectrum, no frequency offset, $\hat{I}_{or}/(I_{oc} + N_o)) = -3dB, |\eta|=0, V_1=50, V=7, K=150$.)

Reuse factor (R)	7	16	19	64	128
Mean acquisition time (ms)	27.8	26.1	32.0	40.8	57.2
Increment due to collision (ms)	2.1	1.3	< 1.0	< 1.0	< 1.0

synchronization and channel estimation process in case the igniter code phases arriving from two co-igniter cells stay coincident for a long time. Table 9.1 exhibits the mean acquisition time performances for different reuse factors when the igniter code phase from the nearest co-igniter cell is kept coincident with that from the current cell. It is assumed that the signals from the two cells arrive at the MS located in the worst-case region of the cell after passing through independent Rayleigh fading channels, and the average fading signal power is proportional to d^{-4}, for the BS-to-MS distance d. The table shows that the reuse factor larger than 16 guarantees nearly perfect blocking of the co-igniter interference, and the performance degradation is very small even for the reuse factor 7.

3.3.2   SYSTEM COMPLEXITY

As an extension of performance comparison between the CDSA and the 3GPP 3-step schemes, we now consider the system complexity. For this, we consider the hardware complexity and operation complexity, as the former is related to the MS chip-set size and the latter to the MS power consumption. Note that the energy consumption during the cell search process is proportional to the product of the acquisition time and the operation complexity. [23]

For the slot boundary acquisition, both schemes employ the same GHG correlator and the same matched filtering operation. So the hardware and operation complexities required for the slot boundary search are the same in both schemes.

In relation to the CPICH spreading, the CDSA scheme employs 16 parallel *short* Gold sequence correlators to identify the igniter code used for the spreading, while the 3GPP 3-step scheme employs 16 parallel *long* Gold sequence correlators to identify the primary scrambling code used for the spreading. Since both schemes use 16 parallel correlators of complex Gold codes and the number of slots used in the CDSA igniter identification process is the same as that of the 3GPP primary scrambling code identification process (i.e, 5 slots), the hardware and operation complexities for the CDSA igniter identification turn out identical to those for the primary scrambling code identification (i.e., the 3rd step) of the 3GPP 3-step scheme, but the CDSA additionally requires a short Gold sequence generator. On the other hand the 3GPP 3-step scheme necessitates the hardware for 16 Golay-Hadamard correlators for the group and frame boundary identification purpose (i.e., the 2nd step). Therefore the overall hardware complexity of the CDSA scheme becomes slightly less than or equal to that of the 3GPP three-step scheme.

Finally, we compare the operation complexity of the remaining operations - - the comparison-correction based synchronization and the state symbol correlation operations of the CDSA scheme in contrast to the group and frame boundary identification operation of the 3GPP 3-step scheme. To collect the state symbol streams for the comparison-correction and the state symbol correlation processes, a single correlation operation is required for the duration of a frame time, and the identified igniter correlator keeps operating to obtain the state symbols. On the contrary, in the 3GPP 3-step case, 16 parallel matched filtering is required for the duration of a frame time to collect the SSC sufficient statistics, with the filtering operation periodically activated (with 10% duty cycle) and performed by a Fast-Hadamard transformer. Once the state symbols are obtained, (15+32) off-line correlations of length 150 and magnitude comparisons are required, *occasionally* (when the state symbol correlation process is triggered), in the CDSA case to determine the cell and frame boundary. In

---

[23] In the complexity comparison, we assume 32 cell groups for the 3GPP scheme. The complexity decreases a little for the 64 cell group cases [110]

contrast, (15×32) off-line correlations of length 15 and magnitude comparisons are required, *always*, in the 3GPP 3-step case to determine the group and frame boundary using the SSC sufficient statistics. It is not difficult to show that the number of arithmetic operations of those off-line processes is of the same order for both schemes, but the number of relevant memory access is smaller for the CDSA case. Note that the operation complexity as well as system complexity for the comparison-correction based synchronization process of the CDSA scheme is negligible (see Section 1 of Chapter 8.). Overall, we may conclude that the operation complexity for the above remaining processes is comparable for the two schemes or slightly less for the CDSA scheme.

Appendix A
Proofs and Lemmas for Theorems in Chapter 8

1. PROOF OF "IF" PART OF THEOREM 8.5

By Eq.(A.10) and Eq.(8.34), we have the relation $\Lambda \cdot \Delta^{-1} \cdot \mathbf{e}_{(W-1)b+(r-1)} = \Lambda_r^{[r]} \cdot (\mathbf{T}^{(W-1)N_I + \beta_{r-1}} \cdot \Delta^{-1} \cdot \mathbf{e}_{(W-1)b+r-1} + \mathbf{c}_{r-1}) = \Lambda_r^{[r]} \cdot (\mathbf{c}_{r-1} + \mathbf{c}_{r-1}) = 0$.
[1] This proves that $\Lambda \cdot \Delta^{-1} \cdot \mathbf{e}_k = 0$ for $k = (W-1)b + (r-1)$. In case $r = 1$, this completes the proof for $k = (W-1)b + j, j = 0, 1, \cdots, r-1$. On the other hand, if $r \geq 2$, we assume that $\Lambda \cdot \Delta^{-1} \cdot \mathbf{e}_k = 0$ for $k = (W-1)b + r - 1, \cdots, (W-1)b + j + 1, 0 \leq j \leq r-2$, and prove that $\Lambda \cdot \Delta^{-1} \cdot \mathbf{e}_k = 0$ for $k = (W-1)b + j$. By Eq.(A.10) and Eq.(8.34), we get $\Lambda \cdot \Delta^{-1} \cdot \mathbf{e}_{(W-1)b+j} = \Lambda_r^{[j+1]} \cdot \mathbf{T}^{(W-1)N_I + \beta_j} \cdot \Delta^{-1} \cdot \sum_{k=(W-1)b+j+1}^{(W-1)b+r-1} u_{j,k} \mathbf{e}_k$. Inserting Eq.(A.8) to this, we get $\Lambda \cdot \Delta^{-1} \cdot \mathbf{e}_{(W-1)b+j} = \sum_{k=(W-1)b+j+1}^{(W-1)b+r-1} u_{j,k} \Lambda_r \cdot \mathbf{T}^{(W-2)N_I + \beta_{b-1}} \cdot \Delta^{-1} \cdot \mathbf{e}_k$. Then, applying Eq.(A.11) to this with $i = W - 1$ and $l = W - 2$, we get the relation $\Lambda \cdot \Delta^{-1} \cdot \mathbf{e}_{(W-1)b+j} = \sum_{k=(W-1)b+j+1}^{(W-1)b+r-1} u_{j,k} \Lambda \cdot \Delta^{-1} \cdot \mathbf{e}_k$, which becomes 0 due to the above assumption. This proves that $\Lambda \cdot \Delta^{-1} \cdot \mathbf{e}_k = 0$ for $k = (W-1)b + r - 1, (W-1)b + r - 2, \cdots, (W-1)b$.

To complete the proof, we now show that $\Lambda \cdot \Delta^{-1} \cdot \mathbf{e}_k = 0$ for $k = 0, 1, \cdots, (W-2)b + b - 1$. In continuation of the above induction process, we assume that $\Lambda \cdot \Delta^{-1} \cdot \mathbf{e}_k = 0$ for $k = (W-1)b + r - 1, (W-1)b + r - 2, \cdots, ib + j + 1 \, 0 \leq i \leq W - 2, 0 \leq j \leq b - 1$, and prove the above relation for $k = ib + j$. When $j = 0, 1, \cdots, r - 1$, we insert Eq.(8.34) to Eq.(A.10) to get $\Lambda \cdot \Delta^{-1} \cdot \mathbf{e}_{ib+j} = \Lambda_r \cdot \Lambda_b^{W-2-i} \cdot \Lambda_b^{[j+1]} \cdot \mathbf{T}^{iN_I + \beta_j} \cdot \Delta^{-1} \cdot \{\mathbf{e}_{ib+j} + \Delta \cdot \mathbf{T}^{(W-1-i)N_I} \cdot \Delta^{-1} \cdot (\mathbf{e}_{(W-1)b+j} + \sum_{k=(W-1)b+j+1}^{(W-1)b+r-1} u_{j,k} \mathbf{e}_k)\}$, which can be written, by Eq.(A.2), as $\Lambda \cdot \Delta^{-1} \cdot \mathbf{e}_{ib+j} = \Lambda_r \cdot \Lambda_b^{W-2-i} \cdot \Lambda_b^{[j+1]} \cdot \mathbf{T}^{iN_I + \beta_j} \cdot \Delta^{-1} \cdot \sum_{k=ib+j+1}^{(W-1)b+r-1} \tilde{u}_{ib+j,k} \mathbf{e}_k$

---

[1] Refer to Lemma A.3 for the definition of $\Lambda_r^{[r]}$.

for some binary numbers $\tilde{u}_{ib+j,k}$'s. When $j = r, r + 1, \cdots, b - 1$, we can get the same result by applying Eq.(A.2) and Eq.(A.1). [2] Now we apply Eq.(A.5) to the result to get $\Lambda \cdot \Delta^{-1} \cdot \mathbf{e}_{ib+j} = \sum_{k=ib+j+1}^{(W-1)b+r-1} \tilde{u}_{ib+j,k} \Lambda_r \cdot \Lambda_b^{W-2-(i-1)} \cdot \mathbf{T}^{(i-1)N_I+\beta_{b-1}} \cdot \Delta^{-1} \cdot \mathbf{e}_k$. By inserting Eq.(A.11) to this, we finally get the relation $\Lambda \cdot \Delta^{-1} \cdot \mathbf{e}_{ib+j} = \sum_{k=ib+j+1}^{(W-1)b+r-1} \tilde{u}_{ib+j,k} \Lambda \cdot \Delta^{-1} \cdot \mathbf{e}_k$, which is zero by the inductive assumption.

2.   PROOF OF "ONLY IF" PART OF THEOREM 8.5

First, we prove the case $j = 0, 1, \cdots, r - 1$. If we insert Eq.(8.35a) to Eq.(A.10) for $i = W - 1$, we get the relation $\Lambda \cdot \Delta^{-1} \cdot \mathbf{e}_{(W-1)b+j} = \Lambda_r^{[j+1]} \cdot \mathbf{T}^{(W-1)N_I+\beta_j} \cdot \Delta^{-1} \cdot \{\mathbf{e}_{(W-1)b+j} + \sum_{k=0}^{(W-1)b+r-1} \mu_{j,k} \mathbf{e}_k\}$, which can be rewritten, by Eq.(A.8), as $\Lambda \cdot \Delta^{-1} \cdot \mathbf{e}_{(W-1)b+j} = \mathbf{T}^{(W-1)N_I+\beta_{r-1}} \cdot \Delta^{-1} \cdot \mathbf{e}_{(W-1)b+j} + \sum_{k=0}^{(W-1)b+j} \mu_{j,k} \mathbf{T}^{(W-1)N_I+\beta_{r-1}} \cdot \Delta^{-1} \cdot \mathbf{e}_k \; + \; \sum_{k=(W-1)b+j+1}^{(W-1)b+r-1} \mu_{j,k} \Lambda_r \cdot \mathbf{T}^{(W-2)N_I+\beta_{b-1}} \cdot \Delta^{-1} \cdot \mathbf{e}_k$. If we apply Eq.(A.11) for $i = W-1$ and $l = W-2$ to the last term of the above equation, it reduces to $\sum_{k=(W-1)b+j+1}^{(W-1)b+r-1} \mu_{j,k} \Lambda \cdot \Delta^{-1} \cdot \mathbf{e}_k$, which is 0 by the assumption that $\Lambda = 0$. So the above equation reduces to $\Lambda \cdot \Delta^{-1} \cdot \mathbf{e}_{(W-1)b+j} = \mathbf{T}^{(W-1)N_I+\beta_{r-1}} \cdot \Delta^{-1} \cdot \{\sum_{k=0}^{(W-1)b+j-1} \mu_{j,k} \mathbf{e}_k + (1 + \mu_{j,(W-1)b+j}) \mathbf{e}_{(W-1)b+j}\} = 0$. But since $\mathbf{T}$ and Δ are both nonsingular, we finally get the results $\mu_{j,k} = 0$ for $k = 0, 1, \cdots, (W-1)b + j - 1$, and $\mu_{j,(W-1)b+j} = 1$.

Now, if $r \leq b - 1$, we prove the case $j = r, r + 1, \cdots, b - 1$. If we insert Eq.(8.35b) to Eq.(A.10) for $i = W-2$, we get the relation $\Lambda \cdot \Delta^{-1} \cdot \mathbf{e}_{(W-2)b+j} = \Lambda_r \cdot \Lambda_b^{[j+1]} \cdot \mathbf{T}^{(W-2)N_I+\beta_j} \cdot \Delta^{-1} \cdot \{\mathbf{e}_{(W-2)b+j} + \sum_{k=0}^{(W-1)b+r-1} \nu_{j,k} \mathbf{e}_k\}$, which can be rewritten, by Eq.(A.5), as $\Lambda \cdot \Delta^{-1} \cdot \mathbf{e}_{(W-2)b+j} = \Lambda_r \cdot \mathbf{T}^{(W-2)N_I+\beta_{b-1}} \cdot \Delta^{-1} \cdot \mathbf{e}_{(W-2)b+j} + \sum_{k=0}^{(W-2)b+j} \nu_{j,k} \Lambda_r \cdot \mathbf{T}^{(W-2)N_I+\beta_{b-1}} \cdot \Delta^{-1} \cdot \mathbf{e}_k + \sum_{k=(W-2)b+j+1}^{(W-1)b+r-1} \nu_{j,k} \Lambda_r \cdot \Lambda_b \cdot \mathbf{T}^{(W-3)N_I+\beta_{b-1}} \cdot \Delta^{-1} \cdot \mathbf{e}_k$. If we apply Eq.(A.11) for $i = W-2$ or $W-1$ and $l = W-3$ to the last term, it reduces to $\sum_{k=(W-2)b+j+1}^{(W-1)b+r-1} \nu_{j,k} \Lambda \cdot \Delta^{-1} \cdot \mathbf{e}_k = 0$. For the other two terms we apply Eq.(A.9) to get the relation $\Lambda \cdot \Delta^{-1} \cdot \mathbf{e}_{(W-2)b+j} = \mathbf{T}^{(W-1)N_I+\beta_{r-1}} \cdot \Delta^{-1} \cdot \{\sum_{k=0}^{(W-2)b+j-1} \nu_{j,k} \mathbf{e}_k + (1 + \nu_{j,(W-2)b+j}) \mathbf{e}_{(W-2)b+j}\} = 0$. Therefore, for the same reason as before, we finally get the results $\nu_{j,k} = 0$ for $k = 0, 1, \cdots, (W-2)b + j - 1$, and $\nu_{j,(W-2)b+j} = 1$. This completes the proof.

---

[2]This part of the proof is needed only when $r \leq b - 1$.

3.   LEMMA A.1

LEMMA A.1 *For $i = 0, 1, \cdots, W - 2, j = 0, 1, \cdots, b - 1$,*

$$
\begin{cases}
\mathbf{e}_l^t \cdot \Delta \cdot \mathbf{T}^{(W-2-i)N_I} \cdot \Delta^{-1} \cdot \mathbf{e}_k = 0, \\
\quad l = 0, 1, \cdots, ib + j, \\
\quad (W-2)b + j < k \le (W-1)b + (r-1), \\[6pt]
\mathbf{e}_l^t \cdot \Delta \cdot \mathbf{T}^{(W-2-i)N_I} \cdot \Delta^{-1} \cdot \mathbf{e}_{(W-2)b+j} \\
\quad = \begin{cases} 0, & l = 0, 1, \cdots, ib + (j-1), \\ 1, & l = ib + j. \end{cases}
\end{cases}
\tag{A.1}
$$

Also, for $i = 0, 1, \cdots, W - 1, j = 0, 1, \cdots, r - 1$,

$$
\begin{cases}
\mathbf{e}_l^t \cdot \Delta \cdot \mathbf{T}^{(W-1-i)N_I} \cdot \Delta^{-1} \cdot \mathbf{e}_k = 0, \\
\quad l = 0, 1, \cdots, ib + j, \\
\quad (W-1)b + j < k \le (W-1)b + (r-1), \\[6pt]
\mathbf{e}_l^t \cdot \Delta \cdot \mathbf{T}^{(W-1-i)N_I} \cdot \Delta^{-1} \cdot \mathbf{e}_{(W-1)b+j} \\
\quad = \begin{cases} 0, & l = 0, 1, \cdots, ib + (j-1), \\ 1, & l = ib + j, \end{cases}
\end{cases}
\tag{A.2}
$$

Proof : To begin with, we expand the equation $\Delta \cdot \Delta^{-1} = \mathbf{I}$ for Δ in Eq.(8.34) to get the relations

$$
\mathbf{h}^t \cdot \mathbf{T}^{iN_I + \alpha_j} \cdot \Delta^{-1} = \mathbf{e}_{ib+j}^t,
\tag{A.3}
$$

$$
\begin{cases}
i = 0, 1, \cdots, W - 2, \ j = 0, 1, \cdots, b - 1, \\
i = W - 1, \ j = 0, 1, \cdots, r - 1.
\end{cases}
$$

We denote by m and n the quotient and the remainder of l divided by b, respectively (i.e., $l = mb + n$, $0 \le n < b$). Then, by Eq.(8.34), $\mathbf{e}_l^t \cdot \Delta \cdot \mathbf{T}^{(W-2-i)N_I} \cdot \Delta^{-1} = \mathbf{e}_{mb+n}^t \cdot \Delta \cdot \mathbf{T}^{(W-2-i)N_I} \cdot \Delta^{-1} = \mathbf{h}^t \cdot \mathbf{T}^{mN_I + \alpha_n} \cdot \mathbf{T}^{(W-2-i)N_I} \cdot \Delta^{-1} = \mathbf{h}^t \cdot \mathbf{T}^{(W-2-(i-m))N_I + \alpha_n} \cdot \Delta^{-1}$, which, by Eq.(A.3), reduces to $\mathbf{e}_{(W-2-(i-m))b+n}^t$. When $l = mb + n \le ib + j$, we have the inequality $\{W-2-(i-m)\}b + n \le (W-2)b + j$, which implies that $\mathbf{e}_l^t \cdot \Delta \cdot \mathbf{T}^{(W-2-i)N_I} \cdot \Delta^{-1} \cdot \mathbf{e}_k = 0$ for $k = (W-2)b + j + 1, (W-2)b + j + 2, \cdots, (W-1)b + r - 1$. Also, when $l < ib + j$, we have $\{W - 2 - (i - m)\}b + n < (W - 2)b + j$, which implies that $\mathbf{e}_l^t \cdot \Delta \cdot \mathbf{T}^{(W-2-i)N_I} \cdot \Delta^{-1} \cdot \mathbf{e}_{(W-2)b+j} = 0$. Finally, when $l = ib + j$, $\{W - 2 - (i - m)\}b + n$ reduces to $(W - 2)b + j$, so we get the relation $\mathbf{e}_{ib+j}^t \cdot \Delta \cdot \mathbf{T}^{(W-2-i)N_I} \cdot \Delta^{-1} \cdot \mathbf{e}_{(W-2)b+j} = 1$. This completes the proof of Eq.(A.1).

By applying Eq.(8.34) and Eq.(A.3), we get the relation $\mathbf{e}_l^t \cdot \Delta \cdot \mathbf{T}^{(W-1-i)N_I} \cdot \Delta^{-1} = \mathbf{e}_{(W-1-(i-m))b+n}^t$. If we repeat a similar reasoning on this, we can get the relations in Eq.(A.2). ∎

4.   LEMMA A.2

LEMMA A.2 *We define* $\Lambda_b^{[b]} \triangleq \mathbf{I}_{L \times L}$ *for* $b \geq 1$, *and* $\Lambda_b^{[j]} \triangleq (\mathbf{T}^{\beta_{b-1}-\beta_{b-2}} + \mathbf{c}_{b-1} \cdot \mathbf{h}^t \cdot \mathbf{T}^{\alpha_{b-1}-\beta_{b-2}}) \cdots (\mathbf{T}^{\beta_j-\beta_{j-1}} + \mathbf{c}_j \cdot \mathbf{h}^t \cdot \mathbf{T}^{\alpha_j-\beta_{j-1}})$, $j=1,2,\cdots,b-1$ *for* $b \geq 2$. *Then*

$$\begin{cases} \Lambda_b = \Lambda_b^{[1]} \cdot (\mathbf{T}^{N_I+\beta_0-\beta_{b-1}} + \mathbf{c}_0 \cdot \mathbf{h}^t \cdot \mathbf{T}^{N_I+\alpha_0-\beta_{b-1}}), \\ \Lambda_b^{[j]} = \Lambda_b^{[j+1]} \cdot (\mathbf{T}^{\beta_j-\beta_{j-1}} + \mathbf{c}_j \cdot \mathbf{h}^t \cdot \mathbf{T}^{\alpha_j-\beta_{j-1}}), \\ \qquad j = 1, 2, \cdots, b-1, \ (b \geq 2), \end{cases} \quad (A.4)$$

$$\Lambda_b^{[j+1]} \cdot \mathbf{T}^{iN_I+\beta_j} \cdot \Delta^{-1} \cdot \mathbf{e}_k =$$
$$\begin{cases} \mathbf{T}^{iN_I+\beta_{b-1}} \cdot \Delta^{-1} \cdot \mathbf{e}_k, \ 0 \leq k \leq ib+j, \\ \Lambda_b \cdot \mathbf{T}^{(i-1)N_I+\beta_{b-1}} \cdot \Delta^{-1} \cdot \mathbf{e}_k, \ ib+j < k \leq L-1, \end{cases} \quad (A.5)$$
$$\text{for } i = 0, 1, \cdots W-2, \ j = 0, 1, \cdots, b-1,$$
$$k = 0, 1, \cdots, L-1,$$

$$\Lambda_b \cdot \mathbf{T}^{(i-1)N_I+\beta_{b-1}} \cdot \Delta^{-1} \cdot \mathbf{e}_k =$$
$$\begin{cases} \Lambda_b^{[j+1]} \cdot (\mathbf{T}^{iN_I+\beta_j} \cdot \Delta^{-1} \cdot \mathbf{e}_{ib+j} + \mathbf{c}_j), \ k = ib+j, \\ \mathbf{T}^{iN_I+\beta_{b-1}} \cdot \Delta^{-1} \cdot \mathbf{e}_k, \ k < ib \text{ or } k \geq (i+1)b, \end{cases} \quad (A.6)$$
$$\text{for } i = 0, 1, \cdots W-2, \ j = 0, 1, \cdots, b-1,$$
$$k = 0, 1, \cdots, L-1.$$

Proof : As Eq.(A.4) can be easily derived from the definition, we prove Eq.(A.5) and Eq.(A.6) only.

To begin with, we prove Eq.(A.5) and Eq.(A.6) for $b = 1$. In this case $W = L$, $j = 0$, $\Lambda_b^{[j+1]} = \Lambda_b^{[b]} = \mathbf{I}_{L \times L}$, and $\Lambda_b = \mathbf{T}^{N_I+\beta_0-\beta_{b-1}} + \mathbf{c}_0 \cdot \mathbf{h}^t \cdot \mathbf{T}^{N_I+\alpha_0-\beta_{b-1}}$. Hence, by Eq.(A.3), we get $\Lambda_b \cdot \mathbf{T}^{(i-1)N_I+\beta_{b-1}} \cdot \Delta^{-1} \cdot \mathbf{e}_k = \mathbf{T}^{iN_I+\beta_0} \cdot \Delta^{-1} \cdot \mathbf{e}_k + \mathbf{c}_0 \cdot \mathbf{h}^t \cdot \mathbf{T}^{iN_I+\alpha_0} \cdot \Delta^{-1} \cdot \mathbf{e}_k = \mathbf{T}^{iN_I+\beta_0} \cdot \Delta^{-1} \cdot \mathbf{e}_k + \mathbf{c}_0 \cdot \mathbf{e}_{ib}^t \cdot \mathbf{e}_k$, which reduces to $\mathbf{T}^{iN_I+\beta_0} \cdot \Delta^{-1} \cdot \mathbf{e}_k$ when $k \neq ib$, and to $\mathbf{T}^{iN_I+\beta_0} \cdot \Delta^{-1} \cdot \mathbf{e}_k + \mathbf{c}_0$ when $k = ib$. If we apply the relation $\Lambda_b^{[j+1]} = \mathbf{I}_{L \times L}$ for $b = 1$ to this, we can obtain the relations in Eq.(A.5) and Eq.(A.6).

Next, we prove Eq.(A.5) for $b \geq 2$. We first consider the case when $0 \leq k \leq ib + j$. By Eq.(A.4) and Eq.(A.3), we can get $\Lambda_b^{[j+1]} \cdot \mathbf{T}^{iN_I+\beta_j} \cdot \Delta^{-1} \cdot \mathbf{e}_k = \Lambda_b^{[j+2]} \cdot (\mathbf{T}^{iN_I+\beta_{j+1}} \cdot \Delta^{-1} \cdot \mathbf{e}_k + \mathbf{c}_{j+1} \cdot \mathbf{e}_{ib+j+1}^t \cdot \mathbf{e}_k) = \Lambda_b^{[j+2]} \cdot \mathbf{T}^{iN_I+\beta_{j+1}} \cdot \Delta^{-1} \cdot \mathbf{e}_k$ for $j = 0, 1, \cdots, b - 2$. Applying this procedure repeatedly, we finally obtain $\Lambda_b^{[j+1]} \cdot \mathbf{T}^{iN_I+\beta_j} \cdot \Delta^{-1} \cdot \mathbf{e}_k = \Lambda_b^{[b]} \cdot \mathbf{T}^{iN_I+\beta_{b-1}} \cdot \Delta^{-1} \cdot \mathbf{e}_k = \mathbf{T}^{iN_I+\beta_{b-1}} \cdot \Delta^{-1} \cdot \mathbf{e}_k$ for $j = 0, 1, \cdots, b-2$, and $\Lambda_b^{[j+1]} \cdot \mathbf{T}^{iN_I+\beta_j} \cdot \Delta^{-1} \cdot \mathbf{e}_k = \mathbf{T}^{iN_I+\beta_{b-1}} \cdot \Delta^{-1} \cdot \mathbf{e}_k$ for $j = b - 1$. Now we consider the case when $ib + j < k \leq L - 1$. By

Eq.(A.4) and Eq.(A.3), we get the relation $\Lambda_b \cdot \mathbf{T}^{(i-1)N_I+\beta_{b-1}} \cdot \Delta^{-1} \cdot \mathbf{e}_k = \Lambda_b^{[1]} \cdot (\mathbf{T}^{iN_I+\beta_0} \cdot \Delta^{-1} \cdot \mathbf{e}_k + \mathbf{c}_0 \cdot \mathbf{e}_{ib}^t \cdot \mathbf{e}_k)$, which reduces to $\Lambda_b^{[1]} \cdot \mathbf{T}^{iN_I+\beta_0} \cdot \Delta^{-1} \cdot \mathbf{e}_k$, for all k's in $ib < k \le L-1$. For mathematical induction, we assume that $\Lambda_b \cdot \mathbf{T}^{(i-1)N_I+\beta_{b-1}} \cdot \Delta^{-1} \cdot \mathbf{e}_k = \Lambda_b^{[j]} \cdot \mathbf{T}^{iN_I+\beta_{j-1}} \cdot \Delta^{-1} \cdot \mathbf{e}_k$, for all k's in $ib+j-1 < k \le L-1$, for a j in $1 \le j \le b-1$. Then, by Eq.(A.4) and Eq.(A.3), we get $\Lambda_b \cdot \mathbf{T}^{(i-1)N_I+\beta_{b-1}} \cdot \Delta^{-1} \cdot \mathbf{e}_k = \Lambda_b^{[j+1]} \cdot (\mathbf{T}^{\beta_j-\beta_{j-1}} + \mathbf{c}_j \cdot \mathbf{h}^t \cdot \mathbf{T}^{\alpha_j-\beta_{j-1}}) \cdot \mathbf{T}^{iN_I+\beta_{j-1}} \cdot \Delta^{-1} \cdot \mathbf{e}_k = \Lambda_b^{[j+1]} \cdot (\mathbf{T}^{iN_I+\beta_j} \cdot \Delta^{-1} \cdot \mathbf{e}_k + \mathbf{c}_j \cdot \mathbf{e}_{ib+j}^t \cdot \mathbf{e}_k)$, which reduces to $\Lambda_b^{[j+1]} \cdot \mathbf{T}^{iN_I+\beta_j} \cdot \Delta^{-1} \cdot \mathbf{e}_k$, for all k's in $ib+j < k \le L-1$. This completes the proof of Eq.(A.5) for $j = 0, 1, \cdots, b-1$.

Finally, we prove Eq.(A.6) for $b \ge 2$. When $j = 0$, for $k = ib$, by Eq.(A.4) and Eq.(A.3), we have the relation $\Lambda_b \cdot \mathbf{T}^{(i-1)N_I+\beta_{b-1}} \cdot \Delta^{-1} \cdot \mathbf{e}_k = \Lambda_b^{[1]} \cdot (\mathbf{T}^{iN_I+\beta_0} \cdot \Delta^{-1} \cdot \mathbf{e}_{ib} + \mathbf{c}_0)$. When $1 \le j \le b-1$, by replacing j in Eq.(A.5) with $j-1$, we get $\Lambda_b \cdot \mathbf{T}^{(i-1)N_I+\beta_{b-1}} \cdot \Delta^{-1} \cdot \mathbf{e}_k = \Lambda_b^{[j]} \cdot \mathbf{T}^{iN_I+\beta_{j-1}} \cdot \Delta^{-1} \cdot \mathbf{e}_k$, $ib+j-1 < k \le L-1$. Therefore, applying Eq.(A.4) and Eq.(A.3), for each $k = ib+j$, $j = 1, 2, \cdots, b-1$, we get $\Lambda_b \cdot \mathbf{T}^{(i-1)N_I+\beta_{b-1}} \cdot \Delta^{-1} \cdot \mathbf{e}_k = \Lambda_b^{[j+1]} \cdot (\mathbf{T}^{iN_I+\beta_j} \cdot \Delta^{-1} \cdot \mathbf{e}_k + \mathbf{c}_j \cdot \mathbf{e}_{ib+j}^t \cdot \mathbf{e}_k) = \Lambda_b^{[j+1]} \cdot (\mathbf{T}^{iN_I+\beta_j} \cdot \Delta^{-1} \cdot \mathbf{e}_{ib+j} + \mathbf{c}_j)$. This proves the upper part of Eq.(A.6). Now, for each $k = mb+n$ with $m \ne i$, $0 \le n < b$ and $0 \le k \le L-1$, we apply Eq.(A.4) and Eq.(A.3) repeatedly. Then we get the relation $\Lambda_b \cdot \mathbf{T}^{(i-1)N_I+\beta_{b-1}} \cdot \Delta^{-1} \cdot \mathbf{e}_k = \Lambda_b^{[1]} \cdot (\mathbf{T}^{iN_I+\beta_0} \cdot \Delta^{-1} \cdot \mathbf{e}_{mb+n} + \mathbf{c}_0 \cdot \mathbf{e}_{ib}^t \cdot \mathbf{e}_{mb+n}) = \Lambda_b^{[1]} \cdot \mathbf{T}^{iN_I+\beta_0} \cdot \Delta^{-1} \cdot \mathbf{e}_{mb+n} = \cdots = \Lambda_b^{[b]} \cdot \mathbf{T}^{iN_I+\beta_{b-1}} \cdot \Delta^{-1} \cdot \mathbf{e}_{mb+n} = \mathbf{T}^{iN_I+\beta_{b-1}} \cdot \Delta^{-1} \cdot \mathbf{e}_k$. This proves the lower part of Eq.(A.6). ∎

5. LEMMA A.3

LEMMA A.3 *We define* $\Lambda_r^{[r]} \triangleq \mathbf{I}_{L \times L}$ *for* $r \ge 1$, *and* $\Lambda_r^{[j]} \triangleq (\mathbf{T}^{\beta_{r-1}-\beta_{r-2}} + \mathbf{c}_{r-1} \cdot \mathbf{h}^t \cdot \mathbf{T}^{\alpha_{r-1}-\beta_{r-2}}) \cdots (\mathbf{T}^{\beta_j-\beta_{j-1}} + \mathbf{c}_j \cdot \mathbf{h}^t \cdot \mathbf{T}^{\alpha_j-\beta_{j-1}})$, $j = 1, 2, \cdots, r-1$ *for* $r \ge 2$. *Then*

$$
\begin{cases}
\Lambda_r = \Lambda_r^{[1]} \cdot (\mathbf{T}^{N_I+\beta_0-\beta_{b-1}} + \mathbf{c}_0 \cdot \mathbf{h}^t \cdot \mathbf{T}^{N_I+\alpha_0-\beta_{b-1}}), \\
\Lambda_r^{[j]} = \Lambda_r^{[j+1]} \cdot (\mathbf{T}^{\beta_j-\beta_{j-1}} + \mathbf{c}_j \cdot \mathbf{h}^t \cdot \mathbf{T}^{\alpha_j-\beta_{j-1}}), \\
\qquad j = 1, 2, \cdots, r-1, \ (r \ge 2),
\end{cases} \tag{A.7}
$$

$$
\Lambda_r^{[j+1]} \cdot \mathbf{T}^{(W-1)N_I+\beta_j} \cdot \Delta^{-1} \cdot \mathbf{e}_k =
$$
$$
\begin{cases}
\mathbf{T}^{(W-1)N_I+\beta_{r-1}} \cdot \Delta^{-1} \cdot \mathbf{e}_k, \\
\qquad 0 \le k \le (W-1)b+j, \\
\Lambda_r \cdot \mathbf{T}^{(W-2)N_I+\beta_{b-1}} \cdot \Delta^{-1} \cdot \mathbf{e}_k, \\
\qquad (W-1)b+j < k \le L-1,
\end{cases}
$$
$$
\text{for } j = 0, 1, \cdots, r-1, \ k = 0, 1, \cdots, L-1, \tag{A.8}
$$

$$\Lambda_r \cdot \mathbf{T}^{(W-2)N_I + \beta_{b-1}} \cdot \Delta^{-1} \cdot \mathbf{e}_k =$$
$$\begin{cases} \Lambda_r^{[j+1]} \cdot (\mathbf{T}^{(W-1)N_I + \beta_j} \cdot \Delta^{-1} \cdot \mathbf{e}_{(W-1)b+j} + \mathbf{c}_j), \\ \qquad\qquad k = (W-1)b + j, \\ \mathbf{T}^{(W-1)N_I + \beta_{r-1}} \cdot \Delta^{-1} \cdot \mathbf{e}_k, \\ \qquad\qquad k < (W-1)b, \end{cases}$$
$$\text{for } j = 0, 1, \cdots, r-1, \ k = 0, 1, \cdots, L-1. \tag{A.9}$$

Proof : The derivation of Eq.(A.7) is trivial. We omit the proof of this lemma as it can be done in a similar way to that for Lemma A.2. The relations in Eq.(A.8) and Eq.(A.9) can be proved in the same procedure as that for Lemma A.2, by applying Eq.(A.7) and Eq.(A.3). Note that the lower line of Eq.(A.8) disappears when $j = r - 1$. ∎

6.   LEMMA A.4

LEMMA A.4 *For the matrices $\Lambda_b^{[j]}$ and $\Lambda_r^{[j]}$ defined in Lemmas A.2 and A.3,*

$$\Lambda \cdot \Delta^{-1} \cdot \mathbf{e}_{ib+j} =$$
$$\begin{cases} \Lambda_r \cdot \Lambda_b^{W-2-i} \cdot \Lambda_b^{[j+1]} \cdot (\mathbf{T}^{iN_I + \beta_j} \cdot \Delta^{-1} \cdot \mathbf{e}_{ib+j} + \mathbf{c}_j), \\ \qquad i = 0, 1, \cdots, W-2, \ j = 0, 1, \cdots, b-1, \\ \\ \Lambda_r^{[j+1]} \cdot (\mathbf{T}^{(W-1)N_I + \beta_j} \cdot \Delta^{-1} \cdot \mathbf{e}_{(W-1)b+j} + \mathbf{c}_j), \\ \qquad i = W-1, \ j = 0, 1, \cdots, r-1, \end{cases} \tag{A.10}$$

$$\Lambda \cdot \Delta^{-1} \cdot \mathbf{e}_{ib+j} = \Lambda_r \cdot \Lambda_b^{W-2-l} \cdot \mathbf{T}^{lN_I + \beta_{b-1}} \cdot \Delta^{-1} \cdot \mathbf{e}_{ib+j},$$
$$l = -1, 0, 1, \cdots, i-1,$$
$$\begin{cases} i = 0, 1, \cdots, W-2, \ j = 0, 1, \cdots, b-1, \\ i = W-1, \ j = 0, 1, \cdots, r-1. \end{cases} \tag{A.11}$$

Proof : We first prove Eq.(A.11) by induction. By the definitions of Λ_b and Λ_o in Eq.(8.32), we have $\Lambda_o = \Lambda_b \cdot \mathbf{T}^{-N_I + \beta_{b-1}}$, and thus we get $\Lambda \cdot \Delta^{-1} \cdot \mathbf{e}_{ib+j} = \Lambda_r \cdot \Lambda_b^{W-1} \cdot \mathbf{T}^{-N_I + \beta_{b-1}} \cdot \Delta^{-1} \cdot \mathbf{e}_{ib+j}$. This proves Eq.(A.11) for $l = -1$. We assume that $\Lambda \cdot \Delta^{-1} \cdot \mathbf{e}_{ib+j} = \Lambda_r \cdot \Lambda_b^{W-2-l} \cdot \mathbf{T}^{lN_I + \beta_{b-1}} \cdot \Delta^{-1} \cdot \mathbf{e}_{ib+j}$ for an l in $-1 \le l \le i-2$. Note that in this interval the relation $l \le W-3$ holds. Then, since by Eq.(A.6) we have $\Lambda_b \cdot \mathbf{T}^{lN_I + \beta_{b-1}} \cdot \Delta^{-1} \cdot \mathbf{e}_k = \mathbf{T}^{(l+1)N_I + \beta_{b-1}} \cdot \Delta^{-1} \cdot \mathbf{e}_k$ for $k < (l+1)b$ or $k \ge (l+2)b$ (for $l = -1, 0, \cdots, W-3$), we get $\Lambda \cdot \Delta^{-1} \cdot \mathbf{e}_{ib+j} = \Lambda_r \cdot \Lambda_b^{W-2-(l+1)} \cdot \mathbf{T}^{(l+1)N_I + \beta_{b-1}} \cdot \Delta^{-1} \cdot \mathbf{e}_{ib+j}$ by noting that $(l+2)b \le ib \le ib+j$. This completes the proof of Eq.(A.11).

Next, we prove Eq.(A.10). For $l = i-1$ Eq.(A.11) takes the expression $\Lambda \cdot \Delta^{-1} \cdot \mathbf{e}_{ib+j} = \Lambda_r \cdot \Lambda_b^{W-1-i} \cdot \mathbf{T}^{(i-1)N_I + \beta_{b-1}} \cdot \Delta^{-1} \cdot \mathbf{e}_{ib+j}$. Applying Eq.(A.6) to this for $i = 0, 1, \cdots, W - 2$, $j = 0, 1, \cdots, b - 1$, we get $\Lambda \cdot \Delta^{-1} \cdot \mathbf{e}_{ib+j} = \Lambda_r \cdot \Lambda_b^{W-2-i} \cdot \Lambda_b \cdot \mathbf{T}^{(i-1)N_I + \beta_{b-1}} \cdot \Delta^{-1} \cdot \mathbf{e}_{ib+j} = \Lambda_r \cdot \Lambda_b^{W-2-i} \cdot \Lambda_b^{[j+1]} \cdot (\mathbf{T}^{iN_I + \beta_j} \cdot \Delta^{-1} \cdot \mathbf{e}_{ib+j} + \mathbf{c}_j)$. This proves the upper line of Eq.(A.10). When $i = W - 1$, $j = 0, 1, \cdots, r - 1$, we insert $l = i - 1 = W - 2$ to Eq.(A.11) to get the relation $\Lambda \cdot \Delta^{-1} \cdot \mathbf{e}_{ib+j} = \Lambda_r \cdot \mathbf{T}^{(W-2)N_I + \beta_{b-1}} \cdot \Delta^{-1} \cdot \mathbf{e}_{(W-1)b+j}$. If we apply Eq.(A.9) to this, we get $\Lambda \cdot \Delta^{-1} \cdot \mathbf{e}_{ib+j} = \Lambda_r^{[j+1]} \cdot (\mathbf{T}^{(W-1)N_I + \beta_j} \cdot \Delta^{-1} \cdot \mathbf{e}_{(W-1)b+j} + \mathbf{c}_j)$. This completes the proof of the lower line of Eq.(A.10).   ∎

7.   LEMMA A.5

LEMMA A.5 *The* (m, n) *element* b_{mn} *of* $\mathbf{B}_1$ *(in the proof of Theorem 8.8) is 0 for* $m = 0, 1, \cdots, \min(r - 1, b - r - 1)$, $n = 0, 1, \cdots, r - 1$, *and* $m = \min(r, b - r), \cdots, (W - 1)b + r - 1$, $n = 0, 1, \cdots, (W - 1)b + r - 1$.

Proof: Since the rank of the corresponding PDSA matrix $\bar{\Delta}$ is L, we can represent the L-vector $\mathbf{h}^t \cdot \mathbf{T}^{\alpha_k}$ by employing its L rows such that

$$\mathbf{h}^t \cdot \mathbf{T}^{\alpha_k} = \sum_{j=0}^{b-r-1} \sum_{i=1}^{W-1} \mu_{j,i}^{(k)} \mathbf{h}^t \cdot \mathbf{T}^{\alpha_j + iN_I} + \sum_{j=b-r}^{b-1} \sum_{i=0}^{W-1} \mu_{j,i}^{(k)} \mathbf{h}^t \cdot \mathbf{T}^{\alpha_j + iN_I}, \quad (A.12)$$

for $k = 0, 1, \cdots, \min(r - 1, b - r - 1)$, where $\mu_{j,i}^{(k)}$'s denote proper binary integers. By post-multiplying $\bar{\Delta}^{-1}$, and then applying the relations in Eq.(8.46), we get $\mathbf{h}^t \cdot \mathbf{T}^{\alpha_k} \cdot \bar{\Delta}^{-1} = \sum_{j=0}^{b-r-1} \sum_{i=1}^{W-1} \mu_{j,i}^{(k)} \mathbf{e}_{b(i-1)+r+j}^t + \sum_{j=b-r}^{b-1} \sum_{i=0}^{W-1} \mu_{j,i}^{(k)} \cdot \mathbf{e}_{b(i-1)+r+j}^t$, and by post-multiplying $\mathbf{e}_{(W-2)b+r+j}$ to each term, in addition, we get $\mu_{j,W-1}^{(k)} = \mathbf{h}^t \cdot \mathbf{T}^{\alpha_k} \cdot \bar{\Delta}^{-1} \cdot \mathbf{e}_{(W-2)b+r+j}$, for $j = 0, 1, \cdots, b - 1$, $k = 0, 1, \cdots, \min(r - 1, b - r - 1)$. Now we insert these final relations to Eq.(A.12) to get [3] $\sum_{j=0}^{b-1} (\mathbf{h}^t \cdot \mathbf{T}^{\alpha_k} \cdot \bar{\Delta}^{-1} \cdot \mathbf{e}_{(W-2)b+r+j}) \mathbf{h}^t \cdot \mathbf{T}^{\alpha_j + (W-1)N_I} = \mathbf{h}^t \cdot \mathbf{T}^{\alpha_k} + \sum_{j=0}^{b-r-1} \sum_{i=1}^{W-2} \mu_{j,i}^{(k)} \mathbf{h}^t \cdot \mathbf{T}^{\alpha_j + iN_I} + \sum_{j=b-r}^{b-1} \sum_{i=0}^{W-2} \mu_{j,i}^{(k)} \mathbf{h}^t \cdot \mathbf{T}^{\alpha_j + iN_I}$, and then post-multiply $\mathbf{T}^{N_I} \cdot \bar{\Delta}^{-1}$ to each term and apply Eq.(8.46) in addition, to obtain $\sum_{j=0}^{b-1} (\mathbf{h}^t \cdot \mathbf{T}^{\alpha_k} \cdot \bar{\Delta}^{-1} \cdot \mathbf{e}_{(W-2)b+r+j}) \mathbf{h}^t \cdot \mathbf{T}^{\bar{\alpha}_j + WN_I} \cdot \bar{\Delta}^{-1} = \mathbf{e}_{r+k}^t + \sum_{j=0}^{b-r-1} \sum_{i=1}^{W-2} \mu_{j,i}^{(k)} \mathbf{e}_{bi+r+j} + \sum_{j=b-r}^{b-1} \sum_{i=0}^{W-2} \mu_{j,i}^{(k)} \mathbf{e}_{bi+r+j}$, for $k = 0, 1, \cdots, \min(r - 1, b - r - 1)$. Then, since $b_{mn} = \mathbf{e}_m^t \cdot \mathbf{B}_1 \cdot \mathbf{e}_n$, we get $b_{mn} = u_m \sum_{j=0}^{b-1} (\mathbf{h}^t \cdot \mathbf{T}^{\alpha_m} \cdot \bar{\Delta}^{-1} \cdot \mathbf{e}_{(W-2)b+r+j}) \mathbf{h}^t \cdot \mathbf{T}^{\alpha_j + WN_I} \cdot \bar{\Delta}^{-1} \cdot \mathbf{e}_n = 0$ for $m = 0, 1, \cdots, \min(r - 1, b - r - 1)$, $n = 0, 1, \cdots, r - 1$. On the other hand, if $\min(r, b - r) \leq m \leq (W - 1)b + r - 1$, we get $b_{mn} = 0$ for $n = 0, 1, \cdots, (W - 1)b + r - 1$, directly from the definition of $\mathbf{B}_1$. This completes the proof.   ∎

---

[3] $\sum_{i=m}^{n} (\cdot)$ is regarded zero if $m > n$.

Appendix B
Derivation of Equations in Chapters 8 and 9

1. DERIVATION OF EQ.(8.80)

The correct symbol detection probability in the orthogonal signaling PDSA scheme $\hat{P}_c(|\eta|)$ can be determined by

$$\hat{P}_c(|\eta|) = \int_0^\infty p_{Y_0|\Omega_1}(y_0) \prod_{j=1}^{2^b-1} \int_0^{y_0} p_{Y_j|\Omega_1}(y_j) dy_j dy_0, \qquad (B.1)$$

where the probability density functions under Ω_1 are given by

$$p_{Y_j|\Omega_1}(y_j) = \begin{cases} \frac{y_j}{\sigma_1^2} e^{-(y_j^2+m_1^2)/2\sigma_1^2} I_0(\frac{y_j m_1}{\sigma_1^2}), & j = 0, \\ \frac{y_j}{\sigma_1^2} e^{-y_j^2/2\sigma_1^2}, & j = 1, 2, \cdots, 2^b - 1, \end{cases} \qquad (B.2)$$

for $\sigma_1^2 \triangleq N_0[1+\gamma_c|\eta|^2]/2$ and $m_1 \triangleq \sqrt{E_s}H(1-|\eta|)$. By inserting Eq.(B.2) to Eq.(B.1), and applying the Laplace transform relation [148] $\int_0^\infty I_0(u\sqrt{\epsilon})e^{-s\epsilon} d\epsilon = \frac{e^{u^2/(4s)}}{s}, s > 0$, we can obtain

$$\hat{P}_c(|\eta|) = \frac{1}{2} e^{-m_1^2/2\sigma_1^2} \int_0^\infty e^{-x/2} I_0(\sqrt{xm_1^2/\sigma_1^2})[1 - e^{-x/2}]^{2^b-1} dx \tag{B.3}$$

$$= \sum_{n=0}^{2^b-1} \frac{(-1)^n}{n+1} \binom{2^b-1}{n} \exp\{-\frac{n}{n+1} \frac{N_I \gamma_c(1-|\eta|)^2}{(1+\gamma_c|\eta|^2)}\}. \tag{B.4}$$

2. DERIVATION OF EQ.(8.89)

To get the correct symbol detection probability in the D^2SA scheme, we first investigate the distribution of E' in Fig. 8.28. Noting $\bar{X}_D = \sqrt{E_s}H(1-|\eta|) \triangleq$

317

m_1 (refer to Eq.(8.83)), and approximating X_D^r to the zero-mean circularly-symmetric complex gaussian random variable with variance $P_s H^2 T_c |\eta|^2 \triangleq 2\sigma_2^2$ (refer to Eq.(8.84)), we get the distribution of $E' = |X_D|^2$

$$p_{E'}(\epsilon) = \frac{1}{2\sigma_2^2} e^{-\frac{\epsilon+m_1^2}{2\sigma_2^2}} I_0(\sqrt{\epsilon} m_1/\sigma_2^2). \tag{B.5}$$

For a given E', the problem of getting the state symbol error probability is the same as the problem of getting the 2^b-ary DPSK symbol error probability in the AWGN channel environment, and the resulting error probability is given by [149, Eq.(7.7)]

$$P(error|E' = \epsilon) = \frac{\sin\frac{\pi}{2^b}}{2\pi} \int_{-\pi/2}^{\pi/2} \frac{\exp\{-\frac{\epsilon}{N_0}(1 - \cos\frac{\pi}{2^b}\cos\theta)\}}{1 - \cos\frac{\pi}{2^b}\cos\theta} d\theta. \tag{B.6}$$

Therefore, the unconditional error probability can be determined by

$$P(error) = \int_0^\infty P(error|E' = \epsilon) p_{E'}(\epsilon) d\epsilon \tag{B.7}$$

$$= \frac{\sin\frac{\pi}{2^b}}{2\pi} \int_{-\pi/2}^{\pi/2} \frac{I(N_0, \sigma_2, m_1, 1 - \cos\frac{\pi}{2^b}\cos\theta)}{1 - \cos\frac{\pi}{2^b}\cos\theta} d\theta, \tag{B.8}$$

where $I(a, b, c, d) \triangleq \frac{e^{-c^2/2b^2}}{2b^2} \int_0^\infty e^{-\epsilon(d/2a^2+1/2b^2)} I_0(\sqrt{\epsilon}c/b^2) d\epsilon$. Now applying the Laplace transform relation $\int_0^\infty I_0(u\sqrt{\epsilon})e^{-s\epsilon} d\epsilon = \frac{e^{u^2/(4s)}}{s}, s > 0$, and carrying out some additional manipulation, we get the relation

$$I(N_0, \sigma_2, m_1, 1 - \cos\frac{\pi}{2^b}\cos\theta) = \frac{\exp\{-\frac{N_I\gamma_c(1-|\eta|)^2(1-\cos(\pi/2^b)\cos\theta)}{1+\gamma_c|\eta|^2(1-\cos(\pi/2^b)\cos\theta)}\}}{[1 + \gamma_c|\eta|^2(1 - \cos\frac{\pi}{2^b}\cos\theta)]}. \tag{B.9}$$

By inserting Eq.(B.9) to Eq.(B.8) and applying the relation $\hat{P}_c(|\eta|) = 1 - P(error)$, we finally get the expression in Eq.(8.89).

3.   DERIVATION OF EQ.(9.4)

Referring to Eq.(B.21) and Fig. 9.5 (b), we can represent the nth matched components $g_n^{[0]}$ by

$$g_n^{[0]} = X_s e^{j(\theta_{r+n,0}+\phi)} + N_n, \quad 0 \le r \le 2^{L-S} - 1, \tag{B.10}$$

where $X_s e^{j(\theta_{r+n,0}+\phi)}$ and N_n respectively denote the matched signal and AWGN components when $\lambda = 0$. Then, the soft state symbol $y_n \equiv g_n g_{n-1}^*$ can be represented by

$$y_n = |X_s|^2 e^{j\Delta\theta_{r+n,0}} + \tilde{N}_n, \tag{B.11}$$

where $\Delta\theta_{r+n,0} \equiv \theta_{r+n,0} - \theta_{r+n-1,0}$ is the phase argument of the nth state symbol and $\tilde{N}_n \equiv X_s N_{n-1}^* e^{j(\theta_{r+n,0}+\phi)} + X_s^* N_n e^{-j(\theta_{r+n-1,0}+\phi)} + N_n N_{n-1}^*$ is the corresponding noise component.

To get the probability P_{TSC}, we consider the magnitude statistics of 2^{L-D} state symbol correlation results

$$Y_{m,n_k} \stackrel{\triangle}{=} \sum_{n=0}^{B-1} y_n e^{-j\Delta\theta_{m+n,n_k}}, \quad m = 0, 1, \cdots, 2^{L-S} - 1,$$

$$n_k = 0, 1, \cdots, 2^{S-D} - 1. \tag{B.12}$$

First, when the correct frame boundary is hypothesized (H_1-hypothesis: $m = r$, $n_k = 0$), the correlation result $Y_{r,0}$ can be represented by the sum of a real random variable $V_1 \stackrel{\triangle}{=} B|X_s|^2 + \sum_{n=0}^{B-2} 2Re[X_s N_n^* e^{j(\theta_{r+n,0}+\phi)}]$ and a complex random variable $V_2 \stackrel{\triangle}{=} \sum_{n=0}^{B-1} N_n N_{n-1}^* e^{-j\Delta\theta_{r+n,0}} + X_s N_{-1}^* e^{j(\theta_{r-1,0}+\phi)} + X_s^* N_{B-1} e^{-j(\theta_{B+r-1,0}+\phi)}$. The two random variables are mutually uncorrelated, and, for a given X_s, V_1 can be approximated to be a real Gaussian random variable with mean $\mu_1 \stackrel{\triangle}{=} B|X_s|^2$ and variance $\sigma_1^2 \stackrel{\triangle}{=} 2(B-1)N_o|X_s|^2$, while V_2 can be approximated to be a zero mean circularly-symmetric complex Gaussian random variable [100] with variance $2\sigma_2^2 \stackrel{\triangle}{=} 2N_o|X_s|^2 + BN_o^2$. Therefore, by [15], we can get the pdf of $R_{r,0} \stackrel{\triangle}{=} |Y_{r,0}|$

$$f_{R_{r,0}}(s) = \int_{-\infty}^{\infty} \frac{s}{\sigma_2^2} I_0(\frac{vs}{\sigma_2^2}) \exp(-\frac{v^2 + s^2}{2\sigma_2^2}) g_{V_1}(v) dv, \tag{B.13}$$

for a Gaussian pdf

$$g_{V_1}(v) = \frac{1}{\sqrt{2\pi}\sigma_1} \exp\{-\frac{(v-\mu_1)^2}{2\sigma_1^2}\}. \tag{B.14}$$

On the other hand, when the correct frame boundary is not hypothesized (H_0-hypothesis: $m \neq r$ or $n_k \neq 0$), regarding the state symbol sequence $\{e^{j\Delta\theta_{m+n,n_k}}\}$ as a random sequence, we can approximate Y_{m,n_k} to be a zero mean circularly-symmetric complex Gaussian random variable with variance $2\sigma_0^2 \stackrel{\triangle}{=} B(|X_s|^2 + N_o)^2$ when $b \geq 2$, and thus get the pdf of the magnitude $R_{m,n_k} \stackrel{\triangle}{=} |Y_{m,n_k}|$

$$f_{R_{m,n_k}}(s) = \frac{s}{\sigma_0^2} \exp(-\frac{s^2}{2\sigma_0^2}), \quad m \neq r \text{ or } n_k \neq 0. \tag{B.15}$$

(The case when $b = 1$ is discussed afterwards.)

Therefore, employing Eqs.(B.13)–(B.15), we can represent the conditional probability of *true acquisition completion through state symbol correlation* for a given X_s by

$$
\begin{aligned}
P_{TSC}(|\eta| \mid X_s) &= \int_0^\infty \int_{-\infty}^\infty \frac{s}{\sigma_2^2} I_0(\frac{vs}{\sigma_2^2}) \exp(-\frac{v^2 + s^2}{2\sigma_2^2}) g_{V_1}(v) dv \\
&\quad \times \ \{\int_0^s \frac{u}{\sigma_0^2} \exp(-\frac{u^2}{2\sigma_0^2}) du\}^{2^{L-D}-1} ds \\
&= \int_{-\infty}^\infty J(v) g_{V_1}(v) dv,
\end{aligned}
\tag{B.16}
$$

where $J(v)$ is determined to be [27, Eq.(19)-Eq.(20)]

$$
\begin{aligned}
J(v) &= \int_0^\infty \frac{s}{\sigma_2^2} I_0(\frac{vs}{\sigma_2^2}) \exp(-\frac{v^2 + s^2}{2\sigma_2^2}) \{\int_0^s \frac{u}{\sigma_0^2} \exp(-\frac{u^2}{2\sigma_0^2}) du\}^{2^{L-D}-1} ds \\
&= \frac{1}{2\sigma_2^2} \exp(-\frac{v^2}{2\sigma_2^2}) \int_0^\infty I_0(\sqrt{\frac{v^2 y}{\sigma_2^4}}) \exp(-\frac{y}{2\sigma_2^2}) \\
&\quad \cdot \{1 - \exp(-\frac{y}{2\sigma_0^2})\}^{2^{L-D}-1} dy \\
&= \sum_{n=0}^{2^{L-D}-1} \frac{\sigma_0^2}{\sigma_0^2 + n\sigma_2^2} (-1)^n \binom{2^{L-D}-1}{n} \exp\{-\frac{nv^2}{2(\sigma_0^2 + n\sigma_2^2)}\} \quad (B.17)
\end{aligned}
$$

By inserting Eq.(B.17) to Eq.(B.16), we can get the conditional probability

$$
\begin{aligned}
P_{TSC}(|\eta| \mid X_s) &= \sum_{n=0}^{2^{L-D}-1} \frac{(-1)^n \binom{2^{L-D}-1}{n} \sigma_0^2}{\sqrt{(\sigma_0^2 + n\sigma_2^2)(\sigma_0^2 + n\sigma_2^2 + n\sigma_1^2)}} \\
&\quad \times \exp\left\{-\frac{n\mu_1^2}{2(\sigma_0^2 + n\sigma_2^2 + n\sigma_1^2)}\right\}.
\end{aligned}
\tag{B.18}
$$

The final step is to remove the dependency of Eq.(B.18) on $|X_s|^2$. When $|\eta| = 0$, the signal component X_s is a constant value $\sqrt{N_o N_I \gamma_c}$ for the chip-SNR $\gamma_c \equiv \frac{P_s H^2 T_c}{N_o}$, and thus Eq.(B.18) reduces to the upper line of Eq.(9.4a). When $|\eta| \neq 0$, X_s has mean $m_s \triangleq \sqrt{N_o N_I \gamma_c}(1 - |\eta|)$, and $(X_s - m_s)$ can be approximated to be a circularly-symmetric complex Gaussian random variable whose variance is $2\sigma_s^2 \triangleq N_o \gamma_c |\eta|^2$. Therefore, the random variable $Z \triangleq |X_s|^2$ has a pdf

$$
f_Z(z) = \frac{1}{N_o \gamma_c |\eta|^2} \exp\{-\frac{N_o N_I \gamma_c(1 - |\eta|)^2 + z}{N_o \gamma_c |\eta|^2}\} I_0(\frac{2\sqrt{z N_o N_I \gamma_c(1 - |\eta|)^2}}{N_o \gamma_c |\eta|^2}),
\tag{B.19}
$$

and by averaging Eq.(B.18) on $|X_s|^2$, we get the lower line of Eq.(9.4a).

When $b = 1$, Y_{m,n_k} (for H_0-hypothesis) is not circularly-symmetric, as the variance of its real part becomes $B|X_s|^4 + BN_o(|X_s|^2 + N_o/2)$ while that of its imaginary part is $BN_o(|X_s|^2 + N_o/2)$. Thus, we shall get a result slightly different from the true P_{TSC} if we employ Eq.(B.15) as the pdf of R_{m,n_k} when $b = 1$. [1] Instead of obtaining the exact value of P_{TSC} for $b = 1$ through a tedious evaluation of the involved multi-dimensional integral, we choose to derive its lower bound by adding to Y_{m,n_k} a new independent imaginary Gaussian random variable of variance $B|X_s|^4$ to get a circularly-symmetric random variable $\tilde{Y}_{m,n_k}$ having a larger variance $2\tilde{\sigma}_0^2 \triangleq B\{|X_s|^4 + (|X_s|^2 + N_o)^2\}$. By replacing σ_0 with $\tilde{\sigma}_0$ in Eq.(B.15), we will get the pdf of the magnitude of $\tilde{Y}_{m,n_k}$ and by repeating the remaining procedures in the above we can get a lower bound of P_{TSC} shown in Eq.(9.4b), which will eventually provide an upper bound of the mean acquisition time of the binary signaling CDSA system.

4.   DERIVATION OF EQ.(9.8)

To determine the branch gains corresponding to the igniter sequence acquisition operation, we investigate the parallel igniter sequence acquisition block in detail, representing the received DPSK state signal by

$$r(t) = \sum_{i=-\infty}^{\infty} \sqrt{P_s} H e^{j(\phi+\theta_{r+i,0})} c^{[0]}(t + [\lambda + \eta]T_c - [i-1]N_I T_c) + N_s(t),$$

(B.20)

where the integer r denotes the frame timing offset satisfying $0 \le r \le 2^{L-S} - 1$, $\theta_{r+i,0} \in \{\frac{2\pi m}{2^b} : m = 0, 1, \cdots, 2^b - 1\}$, and $c^{[l]}(t) \triangleq \sum_{m=0}^{N_I-1} c_m^{[l]} p_{T_c}(t - mT_c)$. In the equation, $c_m^{[0]}$ denotes the igniter sequence of the current cell (or, the 0th cell) normalized such that its absolute value becomes unity, P_s the pilot power, H the channel gain, ϕ the carrier phase, and λ and η the integer and fractional part of the chip alignment offset normalized by T_c respectively. [2] In addition, $N_s(t)$ is the complex white Gaussian noise whose in-phase and quadrature components have the power spectral density of $N_0/2$. Then we can

---

[1] Note that Y_{m,n_k} can be regarded as a random variable formed by adding a circulary-symmetric complex Gaussian random variable of variance $2BN_o(|X_s|^2 + N_o/2)$ to a real (or, one-dimensional) Gaussian random variable of variance $B|X_s|^4$ for $b = 1$ and to a circularly-symmetric complex (or, two-dimensional) Gaussian random variable of the same variance for $b \ge 2$, respectively.
[2] $p_{T_c}(t)$ is the chip pulse shaping filter whose amplitude is 1 in the interval $[0,T_c]$ and 0 elsewhere.

represent the nth matched component of the lth correlator $g_n^{[l]}$ by

$$g_n^{[l]} = \frac{1}{\sqrt{N_I T_c}} \int_{(n-1)N_I T_c}^{n N_I T_c} r(t) \cdot [c^{[l]}(t - [n-1]N_I T_c)]^* dt, \quad l = 0, 1, \cdots, \hat{R} - 1.$$
(B.21)

Since $\{g_n^{[l]} - E[g_n^{[l]}]\}$ can be regarded as circularly-symmetric complex Gaussian random variables [100, 133], independent one another, $T_l(m;n)^2 \triangleq \sum_{i=n}^{n+m-1} |g_i^{[l]}|^2$ is a non-central chi-square random variable when $l = 0$ and $\lambda = 0$, and a central chi-square random variable otherwise, with $2m$ degrees of freedom [71]. In the igniter search stage, the square-root of the maximum value of the $\hat{R}$ correlator output energies (i.e., $\max\{T_0(1;\cdot), T_1(1;\cdot), \cdots, T_{\hat{R}-1}(1;\cdot)\}$) is compared with the threshold R_0, while, in the igniter verification stage, the square-root of the maximum value of the $\hat{R}$ energies noncoherently accumulated for V_I state symbols (i.e., $\max\{T_0(V_I;\cdot), T_1(V_I;\cdot), \cdots, T_{\hat{R}-1}(V_I;\cdot)\}$) is compared with R_1. Now in order to represent the branch gains used in Eq.(9.7), we first define P_d to be *detection probability per H_1 state cell*, P_{fa} the *false acquisition probability per H_0 state cell*, P_{VSI,H_1} the *igniter verification success probability under H_1 state*, and P_{VSI,H_0} the *igniter verification success probability under H_0 state*. Then, referring to [71] and Chapter 8, and noting that $T_l(\cdot;\cdot)^2, l = 0, 1, \cdots, \hat{R}$ are chi-square random variables independent one another, we can determine the probabilities

$$P_d(|\eta|) = 1 - \left\{ 1 - Q\left(\sqrt{\frac{2N_I \gamma_c (1 - |\eta|)^2}{1 + \gamma_c |\eta|^2}}, \sqrt{\frac{\zeta_0}{1 + \gamma_c |\eta|^2}} \right) \right\}$$

$$\times \left\{ 1 - \exp\left(-\frac{\zeta_0}{2[1 + \gamma_c |\eta|^2]}\right) \right\}^{\hat{R}-1},$$
(B.22)

$$P_{fa}(|\eta|) = 1 - \left[1 - \exp\left\{-\frac{\zeta_0}{2[1 + \gamma_c (1 - 2|\eta| + 2|\eta|^2)]}\right\} \right]^{\hat{R}}, \quad (B.23)$$

$$P_{VSI,H_1}(|\eta|) = 1 - \left\{ 1 - Q_{V_I}\left(\sqrt{\frac{2V_I N_I \gamma_c (1 - |\eta|)^2}{1 + \gamma_c |\eta|^2}}, \sqrt{\frac{\zeta_1}{1 + \gamma_c |\eta|^2}} \right) \right\}$$

$$\times \left\{ 1 - \exp\left(-\frac{\zeta_1}{2[1 + \gamma_c |\eta|^2]}\right) \sum_{k=0}^{V_I-1} \frac{1}{k!}\left(\frac{\zeta_1}{2[1 + \gamma_c |\eta|^2]}\right)^k \right\}^{\hat{R}-1},$$
(B.24)

$$P_{VSI,H_0}(|\eta|) = 1 - \left[1 - \exp\left\{-\frac{\zeta_1}{2[1 + \gamma_c (1 - 2|\eta| + 2|\eta|^2)]}\right\} \right.$$

$$\times \left. \sum_{k=0}^{V_I-1} \frac{1}{k!}\left\{\frac{\zeta_1}{2[1 + \gamma_c (1 - 2|\eta| + 2|\eta|^2)]}\right\}^k \right]^{\hat{R}},$$
(B.25)

for the *normalized thresholds* $\zeta_0 \triangleq \frac{2R_0^2}{N_0}$, $\zeta_1 \triangleq \frac{2R_1^2}{N_0}$, and the pilot chip-SNR $\gamma_c \triangleq$
$P_s H^2 T_c / N_0$, where $Q(\alpha, \beta) \triangleq Q_1(\alpha, \beta)$ and $Q_m(\alpha, \beta) \triangleq \int_\beta^\infty x(x/\alpha)^{m-1} \exp$
$\{-\frac{1}{2}(x^2 + \alpha^2)\} I_{m-1}(\alpha x) dx$ for the mth-order modified Bessel function $I_m(x)$.
Furthermore, to specify the igniter sequence identification operation in the verification mode, we define P_G to be the probability that the correlator branch corresponding to the igniter code of the current cell actually produces the maximum accumulated energy in the igniter verification stage and the square-root of the value exceeds the threshold R_1 under the H_1 state. Then, we can determine

$$
\begin{aligned}
P_G(|\eta|) &= \int_{\zeta_1}^\infty \frac{1}{2[1 + \gamma_c|\eta|^2]} \left\{ \frac{y}{2V_I N_I \gamma_c (1 - |\eta|)^2} \right\}^{\frac{V_I - 1}{2}} \\
&\times \exp\left\{ -\frac{y + 2V_I N_I \gamma_c (1 - |\eta|)^2}{2[1 + \gamma_c|\eta|^2]} \right\} I_{V_I - 1}\left(\frac{\sqrt{2V_I N_I \gamma_c (1 - |\eta|)^2 y}}{1 + \gamma_c|\eta|^2} \right) \\
&\times \left[1 - \exp\left\{ -\frac{y}{2[1 + \gamma_c|\eta|^2]} \right\} \sum_{k=0}^{V_I - 1} \frac{1}{k!} \left(\frac{y}{2[1 + \gamma_c|\eta|^2]} \right)^k \right]^{\hat{R} - 1} dy.
\end{aligned}
$$

(B.26)

Using the five probability parameters, we can express the branch gains of the $\hat{R}$-parallel igniter search scheme as given in Eq.(9.8) in Chapter 9.

5.   DERIVATION OF MEAN ACQUISITION TIME FOR $\hat{R}$-PARALLEL SEARCH

On the other hand, by tailoring the mean acquisition time formula derived in [66, Eq.(36)] such that it can be applied to the inter-cell synchronous DS/CDMA systems employing 2^b-ary DPSK-signaling DSA, we can get the mean acquisition time of the DPSK-signaling DSA scheme which does not incorporate the state symbol correlation process

$$
\begin{aligned}
E[T_{DSA}] &\approx \left[E[T_1] \times \left\{ 1 + \frac{1 - (1 - W/2^{L-S})P_c^W}{(1 - W/2^{L-S})P_c^{W+V} 2^{bV+D}} \right\} \right. \\
&\left. + \frac{(W + V)}{(1 - W/2^{L-S})P_c^{W+V}} + \frac{\{1 - (1 - W/2^{L-S})P_c^W\}K}{(1 - W/2^{L-S})P_c^{W+V} 2^{bV+D}} \right] \times N_I T_c,
\end{aligned}
$$

(B.27)

where $E[T_1]$ is equal to $E[T_I]$ in Eq.(9.9) except that the equivalent false acquisition penalty time is [66]

$$
K_I = W + V + K/2^{bV+D}.
$$

(B.28)

Further, we replace $2N_I$ and K_I respectively with $2N_M/\hat{R}$ and K in Eq.(9.9), and modify P_d, P_{VSI,H_1}, and P_G in Eq.(B.22), Eq.(B.24), and Eq.(B.26) to

$$
\begin{aligned}
\hat{P}_d(|\eta|) = {}& 1 - [\, 1 - Q(\, \sqrt{\frac{2N_I\gamma_c(1-|\eta|)^2}{1+\gamma_c|\eta|^2}}\,,\, \sqrt{\frac{\zeta_0}{1+\gamma_c|\eta|^2}}\,)\,] \\
& \times\ [\, 1 - \exp\{-\frac{\zeta_0}{2[1+\gamma_c(1-2|\eta|+2|\eta|^2)]}\}\,]^{\hat{R}-1}, \quad \text{(B.29)}
\end{aligned}
$$

$$
\begin{aligned}
\hat{P}_{VSI,H_1}(|\eta|) = {}& 1 - [\, 1 - Q_{V_I}(\, \sqrt{\frac{2V_I N_I\gamma_c(1-|\eta|)^2}{1+\gamma_c|\eta|^2}}\,,\, \sqrt{\frac{\zeta_1}{1+\gamma_c|\eta|^2}}\,)\,] \\
& \times\ [\, 1 - \exp\{-\frac{\zeta_1}{2[1+\gamma_c(1-2|\eta|+2|\eta|^2)]}\} \\
& \times\ \sum_{k=0}^{V_I-1}\frac{1}{k!}\{\frac{\zeta_1}{2[1+\gamma_c(1-2|\eta|+2|\eta|^2)]}\}^k\,]^{\hat{R}-1}, \quad \text{(B.30)}
\end{aligned}
$$

and

$$
\begin{aligned}
\hat{P}_G(|\eta|) = {}& \int_{\zeta_1}^{\infty} \frac{1}{2[1+\gamma_c|\eta|^2]}\{\frac{y}{2V_I N_I\gamma_c(1-|\eta|)^2}\}^{\frac{V_I-1}{2}} \\
& \times\ \exp\{-\frac{y+2V_I N_I\gamma_c(1-|\eta|)^2}{2[1+\gamma_c|\eta|^2]}\} I_{V_I-1}(\frac{\sqrt{2V_I N_I\gamma_c(1-|\eta|)^2 y}}{1+\gamma_c|\eta|^2}) \\
& \times\ [1 - \exp\{-\frac{y}{2[1+\gamma_c(1-2|\eta|+2|\eta|^2)]}\} \\
& \times\ \sum_{k=0}^{V_I-1}\frac{1}{k!}\{\frac{y}{2[1+\gamma_c(1-2|\eta|+2|\eta|^2)]}\}^k]^{\hat{R}-1}\, dy. \quad \text{(B.31)}
\end{aligned}
$$

Then, we can determine the mean acquisition time of the active correlator-based $\hat{R}$-parallel search acquisition scheme using the legacy unmodulated pilot sequence of period $N_M T_c$ to be

$$
\begin{aligned}
E[T_{PSA}] = {}& \frac{1}{\hat{P}_d\hat{P}_G}\,[\, 1 + V_I\hat{P}_d + K\hat{P}_d(\hat{P}_{VSI,H_1} - \hat{P}_G) \\
& +\ (2N_M/\hat{R}-1)\{1+P_{fa}(V_I+KP_{VSI,H_0})\}(\frac{1}{\hat{F}_d} - \frac{\hat{P}_d\hat{P}_G}{2})\,] \\
& \times\ N_I T_c, \quad\quad\quad\quad\quad\quad\quad\quad\quad\quad\quad\quad\quad\quad \text{(B.32)}
\end{aligned}
$$

$$\text{(B.33)}$$

where $\hat{F}_d=1$ if $|\eta|=0$, and $\hat{F}_d=(2-\hat{P}_d\hat{P}_G)$ if $|\eta|=0.25$.

Appendix C
List of Acronyms

3GPP	3rd Generation Partnership Project
AI	Acquisition Indicator
AICH	Acquisition Indicator CHannel
AMPS	Advanced Mobile Phone System
AP	Access Preamble
AP-AICH	Access Preamble Acquisition Indicator CHannel
API	Access Preamble Indicator
AS	Access Slot
ASC	Access Service Class
ASN	Average Sample Number
ATM	Asynchronous Transfer Mode
BCH	Broadcast CHannel
BDSA	Batch DSA
BER	Bit Error Rate
BLER	BLock Error Rate
BPF	Bandpass Filter
BS	Base Station
BSRG	Basic SRG
CA	Channel Assignment
CAI	Channel Assignment Indicator
CBR	Constant Bit Rate
CCC	CPCH Control Command
CCPCH	Common Control Physical Channel
CCTrCH	Coded Composite Transport Channel
CD	Collision Detection
CD/CA-ICH	Collision Detection/Channel Assignment Indicator Channel

CDI	Collision Detection Indicator
CDMA	Code Division Multiple Access
CD-P	Collision Detection Preamble
CDSA	Correlation-aided DSA
CFN	Connection Frame Number
CPCH	Common Packet CHannel
CPICH	Common PIlot CHannel
CRC	Cyclic Redundancy Check
CSICH	CPCH Status Indicator CHannel
DCA	Dynamic Channel Allocation
D-CDMA	Deterministic CDMA
DCH	Dedicated CHannel
DECT	Digital European Cordless Telephone
DL	Downlink
DLL	Delay-Locked Loop
DPCCH	Dedicated Physical Control CHannel
DPCH	Dedicated Physical CHannel
DPDCH	Dedicated Physical Data CHannel
DS/CDMA	Direct-Sequence CDMA
DSA	Distributed Sample Acquisition
D²SA	Differential DSA
DSCH	Downlink Shared CHannel
DSMA-CD	Digital Sense Multiple Access - Collision Detection
DSS	Distributed Sample Scrambling
DRX	Discontinuous Reception
DTX	Discontinuous Transmission
EDGE	Enhanced Data rates for GSM Evolution
EGC	Efficient Golay Correlator
EW	Expanding Window
FACH	Forward Access CHannel
FAUSCH	FAst Uplink Synchronization CHannel
FBI	Feedback Information
FDMA	Frequency Division Multiple Access
FDD	Frequency Division Duplex
FEC	Forward Error Correction
FER	Frame Error Rate
FH/CDMA	Frequency-Hopping CDMA
FHT	Fast Hadamard Transform
FPC	Full Period Correlation
FSW	Frame Synchronization Word
FTP	File Transfer Protocol
GF	Galois Field

GHG	Generalized Hierarchical Golay
GP	Guard Period
GPS	Global Positioning System
GSM	Global System for Mobile communications
HPSK	Hybrid PSK
IC	Interference Cancellation
ICH	Indicator Channel
ID	Identification
IMT	International Mobile Telecommunications
IS-54	Interim Standard 54
IS-95	Interim Standard 95
ISCP	Interference Signal Code Power
JD	Joint Detection
LPF	Lowpass Filter
L1	Layer 1
L2	Layer 2
MAC	Medium Access Control
MAI	Multiple Access Interference
MAP	Maximum *a posteriori*
MC/CDMA	Multi-Carrier CDMA
MDS	Maximum Distance Separable
ML	Maximum Likelihood
MMSE	Minimum Mean Squared Error
MRC	Maximal-Ratio Combining
MS	Mobile Station
MUI	Mobile User Identifier
m-sequence	Maximal-length Sequence
MSRG	Modular SRG
NMT	Nordic Mobile Telephone
ODMA	Opportunity Driven Multiple Access
OCQPSK	Orthogonal Complex QPSK
OQPSK	Offset QPSK
OVSF	Orthogonal Variable Spreading Factor
PA	Power Amplifier
PAR	Peak-to-Average power Ratio
PCA	Power Control Algorithm
PCCC	Parallel Concatenated Convolutional Code
P-CCPCH	Primary Common Control Physical CHannel
PCH	Paging CHannel
PC-P	Power Control Preamble
PCPCH	Physical Common Packet CHannel
P-CPICH	Primary CPICH

PDC	Personal Digital Cellular
PDSA	Parallel DSA
PDSCH	Physical Downlink Shared CHannel
PhCH	Physical CHannel
PI	Page Indicator
PICH	Page Indicator CHannel
PISO	Parallel Input Serial Output
PLE	Pre-Loop code phase Estimator
PN	Pseudo Noise
PPC	Partial Period Correlation
PRACH	Physical Random Access CHannel
PUSCH	Physical Uplink Shared CHannel
PS	Parallel Sample
PSA	Parallel Search Acquisition
PSC	Primary Synchronization Code
QPSK	Quadrature Phase Shift Keying
RACH	Random Access CHannel
RRASE	Recursion-aided RASE
RASE	Rapid Acquisition by Sequential Estimation
R-CDMA	Random CDMA
RF	Radio Frame
RNC	Radio Network Controller
RRC	Root-Raised Cosine
RS	Reed-Solomon
RSC	Recursive Systematic Convolutional code
RSCP	Received Signal Code Power
RSSI	Received Signal Strength Indicator
RTT	Round Trip Time
RX	Receive
SCCC	Serially Concatenated Convolutional Code
S-CCPCH	Secondary Common Control Physical CHannel
SCH	Synchronization CHannel
S-CPICH	Secondary CPICH
SF	Spreading Factor
SFN	System Frame Number
SI	Status Indicator
SIPO	Serial Input Parallel Output
SIR	Signal-to-Interference Ratio
SNR	Signal-to-Noise Ratio
SRG	Shift Register Generator
SOVA	Soft Output Viterbi Algorithm
SPRT	Sequential Probability Ratio Test

SSA	Serial Search Acquisition
SSC	Secondary Synchronization Code
SSDT	Site Selection Diversity Transmission
SSRG	Simple SRG
SSMA	Spread Spectrum Multiple Access
STD	Selective Transmit Diversity
STTD	Space Time Transmit Diversity
SWI	Sync-Worthiness Indicator
TA	Timing Advance
TACS	Total Access Communication System
TD-CDMA	Time Division CDMA
TDD	Time Division Duplex
TDMA	Time Division Multiple Access
TF	Transport Format
TFC	Transport Format Combination
TFCI	Transport Format Combination Indicator
TFCS	Transport Format Combination Set
TFI	Transport Format Indicator
TFS	Transport Format Set
TPC	Transmit Power Control
TrBK	Transport Block
TrCH	Transport CHannel
TS	Time Slot
TSPRT	Truncated SPRT
TSTD	Time Switched Transmit Diversity
TTI	Transmission Time Interval
TPC	Transmit Power Control
TX	Transmit
TxAA	Transmit Adaptive Antenna
UE	User Equipment
UL	Uplink
UMTS	Universal Mobile Telecommunications System
USCH	Uplink Shared CHannel
USDC	United States Digital Cellular
UTRA	Universal Terrestrial Radio Access
UTRAN	UMTS Terrestrial Radio Access Network
UWC	Universal Wireless Communications consortium
VBR	Variable Bit Rate
VCC	Voltage-Controlled-Clock
W-CDMA	Wideband CDMA

References

[1] T. S. Rappaport, *Wireless communications*. Prentice Hall, 1996.

[2] B. H. Walke, *Mobile radio networks*. Wiley, 2000.

[3] D. E. McDysan and D. L. Spohn, *ATM: Theory and applications*. McGraw-Hill, 1995.

[4] M. K. Simon, J. K. Omura, R. A. Scholtz, and B. K. Levitt, *Spread Spectrum Communiations Handbook*. McGraw-Hill, 1994.

[5] S. Hara and R. Prasad, "Overview of multicarrier CDMA", *IEEE Commun. Mag.*, vol.35, no.12, pp.126-133, Dec. 1997.

[6] W. C. Y. Lee, *Mobile communications design fundamentals*. Wiley, 1993.

[7] A.J.Viterbi, *CDMA: Principles of Spread Spectrum Communication*. Addison-Wesley, 1995.

[8] 3GPP TS 25.213, "Spreading and modulation(FDD)," V3.4.0, Dec. 2000.

[9] TIA/EIA/IS-2000-2-A, "Physical layer standard for cdma2000 spread spectrum systems," Oct. 1999.

[10] TIA/EIA Interim Standard-95, "Mobile station - base station compatibility standard for dual-mode wideband spread spectrum cellular system," July 1993.

[11] H. Holma and A. Toskala, *WCDMA for UMTS - radio access for third generation mobile communications*. Wiley, 2000.

[12] S. S. Rappaport and D. M. Grieco, "Spread-spectrum signal acquisition: Methods and technology," *IEEE Commun. Magazine*, vol. 22, no. 6, pp.6-21, June 1984.

[13] A. Polydoros and S. Glisic, "Code synchronization: A review of principles and techniques," *Proc. ISSSTA'94*, pp.115-136.

[14] A. Polydoros and C. Weber, "A unified approach to serial search spread-spectrum code acquisition - Part I: General theory," *IEEE Trans. Commun.*, vol. COM-32, no. 5, pp.542-549, May 1984.

[15] A. Polydoros and C. Weber, "A unified approach to serial search spread-spectrum code acquisition - Part II: A matched-filter receiver," *IEEE Trans. Commun.*, vol. COM-32, no. 5, pp.550-560, May 1984.

[16] U. Madhow and M. B. Pursley, "Mathematical modeling and performance analysis for a two-stage acquisition scheme for direct-sequence spread-spectrum CDMA," *IEEE Trans. Commun.*, vol. 43, no. 9, pp.2511-2520, Sept. 1995.

[17] A. Polydoros and M. K. Simon, "Generalized serial search code acquisition: The equivalent circular state diagram approach," *IEEE Trans. Commun.*, vol. COM-32, no. 12, pp.1260-1268, Dec. 1984.

[18] S.-M. Pan, D. H. Madill, and D. E. Dodds, "A unified time-domain analysis of serial search with application to spread spectrum receivers," *IEEE Trans. Commun.*, vol. 43, no. 12, pp.3046-3054, Dec. 1995.

[19] S.-M. Pan, H. A. Grant, D. E. Dodds, and S. Kumar, "An offset-Z search strategy for spread spectrum systems," *IEEE Trans. Commun.*, vol. 43, no. 12, pp.2900-2902, Dec. 1995.

[20] S. Davidovici, L. B. Milstein, D. L. Schiling, "A new rapid technique for direct sequence spread-spectrum communications," *IEEE Trans. Commun.*, vol. COM-32, no. 11, pp.1161-1168, Nov. 1984.

[21] Y.-H. Lee and S. Tantaratana, "Sequential acquisition of PN sequences for DS/SS communications: Design and performance," *IEEE J. Select. Areas Commun.*, vol. 10, no. 4, pp.750-759, May 1992.

[22] S. Tantaratana, A. W. Lam, and P. J. Vincent, "Noncoherent sequential acquisition of PN sequences for DS/SS communications with/without channel fading," *IEEE Trans. Commun.*, vol. 43, no. 2/3/4, Feb./Mar./Apr. 1995.

[23] C. W. Helstrom, *Elements of Signal Detection and Estimation*. Prentice-Hall, 1995.

[24] R. L. Pickholtz, D. L. Shilling, and L. B. Milstein, "Theory of spread-spectrum communications - a tutorial," *IEEE Trans. Commun.*, vol. COM-30, no. 5, pp.855-884, May 1982.

[25] L. B. Milstein, J. Gevalgiz, and P. K. Das, "Rapid acquisition for direct sequence spread spectrum communications using parallel SAW convolvers," *IEEE Trans. Commun.*, vol. COM-33, No. 7, pp.593-600, July 1985.

[26] E. Sourour and S. C. Gupta, "Direct-sequence spread-spectrum parallel acquisition in a fading mobile channel," *IEEE Trans. Commun.*, vol. 38, No. 7, pp.992-998, July 1990.

[27] E. Sourour and S. C. Gupta, "Direct-sequence spread-spectrum parallel acquisition in nonselective and frequency-selective Rician fading channels," *IEEE J. Select. Areas Commun.*, vol. 10, no. 3, pp.535-544, Apr. 1992.

[28] H.-R. Park and B.-J. Kang, "On the performance of a maximum-likelihood code-acquisition technique for preamble search in a CDMA reverse link," *IEEE Trans. Veh. Technol.*, vol. 47, no. 1, pp.65-74, Feb. 1998.

[29] R. B. Ward, "Acquisition of pseudonoise signals by sequential estimation," *IEEE Trans. Commun.*, vol. COM-13, pp475-483, Dec. 1965.

[30] R. B. Ward, and K. P. Yiu, "Acquisition of pseudonoise signals by recursion aided sequential estimation," *IEEE Trans. Commun.*, vol. COM-25, pp784-794, Aug. 1977.

[31] C. C. Kilgus, "Pseudonoise code acquisition using majority logic decoding," *IEEE Trans. Commun.*, vol. COM-21, pp.772-774, June 1973.

[32] R. T. Barghouthi and G. L. Stuber, "Rapid sequence acquisition for DS/CDMA systems employing Kasami sequences," *IEEE Trans. Commun.*, vol. 42, no. 2/3/4, pp.1957-1968, Feb./Mar./Apr. 1994.

[33] M. S. Salih and S. Tantaratana, "A closed-loop coherent acquisition scheme for PN sequences using an auxiliary sequence," *IEEE J. Select. Areas Commun.*, vol. 14, no. 8, pp.1653-1659, Oct. 1996.

[34] M. S. Salih and S. Tantaratana, "A closed-loop coherent PN acquisition system with a pre-loop estimator," *IEEE Trans. Commun.*, vol. 47. no. 9, pp.1394-1405, Sept. 1999.

[35] M. H. Zarrabizadeh and E. S. Sousa, "A differentially coherent PN code acquisition receiver for CDMA systems," *IEEE Trans. Commun.*, vol. 45, No. 11, pp.1456-1465, Nov. 1997.

[36] C.-D. Chung, "Differentially coherent detection technique for direct-sequence code acquisition in a rayleigh fading mobile channel," *IEEE Trans. Commun.*, vol. 43, no. 2/3/4, pp.1116-1126, Feb./Mar./Apr. 1995.

[37] U. Cheng, W. J. Hurd, and J. I. Statman, "Spread-spectrum code acquisition in the presence of Doppler shift and data modulation," *IEEE Trans. Commun.*, vol. 38, no. 2, pp.241-250, Feb. 1990.

[38] A. W. Fuxjaeger and R. A. Iltis, "Acquisition of timing and Doppler-shift in a direct-sequence spread-spectrum system," *IEEE Trans. Commun.*, vol. 42, no. 10, pp.2870-2880, Oct. 1994.

[39] S. G. Glisic, T. J. Poutanen, W. W. Wu, G. V. Petrovic, and Z. Stefanovic, "New PN code acquisition scheme for CDMA networks with low signal-to-noise ratios," *IEEE Trans. Commun.*, vol. 47, no. 2, pp.300-310, Feb. 1999.

[40] J. H. J. Iinatti, "On the threshold setting principles in code acquisition of DS-SS signals," *IEEE J. Select. Areas Commun.*, vol. 18, no. 1, pp.62-72, Jan. 2000.

[41] B. B. Ibrahim and A. H. Aghvami, "Direct sequence spread spectrum matched filter acquisition in frequency-selective Rayleigh fading channels," *IEEE J. Select. Areas Commun.*, vol.12, no.5, pp.885-890, June 1994.

[42] W. A. Krzymien, A. Jalali, and P. Mermelstein, "Rapid acquisition for synchronization of bursty transmissions in CDMA microcellular and personal wireless systems," *IEEE J. Select. Areas Commun.*, vol.14, no.3, pp.570-579, Apr. 1996.

[43] E. Brigant and A. Mammela, "Adaptive threshold control scheme for packet acquisition," *IEEE Trans. Commun.*, vol. 46, no. 12, pp.1580-1582, Dec. 1998.

[44] U. Madhow and M. B. Pursley, "Acquisition in direct-sequence spread-spectrum communication networks: An asymptotic analysis," *IEEE Trans. Inform. Theory*, vol. 39, no. 3, pp.903-912, May 1993.

[45] W. R. Braun, "PN acquisition and tracking performance in DS/CDMA systems with symbol-length spreading sequences," *IEEE Trans. Commun.*, vol.45, no.12, pp.1595-1601, Dec. 1997.

[46] M. Katz and S. Glisic, "Modeling of code acquisition process in CDMA networks - Quasi-synchronous systems," *IEEE Trans. Commun.*, vol. 46, no. 12, pp.1564-1568, Dec. 1998.

[47] M. Katz and S. Glisic, "Modeling of code acquisition process in CDMA networks - Asynchronous systems," *IEEE J. Select. Areas Commun.*, vol. 18, no. 1, pp.73-86, Jan. 2000.

[48] H.-R. Park, "Performance anlaysis of a double-dwell serial search technique for cellular CDMA networks in the case of multiple pilot signals," *IEEE Trans. Veh. Technol.*, vol. 48, no. 6, pp.1819-1830, Nov. 1999.

[49] K. Higuchi, M. Sawahashi, and F. Adachi, "Fast cell search algorithm in inter-cell asynchronous DS-CDMA mobile radio," *IEICE Trans. Commun.*, vol.E81, no.7, pp.1527-1534, July 1998.

[50] K.Higuchi, M.Sawahashi, and F.Adachi, "Fast cell search algorithm in DS-CDMA mobile radio using long spreading codes," *Proc. VTC'97*, pp.1430-1434, May 1997.

[51] D. I. Kim and Y. R. Lee, "I/Q multiplexed code assignment for fast cell search in aynchronous DS/CDMA cellular systems," *IEEE Commun. Lett.*, vol.2, no.6, pp.159-161, June 1998.

[52] Siemens and Texas Instruments, "Generalized hierarchical Golay sequence for PSC with low complexity correlation using pruned efficient Golay correlatos," 3GPP TSGR1-554/99, June 1999.

[53] ETSI/SMG/SMG2, "The ETSI UMTS terrestrial radio access(UTRA) ITU-R RTT candidate submission," June 1998.

[54] Texas Instruments, "Comma free codes for fast PN code acquisition in WCDMA systems: A proposal for the UTRA concept-Revised version," Tdoc SMG2 UMTS-L1 72/98, Paris, Apr. 1998.

[55] 3GPP TS 25.214, "Physical layer procedures(FDD)," V3.5.0, Dec. 2000.

[56] B.-H. Kim and B. G. Lee, "DSA: A distributed sample-based fast DS/CDMA acquisition technique," *IEEE Trans. Commun.*, vol.47, no.5, pp.754-765, May 1999.

[57] B.-H. Kim and B. G. Lee, "Performance analysis of DSA-based DS/CDMA acquisition," *IEEE Trans. Commun.*, vol.47, no.6, pp.817-822, June 1999.

[58] B.-H. Kim and B. G. Lee, "A Fast DS/CDMA Acquisition Scheme Based on Igniter Sequence and Distributed Samples," *Proc. ICC'98*, Atlanta, pp.1259-1263, 1998.

[59] B.-H. Kim and B. G. Lee, "Batch distributed sample acquisition for M-ary signaling DS/CDMA systems," *J. Commun. and Networks*, vol.1, no.1, pp.52-62, Mar. 1999.

[60] B.-H. Kim and B. G. Lee, "PDSA: Parallel distributed sample acquisition for M-ary DS/CDMA systems," *IEEE Trans. Commun.*, vol.49, no.4, pp.589-593, Apr. 2001.

[61] B.-H. Kim and B. G. Lee, "BDSA: A batch-type DSA for fast acquisition in M-ary signaling DS/CDMA systems," *Proc. CIC'98*, Seoul, pp.19-23, 1998.

[62] B.-H. Kim and B. G. Lee, "Parallel DSA for fast acquisition in M-ary DS/CDMA systems," *Proc. APCC'99*, Beijing, pp.727-730, Oct. 1999.

[63] B.-H. Kim and B. G Lee, "D^{2}SA: A DPSK-based distributed sample acquisition for coherently-demodulated MPSK/DS-CDMA systems," *IEEE Trans. Commun.*, to be published.

[64] B.-H. Kim and B. G. Lee, "CDSA: A fast and robust PN code acquisition scheme for cellular DS/CDMA applications," *Proc. European Wireless 2000*, Dresden, pp. 301-306, Sept. 2000.

[65] B.-H. Kim and B. G. Lee, "Correlation-aided distributed sample acquisition scheme for fast and robust code synchronization in cellular DS/CDMA systems," submitted to *IEEE J. Select. Areas Commun.* for publication.

[66] B.-H. Kim and B. G. Lee, "Distributed sample acquisition-based fast cell search in inter-cell asynchronous DS/CDMA systems," *IEEE J. Select. Areas Commun.*, W-CDMA special issue, vol.18, no.8, pp.1455-1469, Aug. 2000.

[67] B.-H. Kim, B.-K. Jeong, and B. G. Lee, "Application of correlation-aided DSA(CDSA) technique to fast cell search in IMT-2000 W-CDMA systems," *J. Commun. and Networks*, IMT-2000 special issue, vol.2, no.1, pp.58-68, Mar. 2000.

[68] B.-H. Kim and B. G. Lee, "Distributed sample-based acquisition techniques for fast and robust synchronization of DS/CDMA scrambling codes," *IEEE Commun. Mag.*, vol.38, no.11, pp. 124-131, Nov. 2000.

[69] B. G. Lee and S. C. Kim, *Scrambling Techniques for Digital Transmission*. Springer-verlag, 1994.

[70] B. G. Lee and S. C. Kim, "Recent advances in theory and applications of scrambling techniques for lightwave transmission," *Proc. IEEE*, vol. 83, no. 10, pp. 1399-1428, Oct. 1995.

[71] J. G. Proakis, *Digital Communications*. McGraw-Hill, 1989.

[72] R. L. Peterson, R. E. Ziemer, and D. E. Borth, *Introduction to Spread Spectrum Communications*. Prentice International, Inc, 1995.

[73] R. Gold, "Maximal recursive sequences with 3-valued recursive cross correlation functions," *IEEE Trans. Inform. Theory*, vol. IT-14, pp. 154-156, Jan. 1968.

[74] T. Kasami, "Weight distribution formula for some class of cyclic codes," Coordinated Science Laboratory, Univ. of Illinois, Urbana, Ill., Tech. Report No. R-285, Apr. 1966.

[75] L. R. Welch, "Lower bounds on the maximum cross correlation of signals," *IEEE Trans. Inform. Theory*, vol. IT-20, pp. 397-399, May 1974.

[76] D. V. Sarwate and M. B. Pursley, "Crosscorrelation properties of pseudo-random and related sequences," *Proc. IEEE*, vol. 68, no. 5, pp. 593-619, May 1980.

[77] S. C. Kim and B. G. Lee, "A theory on sequence spaces and shift register generators," *IEEE Trans. Commun.*, vol. 44, no. 5, pp. 609-618, May 1996.

[78] S. C. Kim and B. G. Lee, "Applications of sequence space and SRG theories to distributed sample scrambling," *IEEE Trans. Commun.*, vol. 45, no. 9, pp. 1043-1052, Sep. 1997.

[79] S. W. Golob, *Shift Register Sequences*, 2nd ed. Los Angeles: Aegean Park, 1982.

[80] W. R. Braun, "Performance analysis for the expandign search PN acquisition algorithm," *IEEE Trans. Commun.*, vol. COM-30, pp. 424-435. Mar. 1982.

[81] V. M. Jovanovic, "Analysis of strategies for serial search spread-spectrum code acquisition - direct approach," *IEEE Trans. Commun.*, vol. COM-36, pp. 1208-1220, Nov. 1988.

[82] S. Parkvall, "Variability of user performance in cellular DS-CDMA – long versus short spreading sequences," *IEEE Trans. Commun.*, vol. 48, no. 7, pp. 1178-1187, July 2000.

[83] B. B. Ibrahim and A. H. Aghvami, "Direct sequence spread spectrum code acquisition in mobile fading channel using matched filter with reference filtering," *Proc. GLOBECOMM'93*, Houston, TX, Dec. 1993, pp. 1085-1089.

[84] R. E. Ziemer and R. L. Peterson, *Digital Communications and Spread Spectrum Systems*. New York: MacMillan, 1985.

[85] 3GPP TS 25.211, "Physical channels and mapping of transport channels onto physical channels (FDD)," V3.5.0, Dec. 2000.

[86] Y.-J. Song, "Frame synchronization confirmation technique using pilot pattern," *J. Commun. and Networks*, vol.2, no.1, pp.69-75, Mar. 2000.

[87] S. M. Alamouti, "A simple transmitter diversity scheme for wireless communications," *IEEE J. Select. Areas Commun.*, vol.16, pp.1451-1458, Oct. 1998.

[88] V. Tarokh, H. Jafarkhani, and A. R. Calderbank, "Space-time block coding for wireless communications: Performance results," *IEEE J. Select. Areas Commun.*, vol.17, pp.451-460, Mar. 1999.

[89] 3GPP TS 25.212, "Multiplexing and channel coding (FDD)," V3.5.0, Dec. 2000.

[90] C. Berrou, A. Glavieux, and P. Thitimajshima, "Near Shannon limit error-correcting coding and decoding: Turbo-codes (1)," *Proc. ICC'93*, pp.1064-1070, 1993.

[91] S. Benedetto and G. Montorsi, "Unveiling Turbo Codes: Some results on parallel concatenated coding schemes," *IEEE Trans. Inform. Theory*, vol.42, no.2, pp.409-428, Mar. 1996.

[92] L. R. Bahl, J. Cocke, F. Jelinek, and J. Raviv, "Optimal decoding of linear codes for minimizing symbol error rate," *IEEE Trans. Inform. Theory*, pp.284-287, Mar. 1974.

[93] J. Hagenauer and P. Hoeher, "A Viterbi algorithm with soft-decision outputs and its applications," *Proc. GLOBECOM'89*, pp.47.1.1-47.1.7, 1989.

[94] M. P. C. Fossorier, F. Burkert, S. Lin, and J. Hagenauer, "On the equivalence between SOVA and Max-Log-MAP decodings," *IEEE Commun. Lett.*, vol.2, no.5, pp.137-139, May 1998.

[95] S. Benedetto, G. Montorsi, D. Divsalar, and F. Pollara, "Soft-output decoding algorithms in iterative decoding of Turbo codes," *TDA Progress Report 42- 124*, pp.63-87, Feb. 1996.

[96] S. Benedetto, G. Montorsi, D. Divsalar, and F. Pollara, "A soft-input soft-output maximum a posteriori (MAP) module to decode parallel and serial concatenated codes," *TDA Progress Report 42- 127*, pp.1-20, Nov. 1996.

[97] 3GPP TS 25.302, "Services provided by the physical layer," V3.7.0, Dec. 2000.

[98] F. J. MacWilliams and N. J. A. Sloane, *The Theory of Error-Correcting Codes*, North-Holland, 1977.

[99] 3GPP TS 25.990, "Vocabulary," V3.0.0, Oct. 1999.

[100] E. A. Lee and D. G. Messerschmitt, *Digital Communication.* Kluwer Academic Publishers, 1994.

[101] 3GPP TS 25.101, "UE radio transmission and reception (FDD)," V3.5.0, Dec. 2000.

[102] D. H. Morais and K. Feher, "The effects of filtering and limiting on the performance of QPSK, Offset QPSK, and MSK systems," *IEEE Trans. Commun.*, vol.28, no.12, pp.1999-2009, Dec. 1980.

[103] Y. Akaiwa and Y. Nagata, "Highly efficient digital mobile communications with a linear modulation method," *IEEE J. Select. Areas Commun.*, vol.5, no.5, pp.890-895, June 1987.

[104] K. Laird, N. Whinnett, and S. Buljore, "A peak-to-average reduction method for third generation CDMA reverse links," it Proc. VTC'99, pp.551-555, 1999.

[105] J. Shim and S. Bang, "Spectrally efficient modulation and spreading scheme for CDMA systems," *Electron. Lett.*, vol.34, no.23, pp.2210-2211, Nov. 1998.

[106] J. S. Lee and L. E. Miller, *CDMA Systems Engineering Handbook.* Artech House Publishers, 1998.

[107] Motorola and Texas Instruments, "Proposal for RACH preambles," 3GPP TSGR1-893/99, July 1999.

[108] Nortel Networks, "Golay-Hadamard sequence based RACH preamble design for large cell (Part-I: Algorithm)," 3GPP TSGR1-990/99, July 1999.

[109] Texas Instruments, "Reduced complexity primary and secondary codes with good aperiodic correlation properties for the WCDMA system," 3GPP TSGR1-373/99, Apr. 1999.

[110] Ericsson, "New downlink scrambling code grouping scheme for UTRA/FDD," 3GPP TSGR1-541/99, July 1999.

[111] Y.-P. E. Wang and T. Ottosson, "Cell search in W-CDMA," *IEEE J. Select. Areas Commun.*, vol.18, no.8, pp.1470-1482, Aug. 2000.

[112] 3GPP TS 25.133, "Requirements for support of radio resource management," V.3.4.0, Dec. 2000.

[113] 3GPP TS 25.402, "Synchronization in UTRAN stage 2," V.3.4.0, Dec. 2000.

[114] Siemens, "A new hierarchical correlation sequence with good properties in presence of a frequency error," 3GPP TSGR1-146/99, Mar. 1999.

[115] Texas Instruments, "Secondary SCH structure for OHG harmonization: Simulations and text proposal," TSGR1-923/99, July 1999.

[116] S. Z. Budisin, "New complementary pairs of sequences," *Electron. Lett.*, vol.26, no.13, pp.881-883, June 1990.

[117] S. Z. Budisin, "Efficient pulse compressor for Golay complementary sequences," *Electron. Lett.*, vol.27, no.3, pp.219-220, Jan. 1991.

[118] B. M. Popovic, "Efficient Golay correlator," *Electron. Lett.*, vol.35, no.17, pp.1427-1428, Aug. 1999.

[119] M. J. E. Golay, "Complementary series," *IRE Trans.*, vol.IT-7, pp.82-87, 1961.

[120] H. Meyr, M. Moeneclaey, and S. A. Fechtel, *Digital Communication Receivers - Synchronization, Channel Estimation, and Signal Processing*, Wiley, 1998.

[121] M. Haardt, A. Klein, R. Koehn, S. Oestreich, M. Purat, V. Sommer, and T. Ulrich, "The TD-CDMA based UTRA TDD mode," *IEEE J. Select. Areas Commun.*, vol. 18, no. 8, pp.1375-1385, Aug. 2000.

[122] 3GPP TS 25.221, "Physical channels and mapping of transport channels onto physical channels (TDD)," V.3.5.0, Dec. 2000.

[123] 3GPP TS 25.222, "Multiplexing and channels coding (TDD)," V.3.5.0, Dec. 2000.

[124] 3GPP TS 25.225, "Physical layer - Measurements (TDD)," V3.5.0, Dec. 2000.

[125] 3GPP TS 25.223, "Spreading and modulation (TDD)," V3.4.0, Dec. 2000.

[126] K. S. Gilhousen, I. M. Jacobs, R. Padovani, A. J. Viterbi, L. A. Weaver, and C. E. Wheatley, "On the capacity of a cellular CDMA system," *IEEE Trans. Veh. Technol.*, vol.40, no.2, pp.303-312, May 1991.

[127] 3GPP2 C.S0002-A, "Physical layer standard for cdma2000 spread spectrum systems," Dec. 1999.

[128] S. C. Kim and B. G. Lee, "Synchronization of shift register generators in distributed sample scramblers," *IEEE Trans. Commun.*, vol. 42, no. 2/3/4, pp.1400-1408, Feb./Mar./Apr. 1994.

[129] ARIB, "Japan's proposal for candidate radio transmission technology on IMT-2000: W-CDMA," June 1998.

[130] P. Taaghol, B. G. Evans, E. Buracchini, R. D. Gaudenzi, G. Gallinaro, J. H. Lee, C. G. Kang, "Satellite UMTS/IMT2000 W-CDMA air interfaces," *IEEE Commun. Mag.*, vol.37, no.9, pp.116-126, Sept. 1999.

[131] TIA TR45.5, "The cdma2000 ITU-R RTT candidate submission," June 1998.

[132] U. Fiebig and M. Schnell, "Correlation properties of extended m-sequences," *Electron. Lett.*, vol.29, no.20, pp.1753-1755, 1993.

[133] V. M. Jovanovic and E. S. Sousa, "Analysis of non-coherent correlation in DS/BPSK spread spectrum acquisition," *IEEE Trans. Commun.*, vol. 43, no. 2/3/4, Feb./Mar./Apr. 1995.

[134] S. Haykin, *Adaptive Filter Theory.* Prentice-Hall, 1996.

[135] H. Harashima and H. Miyakawa, "Matched-transmission technique for channels with intersymbol interference," *IEEE Trans. Commun.*, vol.20, no.8, Aug. 1972.

[136] A. K. Aman, R. L. Cupo, and N. A. Zervos, "Combined trellis coding and DFE through Tomlinson precoding," *IEEE J. Select. Areas Commun.*, vol.9, no.6, pp.876-883, Aug. 1991.

[137] W. Y. Chen, G. H. Im, and J. J. Werner, "Design of digital carierrless AM/PM tranceivers," AT&T/Bellcore Contribution T1E1.4/92-149, Aug. 19, 1992.

[138] Joint Technical Committee Standard Contribution, "Technical report on RF channel characterization and system depolyment modeling," JTC(AIR)/94.09.23-065R6, Sep. 23, 1994.

[139] Operators Harmonization Group, "Harmonized global 3G technical framework for ITU IMT-2000 CDMA proposal," Toronto, June 1999.

[140] R. Fantacci and A. Galligani, "An efficient RAKE receiver architecture with pilot signal cancellation for downlink communications in DS-CDMA indoor wireless networks," *IEEE Trans. Commun.*, vol.47, no.6, pp.823-827, June 1999.

[141] Recommendation ITU-R M.1225, "Guidelines for evaluation of radio transmission technologies for IMT-2000."

[142] 3GPP TS 25.201, "Physical layer - General description," V3.1.0, June 2000.

[143] 3GPP TS 25.104, "UTRA (BS) FDD: Radio transmission and reception," V3.5.0, Dec. 2000.

[144] 3GPP TS 25.215, "Physical layer - Measurements (FDD)," V3.5.0, Dec. 2000.

[145] 3GPP TS 25.833, "Physical layer items not for inclusion in Release 99," V3.0.0, Dec. 2000.

[146] 3GPP TS 25.224, "Physical layer procedures (TDD)," V3.5.0, Dec. 2000.

[147] D. Lee, L. B. Milstein, and H. Lee, "Analysis of a multicarrier DS-CDMA code-acquisition system," *IEEE Trans. Commun.*, vol. 47, no. 8, pp.1233-1244, Aug. 1999.

[148] M. K. Simon and M.-S. Alouini, "A unified approach to the probability of error for noncoherent and differentially coherent modulations over generalized fading channels," *IEEE Trans. Commun.*, vol.46, no.12, pp.1625-1638, Dec. 1998.

[149] M. K. Simon, S. M. Hinedi, and W. C. Lindsey, *Digital Communication Techniques*. Prentice-Hall, 1995.

[150] N. Abramson, "Wideband random access for the last mile," *IEEE Personal Commun.*, pp. 29-33, Dec. 1996.

[151] D. C. Cox, "Wireless network access for personal communications," *IEEE Commun. Mag.*, pp.96-115, Dec. 1992.

[152] S. Tantaratana, A. W. Lam, "Noncoherent sequential acquisition for DS/SS systems," *Proc. 29th Annual Allerton Conf. on Comm., Control, and Computing*, Univ. of Illinois, pp.370-379, Oct. 1991.

[153] A. Wald, *Sequential Analysis*. John Wiley and Sons, New York, 1947.

[154] S. C. Kim and B. G. Lee, "Realizations of parallel and multibit-parallel shift register generators," *IEEE Trans. Commun.* vol. 45, no. 9, pp. 1053-1060, Sep. 1997.

[155] Y. Joo, S. C. Kim and B. G. Lee, "Three-state synchronization of distributed sample scramblers in errored environments," *IEEE Trans. Commun.* vol. 45, no. 9, pp. 1021-1024, Sep. 1997.

[156] S. C. Kim and B. G. Lee, "Parallel realization of distributed sample scramblers for application to cell-based ATM transmission," *IEEE Trans. Commun.* vol. 45, no. 10, pp.1245-1252, Oct. 1997.

Index

343

About the Authors

Dr. Byeong Gi Lee received the B.S. and M.E. degrees in 1974 and 1978, respectively, from Seoul National University, Seoul, Korea, and Kyungpook National University, Daegu, Korea, both in electronics engineering, and received the Ph.D. degree in 1982 from the University of California, Los Angeles, in electrical engineering. He was with Electronics Engineering Department of ROK Naval Academy as an Instructor and Naval Officer in active service from 1974 to 1979. He worked for Granger Associates, Santa Clara, CA, from 1982 to 1984 as a Senior Engineer responsible for applications of digital signal processing to digital transmission, and for AT&T Bell Laboratories, North Andover, MA, from 1984 to 1986 as a Member of Technical Staff responsible for optical transmission system development along with the related standard works. He joined the faculty of Seoul National University in 1986, where he is a Professor in School of Electrical Engineering and Vice Chancellor for Research Affairs. He is the Associate Editor-in-Chief of the *Journal of Communications and Networks*, the past Editor of the *IEEE Global Communications Newsletter*, and a past Associate Editor of the *IEEE Transactions on Circuits and Systems for Video Technology*. He is the Director for Membership Programs Development, the past Director of Asia Pacific Region, and a Member-at-Large of the IEEE Communications Society (ComSoc). He was the Chair of the APCC (Asia Pacific Conference on Communications) Steering Committee, and the Chair of the ABEEK (Accreditation Board for Engineering Education of Korea) Founding Committee. His current fields of interest include communication systems, integrated telecommunication networks, and signal processing. He is a coauthor of *Broadband Telecommunication Technology*, 2nd ed., (Artech House, 1996) and *Scrambling Techniques for Digital Transmission* (Springer Verlag, 1994). He holds seven U.S. patents with three more patents pending. Dr. Lee received the 1984 Myril B. Reed Best Paper Award from the Midwest Symposium on Circuits and Systems and Exceptional Contribution Awards from AT&T Bell Laboratories. He is a Fellow of IEEE, a member of National Academy of Engineering of Korea, a member of Board of Governors of IEEE ComSoc, and a member of Sigma Xi.

Dr. Byoung-Hoon Kim received the B.S. and M.E. degrees in electronics engineering, and the Ph.D. degree in electrical engineering and computer science, from Seoul National University, Seoul, Korea, in 1994, 1996, and 2000, respectively. Since August 2000, he has been with the GCT Semiconductor, Inc. developing CDMA communication chip sets. His current research interests include CDMA, digital communications, channel coding, and signal processing for telecommunications. He received the Best Paper Award (on Communications) of Samsung Humantech Paper Contest in 1999, an Excellent Paper Award from Asia Pacific Conference on Communications in 1999, and the Best Paper Award from European Wireless 2000.